IET CONTROL, ROBOTICS AND SENSORS SERIES 124

Energy Harvesting in Wireless Sensor Networks and Internet of Things

Other related titles:

Volume 68	**Flexible Robot Manipulators: Modelling, simulation and control** M.O. Tokhi and A.K.M. Azad (Editors)
Volume 69	**Advances in Unmanned Marine Vehicles** G.N. Roberts and R. Sutton (Editors)
Volume 77	**Further Advances in Unmanned Marine Vehicles** G.N. Roberts and R. Sutton (Editors)
Volume 86	**Modeling and Control of Flexible Robot Manipulators** M.O. Tokhi and A.K.M. Azad.
Volume 96	**Cyber-Physical System Design with Sensor Networking Technologies** S. Zeadally and N. Jabeur (Editors)
Volume 97	**Sensory Systems for Robotic Applications** R. Dahiya (Editor)
Volume 99	**Practical Robotics and Mechatronics: Marine, space and medical applications** I. Yamamoto
Volume 102	**Recent Trends in Sliding Mode Control** L. Fridman, J.P. Barbot and F. Plestan (Editors)
Volume 104	**Control of Mechatronic Systems** L. Guvenc, A. Guvenc, B. Demirel and M.T. Emirler
Volume 105	**Mechatronic Hands: Prosthetic and Robotic Design** P.H. Chappell

Energy Harvesting in Wireless Sensor Networks and Internet of Things

Edited by
Faisal Karim Shaikh and Sherali Zeadally

The Institution of Engineering and Technology

Published by The Institution of Engineering and Technology, London, United Kingdom

The Institution of Engineering and Technology is registered as a Charity in England & Wales (no. 211014) and Scotland (no. SC038698).

First published 2021

The Institution of Engineering and Technology
Michael Faraday House
Six Hills Way, Stevenage
Herts, SG1 2AY, United Kingdom

www.theiet.org

British Library Cataloguing in Publication Data
A catalogue record for this product is available from the British Library

ISBN 978-1-78561-736-2 (Hardback)
ISBN 978-1-78561-737-9 (PDF)

Typeset in India by MPS Limited
Printed in the UK by CPI Group (UK) Ltd, Croydon

To my wife Shumaila, my children Eshaal, Memoona, Faseeh, and Hassan, and my parents
– Faisal Karim Shaikh

To my wife Borrara, my two daughters Zobia and Zofia, and my parents
– Sherali Zeadally

Contents

About the editors

Faisal Karim Shaikh is a professor in the Department of Telecommunication Engineering at the Mehran University of Engineering and Technology, Jamshoro, Pakistan. His research is focused on investigating energy-efficient communication protocols in wireless sensor networks (WSNs) for mobile, ubiquitous, and pervasive applications. He is interested in environmental monitoring, vehicular ad hoc networks, smart homes, telehealth (body area networks), and Internet of Things. He is a member of PEC, IEEE, and ACM.

Sherali Zeadally is a professor in the College of Communication and Information at the University of Kentucky, USA. He is the editor-in-chief of the IET *Wireless Sensor Systems* Journal, a board member of the IET International Book Series on Sensors and he currently serves as an associate editor or editorial board member for over 30 international refereed journals. He is a fellow of the British Computer Society and a fellow of the IET.

Preface

Recently, we have seen massive deployments of wireless sensor networks (WSNs) in the physical environment to gather data to better understand the environment. WSNs are composed of a distributed network of similar, low powered, low storage, low processing, and low wireless communication range nodes. The nodes are battery operated with a limited lifespan and therefore replacing the batteries becomes a cumbersome process. Typically, the data in WSNs is gathered at the base station for onward transmission and processing by the users. Generally, the major source of power drain is associated with the wireless communication between the nodes.

On the other hand, an increasing number of nodes (things) are being connected to the Internet as they become more advanced, compact, and affordable. The network of Internet-connected things is generally referred to as the Internet of Things (IoT). The IoT ecosystem includes a wide range of devices with heterogeneous networking, computing, and storage capabilities. IoT enables the exchange of various types of information (e.g., sensor data, multimedia data) between the things. The things associated with IoT range from sensors, actuators, and mobile phones to humans, vehicles, and so on. Again, the main limiting factor for IoT is battery-operated things. The batteries need to be replaced after some time.

Therefore, energy harvesting (also known as energy scavenging) from the environment to power the nodes is a better option compared to battery-powered nodes/things along with effective energy management schemes. There are many natural and human-made resources present in the environment that can be utilized for energy harvesting. The various resources in the environment which can be used to convert into electrical power include motion, light, heat, electromagnetics, radio frequency, chemical, mechanical, and the human body.

This book covers various energy harvesting aspects for WSNs and IoT. Specifically, the book is divided into four sections, which cover the broad spectrum of audience interested in energy harvesting mechanisms for WSNs and IoT. The first section deals with energy harvesting from ambient environments such as thermal, vibrational, and the human body. The second section specifically covers radio frequency to be utilized for energy harvesting purposes for WSNs and IoT. Radio frequency is a suitable choice for energy harvesting because WSNs and IoT inherently utilize radio frequency for information transfer. The third section covers recent trends in energy harvesting that have been recently explored such as UAV-enabled networks, simultaneous information and power transfer, efficient computing mechanisms, and utilization of software models. The last section focuses on security aspects for energy harvesting enabled WSN and IoT.

Chapter 1 describes thermal energy harvesting in WSN. The chapter provides an overview of energy harvesting sources, especially thermoelectric harvesters in smart cities' scenarios. The authors focus on electric vehicles, smart grids, medical and environmental aspects of smart cities, and the usage of thermal energy harvesters in these applications. The chapter also investigated the challenges associated with the application of WSNs in smart cities.

Chapter 2 focuses on auxetic designs to improve ambient strain energy harvesting in WSN and IoT. The power generated from vibration energy can be increased by using an auxetic substrate under a piezoelectric layer. The chapter discusses how the auxetic designs impact the energy harvesting rate. The authors have considered the following auxetic designs, i.e., re-entrant honeycomb, rotating triangle, parallelogram, and hexagram. The auxetic energy harvesters open up new avenues for wireless self-powered structural health monitoring.

Chapter 3 reviews the energy harvesting approaches inside the human body and compares the energy harvesting rate of various harvesters. Generally, the implantable energy harvesters utilize internal human body energy for energy harvesting such as expanding and contracting arteries, blood flow and pressure in the veins and arteries, vibrating organs, body heat, and chemical reactions. Accordingly, the chapter categorizes the sources of energy inside the human body and classifies the literature based on various energy harvesting concepts, i.e., electromagnetic, piezoelectric, fuel cell, and electrostatic. The chapter also proposed an energy harvesting mechanism using piezoelectric inside a vein. The blood flow and pressure inside a vein deform the piezoelectric to generate the power.

Chapter 4 uses a statistical approach based on stochastic geometry to model and evaluate the energy harvesting performance of a cognitive device-to-device wireless multimedia sensor network. The chapter considered a multi-tier multi-channel cellular network for energy harvesting. The proposed model shows that nodes can be efficiently powered to enable cognitive device-to-device communications. The authors also discussed future aspects of energy harvesting in cellular networks.

Chapter 5 studies the metamaterial microstrip antenna design for energy harvesting in wireless systems. Specifically, the chapter discusses the design of a low-profile metamaterial (MM) and electromagnetic bandgap (EBG) structures for energy harvesting. The authors considered a Hilbert-shaped MM antenna mounted on a flexible solar panel substrate for wearable IoT applications. Accordingly, they used solar and RF energy harvesting techniques in the proposed design. The key design points considered by the authors include: eliminating the shadowing effects of the antenna on the solar panel and increasing the antenna bandwidth to harvest the RF energy. The chapter also identifies the major advances for each design step to develop cost-effective, efficient, and reliable energy harvesting systems for IoT.

Chapter 6 explores an optimal mode selection strategy for RF-powered IoT devices based on cognitive radios (CR) using the fuzzy logic technique. The authors assumed that it is not possible for the IoT nodes to harvest RF energy obtained from the primary user and simultaneously provide opportunistic spectrum access. Using a fuzzy logic-based decision-making process, the IoT nodes determine whether to access the spectrum or harvest RF energy at each time slot to optimize the expected total throughput.

Chapter 7 discusses the domain of unmanned aerial vehicle (UAV) assisted energy harvesting for WSNs/IoT networks. A UAV network is a promising technology that is already being used for several IoT applications, especially in remote areas which are hard to access. The chapter comprehensively reviews the various energy harvesting schemes in a UAV-assisted WSNs/IoT network.

Chapter 8 reviews resonant beam charging (RBC) with simultaneous data transmission and energy harvesting techniques in WSNs and IoT. The chapter provides a detailed description of the system model, working mechanism, and mathematical analysis of the RBC system. The chapter also proposed an adaptive control system and the scheduling algorithm for multiple users' charging to improve the overall system efficiency.

Chapter 9 surveys the recent developments in simultaneous wireless information and power transfer (SWIPT) in IoT. The chapter describes various SWIPT architectures and enabling technologies in IoT. The chapter also discusses future design challenges of SWIPT that need to be addressed to continuously and reliably deliver the information and energy to IoT devices.

Chapter 10 focuses on energy-efficient computing for future IoT applications. The authors focused on energy-efficient computing at the fog nodes and energy-efficient transmission between IoT devices and fog nodes. The chapter presents techniques and algorithms related to energy-efficient task offloading, device processing parameters control, content caching, and intelligent surfaces-based transmissions.

Chapter 11 uses an intelligent programming model called MapReduce for energy-harvesting by exploiting IoT nodes. MapReduce allows IoT devices to produce data analytics within the device and prepare a complex analytics dataset. The chapter proposes an energy harvesting technique based on the data's production pattern at the nodes.

Chapter 12 discusses the denial-of-service (DoS) attack because of the shared medium, lack of resources, centralized coordination, limited computing, and communication capabilities in WSN and IoT networks. The chapter proposes a solution (called EYES) to mitigate the attack, by forwarding the misbehavior of single and multiple conspiring nodes. In the proposed solution, each node disguises itself as an energy harvesting node and monitors the forwarding of adjacent nodes. The uncertain forwarding operations are validated by periodically collecting the information of overheard packets from adjacent nodes. The authors used the OMNET++ simulator for performance evaluation and comparison with existing schemes. The results obtained show the effectiveness of the proposed scheme in terms of higher detection rate, lower detection latency, and packet delivery ratio.

We would like to thank all the authors for their contributed chapters. We take this opportunity to thank all the reviewers who kindly reviewed all chapters submitted for consideration for inclusion in the book. Based on the reviews obtained, we finalized the list of accepted chapters. We would also like to thank all the staff members at IET for their patience with us over the years as we developed the book. We thank them for their advice and administrative support throughout the preparation of this book. Finally, we hope all readers will enjoy reading this book as much as we did!

Part I

Energy harvesting from ambient environments

Chapter 1

Thermal energy harvesting for wireless sensor networks

Sahar Ahmadzadeh[1]

Abstract

Development of a self-powered electronic module that does not need to replace power supply has been considered in the context of energy harvesting. Energy harvesting, which refers to the process of storing derived energy from ambient or external sources, is widely investigated, particularly for providing a small amount of power for low-energy electronics. The sources might be kinetic, solar power, or thermal, also known as ambient energy. The interest in wireless sensor networks (WSNs) which are basically a network of smart nodes that are connected wirelessly led to rapid progress in this area for various applications, more specifically in industrial and medical sections. With progressing the industrial internet of things (IIoT), the need for energy harvesting in smart cities is increased in recent years. With respect to the importance of energy harvesting in WSNs, this chapter provides an overview of energy harvesting sources, especially thermoelectric harvesters in smart cities with more emphasis on electric vehicles (EVs), smart grids, and medical and environmental aspects. Also, challenges regarding the application of WSNs have been investigated.

1.1 Introduction

With respect to augmented medical needs for aging populations, because of longer life expectations in the industrial world, the demand for integrated medical devices and home healthcare systems has been increased widely. However, the energy consumption and battery lifetime of implantable devices and the cost of additional surgery for battery replacement caused challenges for engineers and scientists [1–3].

Energy which is harvested from Foot impact during the movement or cardiac muscle contraction vibration or thermal energy as the most abundant source of

[1]School of Computing Sciences, University of East Anglia, Norwich, UK

energy in the human body can be used for implantable or wearable devices. The abundance of heat in the human body makes it an excellent choice for thermoelectric generators for energy harvesting in the body.

Also, in the industrial world and smart cities with the implementation of WSNs in smart grids, the importance of energy harvesting for continuous real-time applications is even more highlighted to monitor different parts of the grid consistently. Energy harvesting for electric vehicles (EVs) in smart cities is a challenging issue in recent years. For example, a small squre photovoltaic panel can harvest energy for EVs to come up with reasonable scheduling charge/discharge programs that not only has advantage for consumers but also generating and distribution companies may take advantage from it [4–7]. These reasons motivated scientists to study about gaining energy from ambient resources.

As represented, thermal energy, which is gained from the environment, can be used in smart cities for applications in generation, distribution, and transmission lines in the smart grid. In another application, it can be used to provide the required source of energy for EVs as well as unmanned aerial vehicles (UAVs). In recent years, UAVs have been widely used for several purposes such as military, search and rescue, agriculture, commercial, and maintenance applications. Therefore, providing the required energy for a long time flight without a frequent need to change the battery is a challenging issue that highlights the importance of the application of energy harvesting. Also, it can be considered for medical purposes in wearables and implantable devices that are used in the body to provide required energy without the need for battery replacement. Another application of energy harvesting is for WSNs which are used for environmental monitoring purposes. Environmental applications for WSNs have included agriculture, irrigation, soil monitoring, and so on.

The rest of the chapter is organised as follows: in section 1.2, motivation and an overview on application scenarios of thermal energy harvesting of WSNs are explained. Also, the concept of primary/secondary batteries, as well as supercapacitors and hybrid combination of them, has been investigated. In section 1.3, the classification of energy harvesting sources in WSNs is considered. Section 1.4 reviews the technologies that are used for energy harvesting in smart cities. We divide the section into subsections related to the application of thermal energy harvesting in WSNs which are used in the smart grid. Then, we discuss energy harvesting technologies that are used in EVs/roads/UAVs and then we discuss medical applications of energy harvesting in wearable and implantable medical devices and finally the application of energy harvesting in WSNs used for environmental planning have been investigated. In section 1.4, challenges in WSNs have been discussed, and finally, section 1.5 is the conclusion.

1.2 Thermal energy harvesting in WSNs: motivation and application scenarios

In this section, we will provide motivation, background, and application scenarios of energy harvesting in WSNs with more focus on thermal resources.

1.2.1 Requirement of thermal energy harvesting

To answer the question of why energy harvesting is required, we discuss its importance in WSNs. The importance of energy harvesting in WSNs is because of its wide range of industrial and medical applications. However, the most challenging issue would be the consumed energy by the sensor nodes that must work continuously for months or years. For example, the battery which is used in sensor nodes might be affected because of leakage currents, or even extreme weather conditions. Therefore, harvesting energy from different sources has been a challenging issue for many years and a lot of literature has already paid attention to this challenging area [7]. There are two ways to produce energy in WSNs [8]: (1) energy which is stored in battery, supercapacitors or a combination of both and (2) energy harvesting without storage which will be explained in the following section. According to [9], storage devices can be categorised as follows:

- Primary or secondary batteries;

Based on their application, batteries have different characteristics which include lifetime, size, weight, etc. Batteries' main functionality is not only limited to providing supply for the system but also to store energy that is harvested from the environment. Factors such as storage technology, amount of energy stored in the system, internal resistance, tolerance to overcharging as well as the depth of discharge play an important role to determine battery specifications. Batteries are divided into two categories based on their rechargeability, called primary or secondary, in which primary batteries are non-rechargeable with high capacity and temperature stability. On the other hand, secondary batteries are rechargeable. However, the number of times for charge/recharge is limited. Figure 1.2 shows the difference between primary and secondary batteries.

- Supercapacitors High power density is considered as the main characteristic of supercapacitors in comparison with batteries. They might work based on electrochemical principle or based on generating a charge from a reaction that occurs on an electrode. Supercapacitors have some advantages compared to batteries such as the capability for charge/discharge after many times and in a shorter timing duration. Also, supercapacitors operating temperature is higher than batteries. However, self-discharge is considered as a disadvantage for it. Figure 1.3 illustrates the difference between conventional capacitors and supercapacitors.
- Hybrid combinations of supercapacitors and rechargeable batteries. which are considered to store energy for applications that a battery is used for a certain amount of time.

1.2.2 Application scenarios

As is shown in Figure 1.1, there are different sources of environmental energy including RF, solar, flow, and thermal. These sources are considered to harvest electricity for applications that require longtime available energy without the need to change the battery. In the next section, application of these sources for energy harvesting in WSNs will be explained in detail.

1.3 Energy harvesting techniques in WSNs

In this section, we review energy harvesting techniques in WSNs. As is illustrated in Figure 1.1, the harvested energy in WSNs might be gained directly from the ambient or indirectly from the human body or external resources. In this section, we discuss ambient energy harvesting resources such as RF, flow, solar, and highlight the thermal resources for various applications in smart cities which will be explained further.

1.3.1 Solar-based energy harvesting

Sun is an affordable and clean source for energy harvesting. Although solar energy cannot be controlled, it is predictable based on its daily and seasonal patterns. To transfer solar energy to electricity, photovoltaic cells are used to produce electricity which depends on the intensity of light as well as the cell size and also ambient temperature. In order to increase the output power, multiple solar cells are gathered to form a solar panel.

As is illustrated in Figure 1.2, there are different types of solar panels based on direction, curvature, and shape of the mirrors, which reflect the sunlight with different applications that are explained in detail in [11] based on their commercial applications.

1.3.2 Thermal-based sources

Thermoelectric transducers convert thermal energy to electricity. The principle of working is based on the Sedbeck effect which is a phenomenon in which a temperature difference between two different electrical conductors or semiconductors with p- and n-type blocks, which are ordered in parallel produces a voltage difference depending on temperature and material properties. Figure 1.3 represents the schematic diagram of the Seebeck effect which is explained in [10].

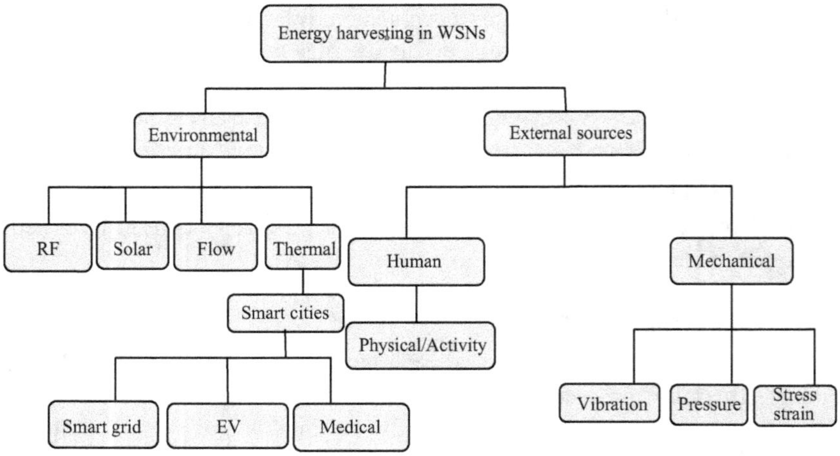

Figure 1.1 Difference between primary and secondary batteries

Primary cell	Secondary cell
Have high energy density and slow in discharge and easy to use	They are smaller energy density
There are no fluids in the cells hence it is also called as dry cells	There are made up of wet cells (flooded and liquid cells) and molten salt (liquid cells with different composition)
It has high internal resistance	It has low internal resistance
It has an irreversible chemical reaction	It has a reversible chemical reaction
Its design is smaller and lighter and the initial cost is cheaper	Its design is more complex and heavier and the initial cost is high

Figure 1.2 Difference between primary and secondary batteries

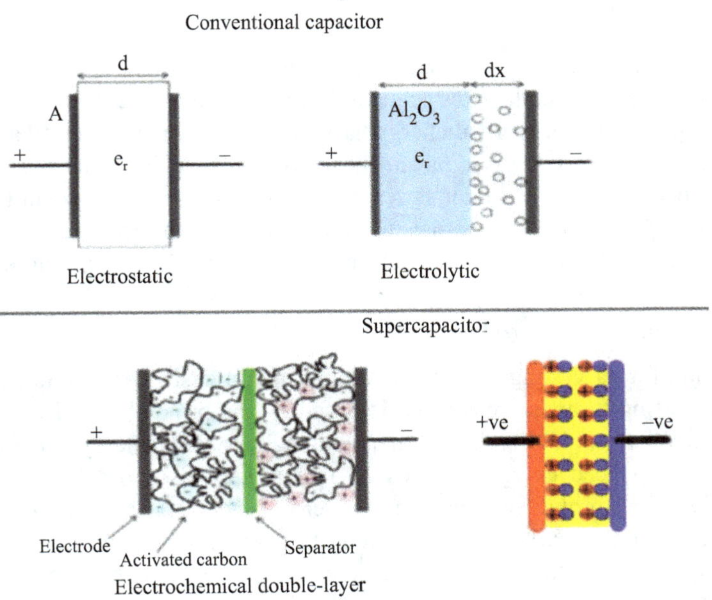

Figure 1.3 Difference between conventional and supercapacitors

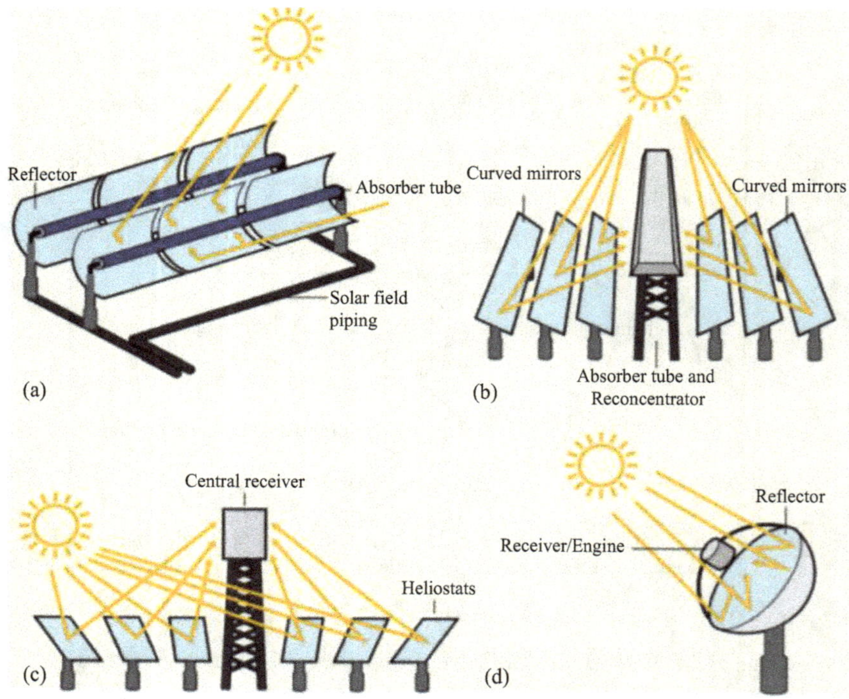

Figure 1.4 Schematic diagram of four basic concentrated solar power configurations as explained in [11] including (a) parabolic trough, (b) linear Fresnel reflector, (c) central receiver/power tower, and (d) dish system

The disadvantage of thermoelectric generators is stopped generating power after some time when both plates achieve the same temperature. However, there are several applications for thermal energy harvesting. For instance, in [10] application of thermoelectric harvesting in pavement roadways is explained in detail.

The thermoelectric generator system for a pipe-pavement is shown in Figure 1.4.

As is represented, the thermal difference between the road pavement as hot surface and river water as the cold surface is considered for generating electricity.

1.3.3 Wind-based harvesting

Using wind for harvesting is basically a kind of conversion of kinetic energy into electricity using turbines, rotors and is based on the principle of electromagnetic induction. Wind turbines are a common approach for wind energy harvesting. The amount of produced energy depends on factors such as chord length and wind speed. For wind turbine systems, the number of blades plays an important role in energy production in a way that according to [9] a six-bladed wind turbine compromises among efficiency, size, and delivered power. Another way to produce electricity using wind for energy harvesting is wind fluter generators that work based on the aeroelastic flutter effect.

Figure 1.5 illustrates a schematic of a wind energy harvester with a stator claimed with six piezoelectric cantilever beams on top and four-blade rotor on the bottom, as an example for the wind-based harvesting system.

1.3.4 RF-based sources

Transmission of radio waves in the range between 3 and 300 kHz and its conversion to direct current is the principle of RF-based energy harvesting. The important parameters for RF-based energy harvesting consist of a source of power, antenna gain, and the distance from the RF source. There are some advantages and disadvantages to RF-based sources. For instance, the availability of RF-based energy harvesting in comparison with thermal, solar, and wind in indoor applications such as smart home, health, and environmental monitoring can be considered as superior of RF sources compared to others.

Figure 1.6 illustrates a wireless charging pad based on inductive coupling which is described in [13].

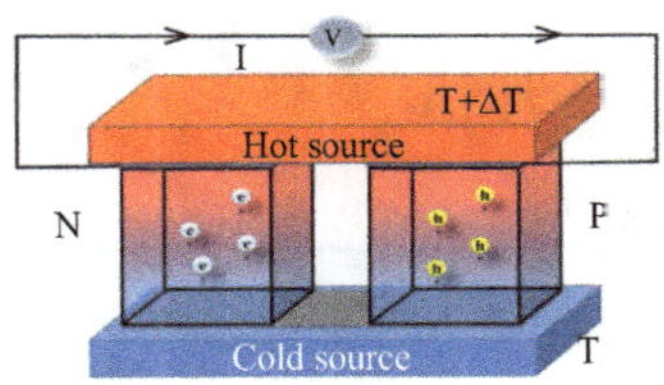

Figure 1.5 Schematic diagram of the Seebeck effect [10]

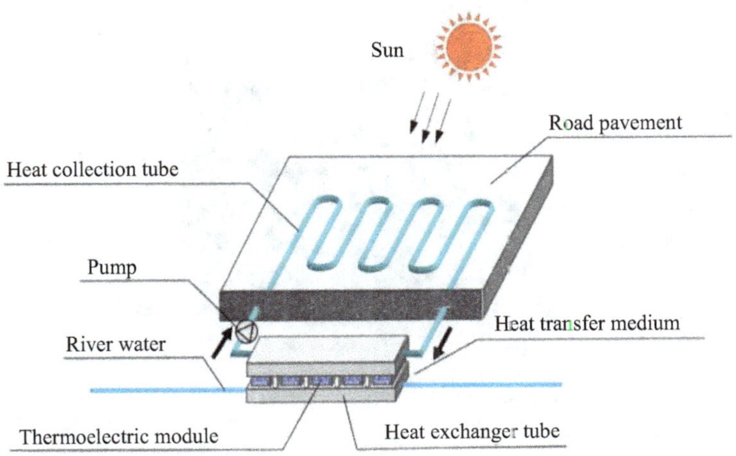

Figure 1.6 Schematic diagram of the pipe-pavement-thermoelectric generator system [10]

Table 1.1 illustrates type, nominal power, and transducer type for various energy harvesting sources. As shown, wind has the highest power density than RF which has the lowest power density compared to other solar and thermal energy, while solar power density is higher than thermal as well.

Figure 1.7 represents the design process for environmental wireless sensor networks. As is shown after considering the energy source based on its application, the amount and time distribution of available energy is estimated. The next step is considering the energy storage method which is autonomous without needing store energy, hybrid which the energy might be stored in supercapacitors, or secondary batteries which are chargeable. Also, the energy for WSN might be provided by the battery supplemented which is a combination of chargeable and nonchargeable batteries or supercapacitors.

Table 1.1 Comparison of energy harvesting sources

Power source	Type	Embedded nominal power	Transducer
Wind	Mechanical	50 W	Wind turbine
Solar	Electromagnetic	15 W	Solar panels
Thermal	Thermal	150 mW	Thermoelectric generator
Ambient RF	Electromagnetic	5 μm	Antenna

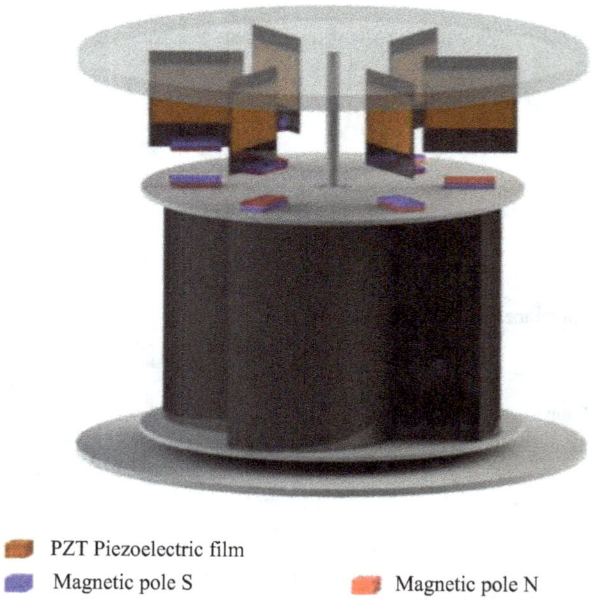

Figure 1.7 Schematic illustration of the proposed vertical axis wind energy harvester in [12]

1.4 Application of thermal energy harvesting in smart cities

Nowadays, with progress in IoT, the concept of smart cities is a hot topic for investigation. A smart city consists of different parts such as a smart environment, smart houses, and smart hospitals. However, the most important characteristics that make a city to be defined as a smart city can be summarized as providing city services with the use of real-time data, analysis of all data, and seamless access to data from various interlinked domains [14].

Figure 1.8 represents different components of smart cities.

As is represented in Figure 1.8, one of the important components of smart cities is the smart grid. WSNs are widely used in smart grids. The smart grids are electrical grid that consists of different parts including smart meters, smart appliances, renewable energy, and energy-efficient resources. According to [15], electronic power conditioning and control of the production and distribution of electricity are important aspects of the smart grid. Smart grid involves cities and offices as is represented in Figure 1.9.

With respect to the concept of smart grid, energy harvesting plays an important role in such systems. In the next section, the application of energy harvesting in smart grids is explained in more detail.

1.4.1 Application of WSNs in smart grids

Smart grid consists of different layers including generation, transmission, distribution, and end-users. The application of renewable energy such as solar and wind with self-production is a hot topic which is taken into consideration in recent

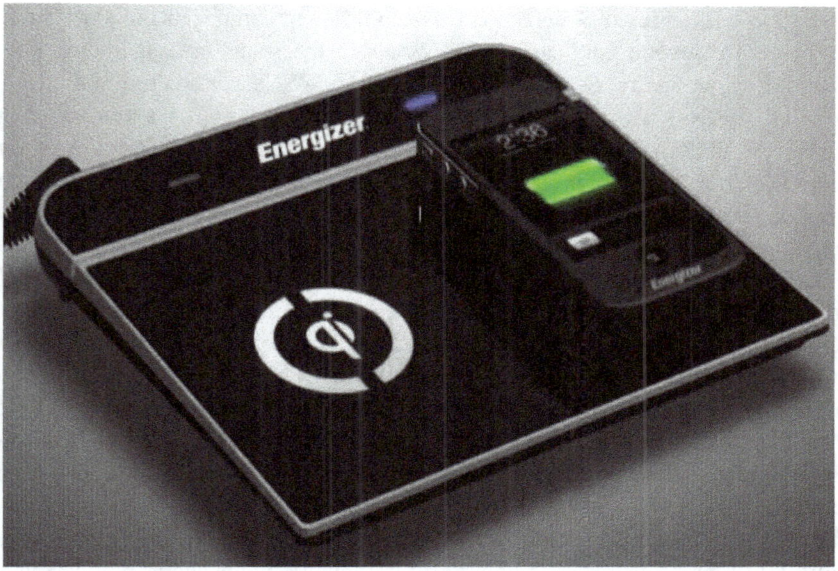

Figure 1.8 Wireless power inductive charging [13]

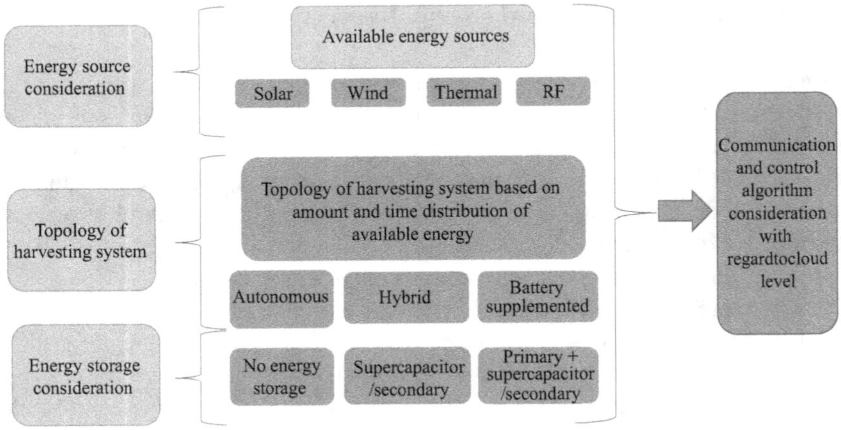

Figure 1.9 Design process of environmental wireless sensor network based on energy harvesting [9]

years. Therefore, energy management based on required demands which is called demand response management in smart grid is discussed widely. In that way, different management methods and algorithms have been introduced [16,17].

In a smart grid, the smart meters are connected to a local area network. There might be WSNs in connection with smart meters used in different parts of the network, including generation companies, transmission lines, service providers, distribution networks, and end-users for providing two-way information between end-users and energy providers.

Figure 1.10 represents interaction between different components of smart grid.

There are several applications for WSNs in the smart grid, including power utilization, power delivery, and power generation. WSNs can be used in home area networks (HANs). A typical application of WSN is wireless automatic meter reading (WAMR), which can determine real-time energy consumption of end-users. The data which is provided by WAMR can be used to optimize the consumption level of end-users compared with energy which is produced by generating companies. The utility companies can identify more valuable consumers. There are some demand response management programs that provide some incentives for end-users to control consumers' demand in peak time periods. Application of energy harvesting technologies such as wind or solar on users' side can decrease the gap between users' actual demand and energy which is produced by bulk generation. WSNs can be used for operations and substation automation. Wireless sensors can be used to monitor the delivery system and being classified based on their locations. In transmission lines, WSNs can be widely used for remote system monitoring, especially where it is difficult to check the lines in person. This is important to protect the system from faults and is essential for grid maintenance. Figure 1.11 represents the NIST smart grid framework which shows how WSNs are connected in the smart grid through the communication path.

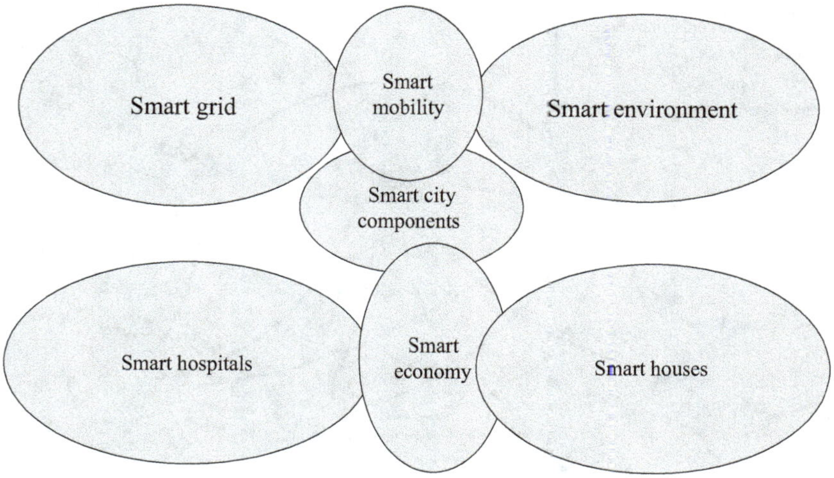

Figure 1.10 Illustration of the components of smart city

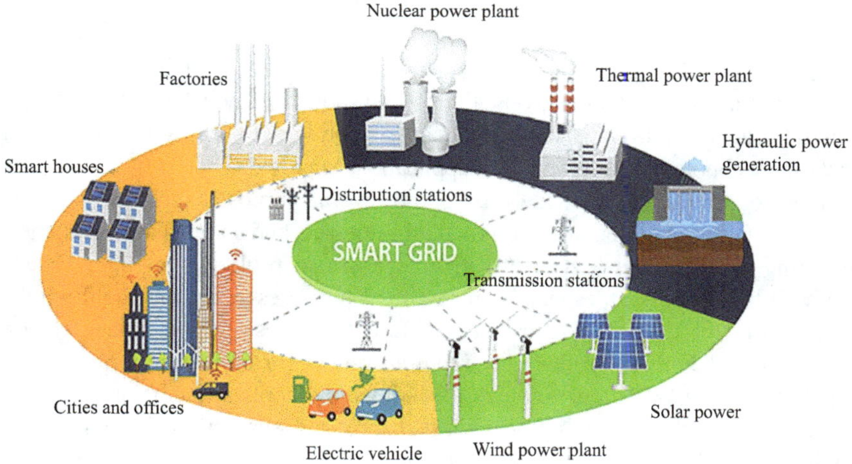

Figure 1.11 Illustration of the components of a smart grid

1.4.2 Energy harvesting in electrical vehicles and roads

In the smart grid, demand response management programs are considered to control the EVs consumption, especially during peak time periods using smart meters. Energy harvesting for EVs is considered to manage the consumers' demand by converting the wasted heat and movement in the vehicles into electricity, as with the turbine in the exhaust. EVs can use piezoelectric effect to generate electricity. The piezoelectricity is defined as the electric charge that accumulates in certain materials in response to applied mechanical stress as a result of surface heat or pressure. This characteristic is used in EVs to generate electricity. In the roads, energy harvesting can be used to produce electricity.

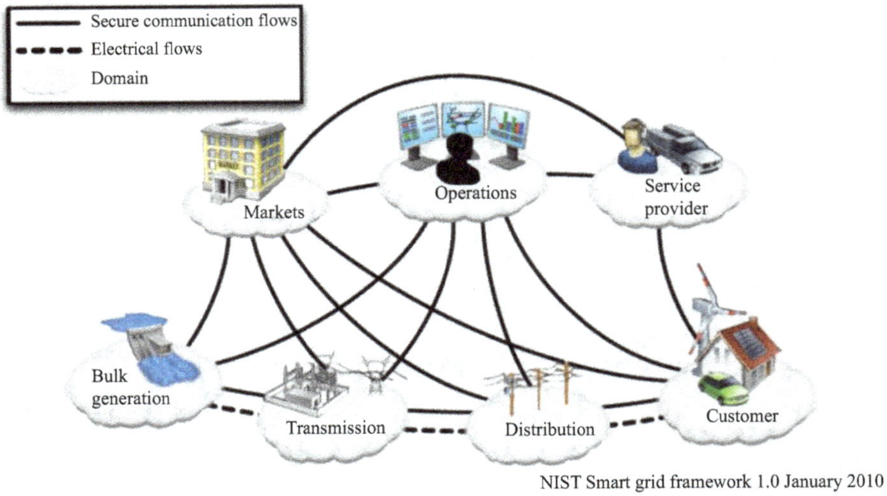

Figure 1.12 Interaction of different smart grid sectors throughout secure communication and electrical follows [18]

Figure 1.12 and 1.13 illustrate smart grid sectors regarding electrical fellows as well as a framework for smart grid respectively. Figure 1.14 shows different technologies that can be used to capture and store energy from roads and bridges. In [22], a review on energy harvesting technologies for biomedical applications is provided.

Nowadays, UAVs are widely used for several purposes such as search and rescue, irrigation, maintenance, commercial purposes, and so on. Therefore, energy harvesting from ambient is an effective measure to prolong the flight time for UAVs. This is important, especially for military purposes to support the flight for a longer period of time. UAVs can use different sources such as solar, RF, or a combination of both to produce the required power for their long-term flight. In [22], a review on WSN technologies for biomedical applications is provided. The authors in [23] propose a hybrid energy harvesting system for UAVs that can simultaneously harvest energy from solar and RF. Figure 1.15 illustrates the block diagram of the proposed energy harvesting system in [23].

1.4.3 Smart health

In this section, importance of WSNs for biomedical applications will be explained briefly and then different ways of energy harvesting for wearables and implantable devices will be studied. Energy harvesting-based body area networks (BANs) and the IoT have been studied in the literature [19,20]. Figure 1.10 illustrates a wireless BAN which is considered for health monitoring.

Harvesting energy from the body during exercise with piezoelectric materials is studied in [20] based on the proposed architecture.

Energy harvesting in IoT, especially in health monitoring sensors which are needed to work for a long time period without the need to change the battery, is a crucial issue that should be taken into consideration. In that way, piezoelectric sensors can be

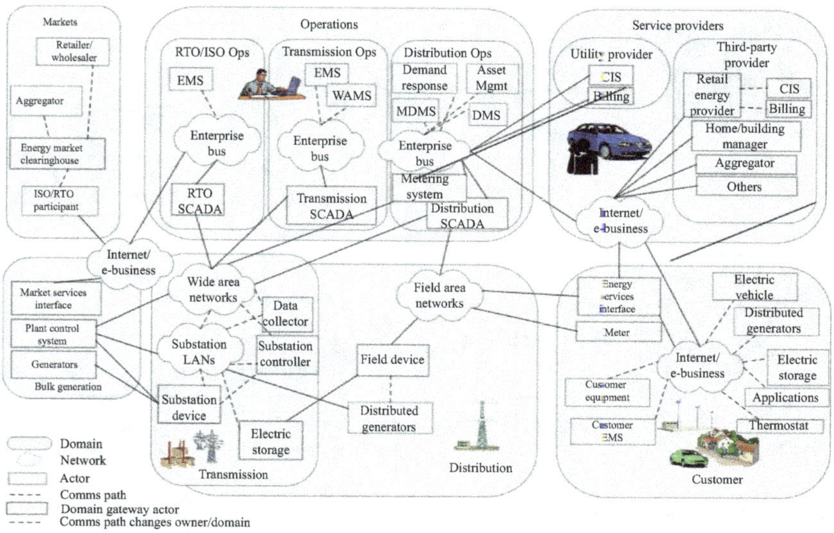

Figure 1.13 NIST smart grid framework [18]

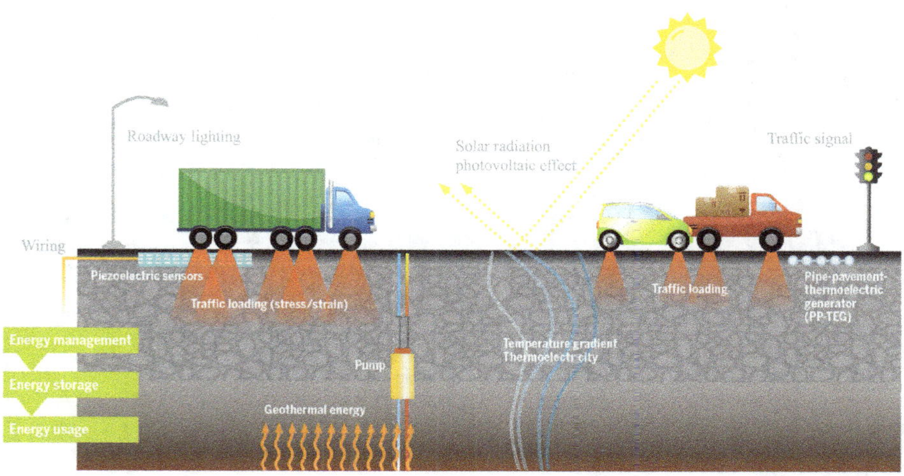

Figure 1.14 Different technologies to capture and store energy in roadways

attached to different parts of the human body which are under certain pressure areas. The pressure area causes piezoelectric characteristics to have appeared and as a result, electric energy being provided to be supplied to wearable health monitoring sensors.

1.4.3.1 Wearable

Nowadays, wearable devices are widely used for self-health management applications. Smart rings, smartwatches, and many other smart wearable devices require energy harvesting to overcome the challenges related to battery lifetime.

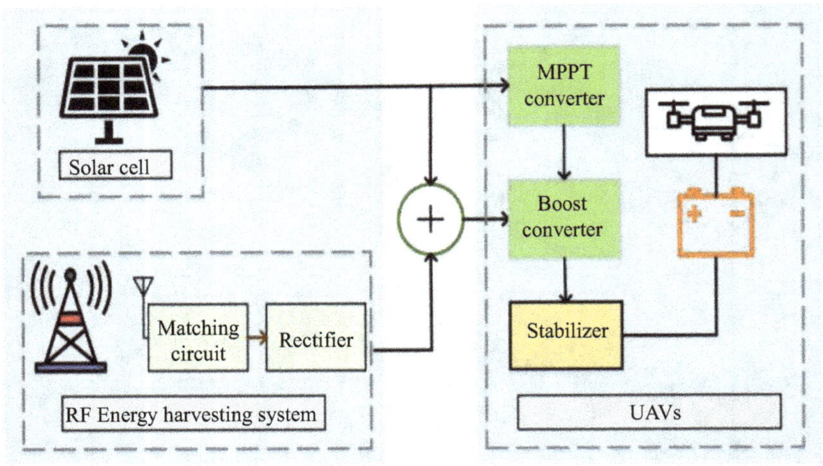

Figure 1.15 A block diagram of hybrid RF-solar system for UAVs proposed in [23]

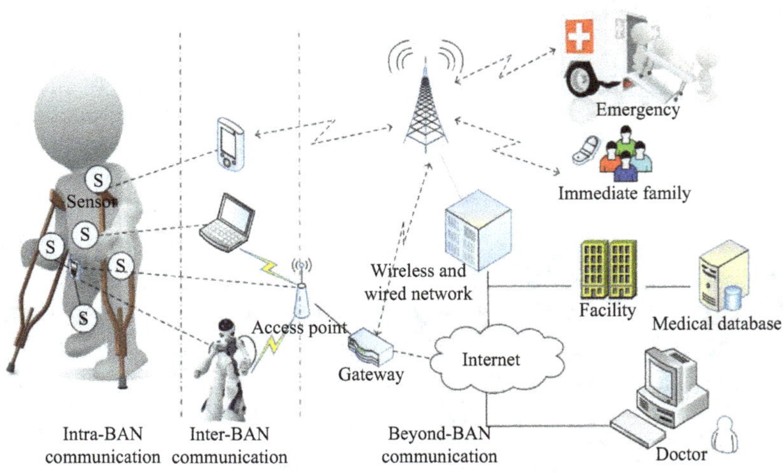

Figure 1.16 A health monitoring system based on wireless BAN

Figures 1.11 and 1.12 show the application of solar and thermal energy harvesting in wearables. Photovoltaic (PV) cells, which are nonlinear semiconductor devices, generate electricity when the chemicals in them are exposed to solar radiation. The performance and efficient output of solar cells depend on the material which is used in them. Three materials that are typically used in PVs consist of amorphous, monocrystalline, and polycrystalline silicon. Wearable devices have their own advantages and disadvantages [21]. The high conversion efficiency and long-term availability of solar energy over an extended duration make it attractive for wearable devices that need continuous monitoring. One

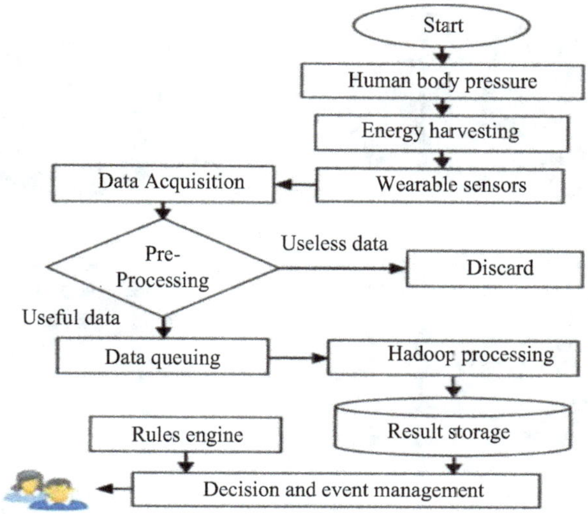

Figure 1.17 A health monitoring system based on wireless BAN

example is Soliband which is represented in Figure 1.11. Soliband is a wearable device that works uninterupted as long as a light source is available. It consists of monocrystalline silicon solar cells, a lithium polymer (LiPo) battery, a supercapacitor, a power management unit, and a photoplethysmogram (PPG) sensor that monitors heart rate and blood oxygenation. The disadvantage of solar wearable devices is that they are uncontrollable and only partially predictable. The magnitude of energy harvested is influenced by weather patterns and the time of the day. Because PV cells are non-linear devices, they have an optimal operating point at which maximum power output is generated [1,21]. Energy can be harvested from thermal sources by exploiting the Seebeck effect to convert the temperature differential between the body skin and the ambient into voltage. As is illustrated in Figure 1.12, a thermoelectric energy generator (TEG) consists of an n-type material placed in series with a p-type, in a way that with the occurrence of temperature variation between materials' heat electrons moves from the hotter part to the cooler part. In that case, formed electric potential causes current to flow in a closed loop. In order to increase the voltage output, multiple p- and n-type semiconductors can be connected to form a thermopile. The reason TEG is suitable for wearable devices is that because the human body produces thermal energy which is dissipated to the environment. Usually, a temperature difference of 5–10 °C between the body and the surrounding environment makes TEG suitable for wearable devices.

1.4.4 WSNs for environmental monitoring

WSNs are widely used for surveillance and monitoring the environment. For instance, Zhang [24] represents an integrated environmental monitoring system for underground coal mines with multi-parameter monitoring capabilities. Rossi *et al.* [25] use hybrid data and decision fusion techniques for environmental monitoring

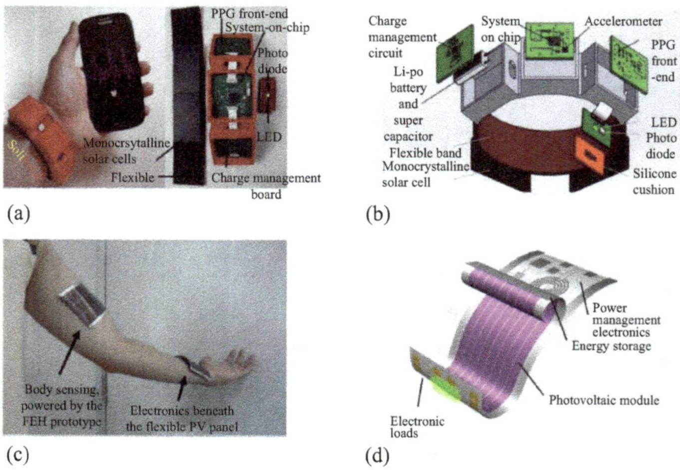

Figure 1.18 Overview of the proposed architecture in [20]

applications using WSNs. Dziadak *et al.* [26] study three groups of WSNs which are used for environmental monitoring as follows: air and atmosphere quality monitoring, water/aquatic environment quality monitoring and cultivated areas, and agriculture monitoring. Xi *et al.* [27] discuss the application of WSNs for soil environmental monitoring purposes. Figure 1.18 illustrates a system architecture for overhead/underground storage reservoirs (OSHR/USHR) based on WSNs, which is represented in [28].

Monitoring and control of water networks require instrumentation, automation of manual valves, and proper communication to cover large distances. The proposed network in [28] consists of remote/actuator nodes, relay nodes, gateway nodes, web servers, and end devices, as shown in Figure 1.18. The remote nodes are installed in storage reservoirs and electrically actuated ON/OFF valves. There is no line of sight between UGSR nodes and gateway. Hence, nearby OHSR nodes function as repeater nodes (relay node). A relay node not only sends its own data but also sends the data of other remote nodes to the gateway. The gateway collects data from all remote nodes to process the data and sends the data to the local/cloud server. It also sends the control signal to the actuator. GSM is used for redundancy. Web and mobile API is developed to monitor the water level and to control the actuator.

As is shown in Figure 1.19, solar energy harvesting can be considered for electricity generation in the proposed system.

1.5 Challenges and issues in WSNs

The challenges related to WSNs can be summarised as follows:

- Energy

  Sensor nodes require energy to operate and process several applications. Energy management is an essential issue in WSNs and therefore energy

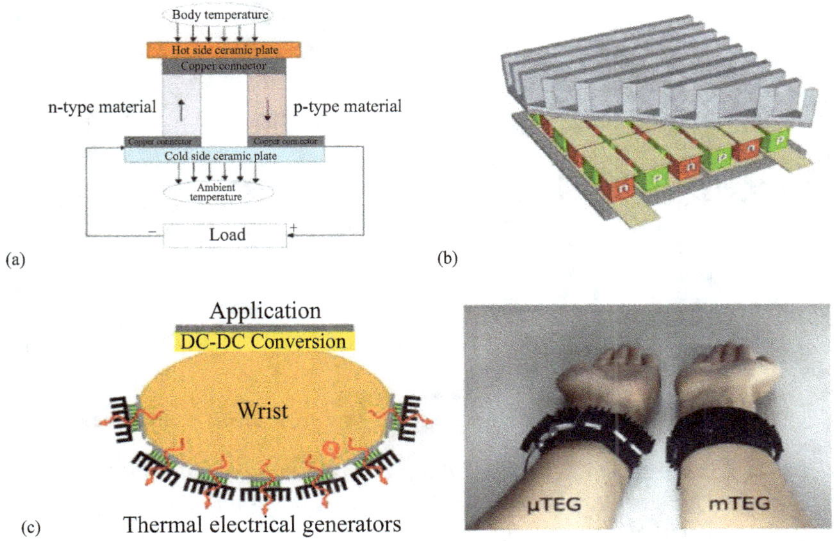

Figure 1.19 Overview of the proposed architecture in [21]

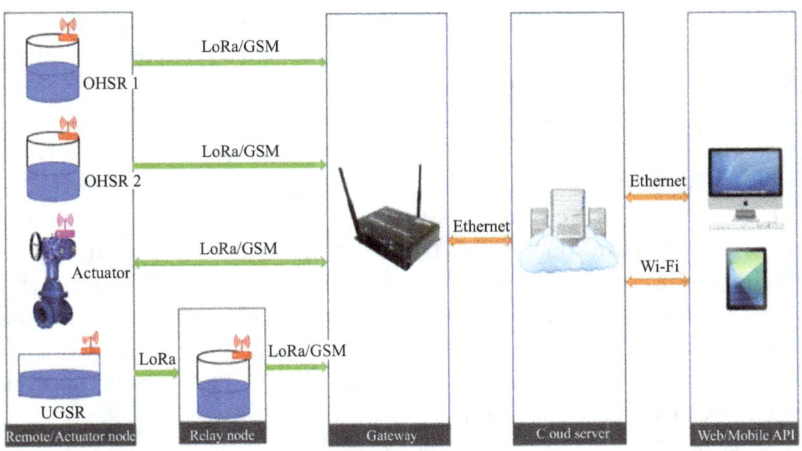

Figure 1.20 System architecture for application of WSNs in environmental monitoring [28]

harvesting is an important issue that should be considered for sensor nodes. The provided energy in WSNs is used for several applications such as sensing, data collection, processing, and data transmission. There are other ways to control and manage the provided energy in sensor nodes such as appropriate routing protocols, or sleep and wakeup microcontroller programming. In [29], transport properties and structural characterization of

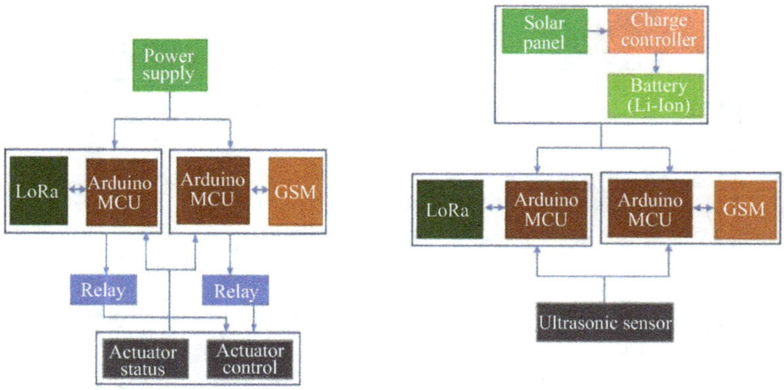

Block diagram of actuator node   Block diagram of remote node

Figure 1.21 Illustration of the block diagrams of actuator and remote nodes which are explained in [30] for monitoring and control of water distribution network

some nanomaterials for fabrication of hybrid supercapacitors are investigated. Desai *et al.* [30] provide a distributed algorithm for node energy management for self-organizing wireless networks. Also authors in [33,34] discuss energy management in WSNs with regard to energy harvesting.

- Communication and routing

  In WSNs, communication with a base station is a challenging issue as nodes need to send packets of data through other nodes which may cause failed packets. Route discovery in WSNs may take a longer period of time due to network traffic or congestion. This is an effective factor in battery energy consumption in WSNs. Therefore, using appropriate protocols for certain applications in WSNs is an important issue that should be taken into consideration. There are different approaches for routing and communication in WSNs. Pasupuleti and Balaswamy [35] discuss an optimized sensing-based data communication approach. There ishuge literature in this context which we briefly refer [36–38].

- Security issues

  In WSNs, security is a challenging issue as data is transferring through the nodes wirelessly. In WSNs, most nodes are operating through the ISB band which is license-free. Security is very important in WSNs for commercial or military applications due to the possibility of malicious attacks. There are several security requirements for WSNs such as data confidentiality to protect the radio spectrum protected to be monitored by everyone, data authenticity to prevent new misleading packets to be injected between nodes by the attacker, and data integrity to protect the signal from fading, reflection signal diffraction, scattering, and various kind of noises. The authors in [39–42] discuss security issues in WSNs.

- Operating System

  For WSNs, a small size and efficient operating system are needed to provide basic resource and memory management. The operating system in WSNs should operate to consume less energy to keep the device working for a longer period of time. The authors in [41–46] discuss the impact of operating systems on WSNs.
- Hardware and software

  With respect to the small size of WSNs, the limited amount of memory, processing speed, and energy are issues that should be taken into consideration. Therefore, by appropriate hardware design, the size of the sensor can be optimized. Also, energy harvesting using ambient resources such as solar can be applied to use the sensor for a longer period of time without the need to change the battery. Also, with programming, it is possible to manage the on/off timing period when the device is active or not to control the power consumption [47]. Galos.
- Localization

  As sensor nodes are deployed randomly in different senarios, it is difficult to manage them in the absence of supporting infrastructure. Wang *et al.* [48] design a localization platform for WSNs.
- Time synchronization

As sensor nodes are operated independently, synchronization of the local clocks is a challenging issue that should be considered. Scheible *et al.* [49] address this issue throughout the design and implementation of real-time sensor interfaces for manufacturing automation purposes.

1.6 Conclusion

With respect to the long-time continuous monitoring requirements for certain medical or industrial applications and the limited capacities of batteries that are used in wireless sensor nodes, investigation about energy harvesting in WSNs is an inevitable issue that is studied from different perspectives. With regard to the importance of thermal resources due to their abundant availability, we reviewed thermal energy harvesting for several applications.

This chapter provides a broad understanding of the concept of energy harvesting, energy harvesting resources, application of energy harvesting in smart cities with more emphasis on smart grids, smart health and EVs, UAVs and roads, and also environmental applications of energy harvesting in WS nodes. Then, the general challenges in WSNs are explained.

Reading the sections reveals a general understanding of the application of energy harvesting for WSNs with more emphasis on thermal energy harvesting. Investigation about other sources of energy harvesting in more detail is suggested for further investigation.

List of acronyms

Acronyms	Definitions
BAN	Body area network
EV	Electric vehicles
HAN	Home area network
IoT	Internet of Things
Lip	Lithium polymer
OSHR	Overhead storage reservoirs
PPG	Photoplethysmogram
PV	Photovoltaic
TEG	Thermoelectric energy generator
UAV	Unmanned aerial vehicles
USHR	Underground storage reservoirs
WAMR	Wireless automatic meter reading
WSN	Wireless sensor networks

References

[1] Y.-W. Chong, W. Ismail, K. Ko, and C.-Y. Lee. “Energy harvesting for wearable devices: A review.” *IEEE Sensors Journal* 19(20) (2019): 9047–9062.

[2] A. Chen. “Thermal Energy Harvesting with Thermoelectrics for Self-powered Sensors: With Applications to Implantable Medical Devices, Body Sensor Networks and Aging in Place.” *UC Berkeley* (2011). Retrieved from https://escholarship.org/uc/item/5qj8d707.

[3] S. Dalola, M. Ferrari, V. Ferrari, M. Guizzetti, D. Marioli, and A. Taroni. “Characterization of thermoelectric modules for powering autonomous sensors.” *IEEE Transactions on Instrumentation and Measurement* 58(1) (2009): 99–107.

[4] L. Hou, S. Tan, Z. Zhang, and N.W. Bergmann. “Thermal energy harvesting WSNs node for temperature monitoring in IIoT.” *IEEE Access* 6 (2018): 35243–35249.

[5] X. Lu and S.-H. Yang. “Thermal energy harvesting for WSNs.” In 2010 IEEE International Conference on Systems, Man and Cybernetics, pp. 3045–3052. IEEE, 2010.

[6] F. Deng, X. Yue, X. Fan, S. Guan, Y. Xu and J. Chen. “Multisource energy harvesting system for a wireless sensor network node in the field environment.” *IEEE Internet of Things Journal* 6(1) (2019): 918–927.

[7] F. K. Shaikh and S. Zeadally. “Energy harvesting in wireless sensor networks: A comprehensive review.” *Renewable and Sustainable Energy Reviews* 55 (2016): 1041–1054.

[8] I. F. Akyildiz, T. Melodia, and K. R. Chowdhury. “A survey on wireless multimedia sensor networks.” *Computer networks* 51(4) (2007): 921–960.

[9] M. Prauzek, J. Konecny, M. Borova, K. Janosova, J. Hlavica, and P. Musilek. “Energy harvesting sources, storage devices and system topologies

for environmental wireless sensor networks: A review." *Sensors* 18(8) (2018): 2446.

[10] U. Datta, S. Dessouky, and A. T. Papagiannakis. "Harvesting thermoelectric energy from asphalt pavements." *Transportation Research Record* 2628(1) (2017): 12–22.

[11] J. Kauffman and K.-M. Lee, eds. *Handbook of Sustainable Engineering*. Dordrecht: Springer Science+Business Media, 2013.

[12] Y. Na, M.-S. Lee, J.W. Lee, and Y.H. Jeong. "Wind energy harvesting from a magnetically coupled piezoelectric bimorph cantilever array based on a dynamic magneto-piezo-elastic structure." *Applied Energy* 264 (2020): 114710.

[13] C. Song. *Broadband Rectifying-Antennas for Ambient RF Energy Harvesting and Wireless Power Transfer*. PhD thesis, University of Liverpool, 2017.

[14] M.S. Obaidat and P. Nicopolitidis, eds. *Smart Cities and Homes: Key Enabling Technologies*. Cambridge, MA: Morgan Kaufmann, 2016.

[15] R.K. Yadav, V. Kumar, H. Shriwastava. "Smart grid: Market opportunities and challenges," *International Journal of Technical Research and Science* IV(III), (2019): 33–36.

[16] S. Ahmadzadeh and K. Yang. "An optimal real-time pricing algorithm under load uncertainty and user number variation in smart grid." *International Journal of Communication Systems* 30(11) (2017): e3246.

[17] A.A.G. Agung and R. Handayani. "Blockchain for Smart Grid." *Journal of King Saud University-Computer and Information Sciences* (2020).

[18] Office of the National Coordinator for Smart Grid Interoperability. NIST Framework and Roadmap for Smart Grid Interoperability Standards, Release 1.0. No. Special Publication (NIST SP)-1108. U.S. Department of Commerce, 2010.

[19] Y. Hao, L. Peng, H. Lu, M.M. Hassan, and A. Alamri. "Energy harvesting based body area networks for smart health." *Sensors* 17(7) (2017): 1602.

[20] M. Babar, A. Rahman, F. Arif, and G. Jeon. "Energy-harvesting based on internet of things and big data analytics for smart health monitoring." *Sustainable Computing: Informatics and Systems* 20 (2018): 155–164.

[21] X. Xu, L. Shu, M. Guizani, M. Liu, and J. Lu. "A survey on energy harvesting and integrated data sharing in wireless body area networks." *International Journal of Distributed Sensor Networks* 11(10) (2015): 438695.

[22] A. Nozariasbmarz, H. Collins, K. Dsouza, *et al.* "Review of wearable thermoelectric energy harvesting: From body temperature to electronic systems." *Applied Energy* 258 (2020): 114069.

[23] C.V. Nguyen, T.V. Quyen, A.M. Le, L.H. Truong, and M.T. Nguyen. "Advanced Hybrid Energy Harvesting Systems for Unmanned Aerial Vehicles (UAVs)." *Advances in Science, Technology and Engineering Systems Journal* 5(1) (2020): 34–39.

[24] Y. Zhang, W. Yang, D. Han, and Y.-I. Kim. "An integrated environment monitoring system for underground coal mines – Wireless sensor network

subsystem with multi-parameter monitoring." *Sensors* 14(7) (2014): 13149–13170.

[25] L. A. Rossi, B. Krishnamachari and C.-C.J. Kuo. "Hybrid data and decision fusion techniques for model-based data gathering in wireless sensor networks [environmental monitoring applications]." In IEEE 60th Vehicular Technology Conference, 2004. VTC2004-Fall. 2004. Vol. 7, pp. 4616–4620. IEEE, 2004.

[26] B. Dziadak, Ł. Makowski, and A. Michalski. "Survey of energy harvesting systems for wireless sensor networks in environmental monitoring." *Metrology and Measurement Systems* 23(4) (2016): 495–512.

[27] X. Xi, X. Xie, and Z. Hai. "Application of theory and technology of wireless sensor network system for soil environmental monitoring." *Sensors & Transducers* 21(5) (2013): 78–84.

[28] S. Chinnusamy, P. Mohandoss, P. Paul, *et al.* "IoT Enabled Monitoring and Control of Water Distribution Network." *1st International WDSA / CCWI 2018 Joint Conference*, Kingston, Ontario, Canada. July 23–25, 2018.

[29] N. K. Singh. *Investigation on Transport Properties and Structural Characterization of Some Nanomaterials for Fabrication of Hybrid Supercapacitors*. Thesis, Chhattisgarh Swami Vivekanand Technical University, 2014.

[30] S.S. Desai and M. J. Nene. "DANES—Distributed algorithm for node energy-management for self-organizing wireless sensor networks." In 2016 IEEE International Conference on Recent Trends in Electronics, Information & Communication Technology (RTEICT), pp. 1296–1301. IEEE, 2016.

[31] O. Kanoun, ed. *Energy Harvesting for Wireless Sensor Networks: Technology, Components and System Design*. Berlin, Boston: De Gruyter Oldenbourg, 2018.

[32] N.G. Pavithra, M. G. Sumithra, and E. Shalini. "Efficient energy management in wireless sensor networks using node rotation." In 2016 Online International Conference on Green Engineering and Technologies (IC-GET), pp. 1–5. IEEE, 2016.

[33] V.R. Pasupuleti, and C. Balaswamy. "Optimised routing and compressive sensing based data communication in wireless sensor network." *IET Communications* 14(6) (2020): 982–993.

[34] M. Jayekumar, and V. Nagarajan. "Route Failure Recovery Algorithm for Dense Wireless Sensor Networks." In 2018 IEEE International Conference on System, Computation, Automation and Networking (ICSCA), pp. 1–6. IEEE, 2018.

[35] V. Mythili, A. Suresh, M.M. Devasagayam, and R. Dhanasekaran. "SEAT-DSR: Spatial and energy aware trusted dynamic distance source routing algorithm for secure data communications in wireless sensor networks." *Cognitive Systems Research* 58 (2019): 143–155.

[36] H. S. Ramos, E. M. R. Oliveira, A. Boukerche, A. C. Frery, and A. A. F. Loureiro. "Characterization and mitigation of the energy hole problem of many-to-one communication in wireless sensor networks." In 2012

International Conference on Computing, Networking and Communications (ICNC), pp. 954–958. IEEE, 2012.

[37] R. Kumar, S. Tripathi, and R. Agrawal. "A review on security in wireless sensor network." In 2020 International Conference on Emerging Smart Computing and Informatics (ESCI), pp. 304–308. IEEE, 2020.

[38] S. A. Jilani, C. Koner, and S. Nandi. "Security in Wireless Sensor Networks: Attacks and Evasion." In 2020 National Conference on Emerging Trends on Sustainable Technology and Engineering Applications (NCETSTEA), pp. 1–5. IEEE, 2020.

[39] A. Tyagi, J. Kushwah, and M. Bhalla. "Threats to security of wireless sensor networks." In 2017 7th International Conference on Cloud Computing, Data Science & Engineering-Confluence, pp. 402–405. IEEE, 2017.

[40] G.S. Oreku and T. Pazynyuk. Security in Wireless Sensor Networks. Cham: Springer, 2016.

[41] A. Kurniawan. *Practical Contiki-NG: Programming for Wireless Sensor Networks*. New York: Apress, 2018.

[42] A. Förster. *Introduction to Wireless Sensor Networks*. Hoboken, NJ: John Wiley & Sons, Inc., 2016.

[43] C. B. Margi, M. A. Simplicio Jr., M. Näslund, *et al*. "Impact of operating systems on wireless sensor networks (security) applications and testbeds." In 2010 Proceedings of 19th International Conference on Computer Communications and Networks, pp. 1–6. IEEE, 2010.

[44] C. Wang and W. Wu. "A light-weighted operating system with deadlock prevention strategy for wireless sensor nodes." In 2009 WRI International Conference on Communications and Mobile Computing, vol. 1, pp. 578–583. IEEE, 2009.

[45] M. Healy, T. Newe, and E. Lewis. "Power management in operating systems for wireless sensor nodes." In 2007 IEEE Sensors Applications Symposium, pp. 1–6. IEEE, 2007.

[46] H. Zhou and K. Hou. "Limos: A lightweight multi-threading operating system dedicated to wireless sensor networks." In 2007 International Conference on Wireless Communications, Networking and Mobile Computing, pp. 3051–3054. IEEE, 2007.

[47] M. Galos, F. Mieyeville, D. Navarro, and I. O'Connor. "Reprogramming hardware-software heterogeneous wireless sensor networks." In 2011 The 14th International Symposium on Wireless Personal Multimedia Communications (WPMC), pp. 1–5. IEEE, 2011.

[48] Y. Wang, F. Yao, S. Chai, Z. Wang, and X. Liu. "Localization platform design of wireless sensor network." In 2020 39th Chinese Control Conference (CCC), pp. 5209–5214. IEEE, 2020.

[49] G. Scheible, D. Dzung, J. Endresen, and J.-E. Frey. "Design and implementation of a truly-wireless real-time sensor/actuator interface for discrete manufacturing automation." In *Embedded Systems Handbook*, pp. 28–1. CRC Press, 2017.

Chapter 2

Auxetic designs to improve ambient strain energy harvesting for WSN/IoT

William J.G. Ferguson[1]

Abstract

Using an auxetic (material possessing an effective negative Poisson's ratio, such that it expands laterally under tension and vice versa) substrate under a piezoelectric layer can greatly increase the power output of a vibration energy harvester. This chapter discusses the testing and optimisation simulations for a selection of auxetic designs: re-entrant honeycomb, and rotating triangle, parallelogram and hexagram. The four best auxetic designs from this process were taken forward to the experimental stage to compare them empirically. All the auxetic samples had outputs more than triple the baseline, with the honeycomb having the highest gain of 5.7 times (570 versus 101 μW from an input excitation of 10 Hz, 100 με). These auxetic energy harvesters open up new avenues for wireless self-powered structural health monitoring sensor nodes in infrastructure, buildings and vehicles, where the ambient vibration energy might otherwise be too diffuse to harvest from.

2.1 Introduction

There are many scenarios in which it may be desirable to acquire sensor data about a local environment, e.g. structural health monitoring of a building or machine [1–4], a pervasive sensor network or wearable medical instruments [5–9], which may not have a suitable power supply. Both wired and battery-powered sensors might be undesirable or impractical in many cases; for example, in the former instance, retrofitting sensors and wires can be difficult and expensive, or a grid supply may be unavailable, while in the latter case, there would be a requirement for regular replacement or perhaps a risk due to chemical reactivity. In many cases, the current generation of energy harvesting devices may be challenged to extract sufficient ambient energy to power most sensors of interest. Motivated by this deficiency, this

[1]College of Engineering, Mathematics and Physical Sciences, University of Exeter, Exeter, UK

chapter presents work designing a new generation of energy harvesting devices, which concentrate the environmental strain energy into a more potent power supply.

The main application of these energy harvesters would be to power systems for remotely monitoring an area for changes over a long timescale [1–4] typified by structural health monitoring. Using sensors to measure environmental conditions can notify maintenance engineers of changes, which could indicate structural damage, along with its location and severity [1–4,10–13]. This has been trailed in buildings and bridges to give advance warning of cracks before they become visible [10] under a train to monitor the condition of the train's undercarriage as well as of the tracks [4,14], detecting leaks in a pipeline [11,12], and in an aircraft to detect growing fatigue in the fuselage, wings, or fuel tanks [13]. Many companies find this kind of asset management or predictive maintenance preferable to annual checks or scheduled replacements [13].

Piezoelectricity allows for the mechanical strain of a host structure to be converted directly into electricity [15–20]. This produces higher voltages, but lower currents, than an equivalent electromagnetic induction harvester [8,21]. It is a property of certain crystal structures and found most strongly in specifically designed ceramic perovskite materials such as PZT (lead zirconate titanate) [17–20]. These materials tend to be very brittle, so piezo-composites were developed [18,19,22,23]. MFC (macro-fibre composite) is one such material, which embeds parallel fibres of PZT in the polymer [15,18,24–26]. This is much more flexible than bulk ceramic while retaining much of its piezoelectric properties [15,18,19,27] and is already used for energy harvesting applications in the aerospace sector [15,18,25–30].

Much of the current work on vibration energy harvesting targets machinery and vehicles, especially aircraft, where this energy is most accessible [13]. Vibrations in the built environment are often generated by wind or other variable loading and are generally stochastic, low frequency, and of small amplitudes [8,10,29,31]. In order to successfully harvest these ambient vibrations, they will need to first be concentrated into the piezoelectric material.

This can be achieved with the use of auxetic materials, i.e. those with an effectively negative Poisson's ratio (meaning that they expand laterally under tension, and vice versa) [32–36]. This can be achieved by the material having a re-entrant or self-infolding internal structure at some scale [36,37], or from internal holes between hinged areas that are able to rotate and thus push outwards [36,38,39]. Combining auxetic and piezoelectric materials could strain the latter much more from the same mechanical input and thereby increase their power output dramatically [40–45].

2.2 Review of related works

The use of auxetic materials in energy harvesting could lead to great improvements in power output over currently available harvesters and potentially open up untapped sources that would otherwise be insufficient to power a sensor. This research is still at an early stage, often based on simulations rather than experiments, and is summarised in Table 2.1. Li *et al.* [40] modelled a bimorph energy harvester with an auxetic rotating squares substrate which they expect could generate 2.76 times the power of an equivalent plain, solid substrate. Their results show dramatic increases of stress in the polyvinylidene difluoride (PVDF) piezoelectric layers over the slits in the auxetic substrate compared to a plain substrate. The regions bonded over the solid squares of the substrate show a minimal stress increase, indicating that there is still much room for improvement on this principle.

Building on this, Ferguson *et al.* [41] designed and built a proof-of-concept auxetic energy harvester, using a single re-entrant hexagon to focus the applied strain into the attached PZT. This resulted in a power output 9.1–14.5 times that of the baseline plain structure under the same excitation (from 14 to 127–202 μW at a 250 με, 10 Hz excitation). This gain factor was constant for a range of input frequencies and strain amplitudes. This improvement was achieved with the aid of a finite-element model to optimise the prospective design for power output, within material stress limits. The modelling process required thin elastic layers (TELs) to account for the imperfect bonding between the epoxy and other components, which served as a consistent fitting parameter for mapping the model onto each sample. This method proved an effective tool, used for the work discussed below.

Farhangdoust [46] took this model and applied it to a cantilever beam harvesting from a bridge cable's oscillations and found a 2.51 times power gain over an equivalent plain harvester. Farhangdoust *et al.* [47] also modified Ferguson's model to have thicker beams with a cut hinge and found a simulated power gain of up to 19.1 times a plain harvester. Eghbali *et al.* [48] made a similar cantilever harvester utilising the widely available circular PZT buzzer and compared two radial auxetic designs to a rectangular plain substrate. Both designs improved the power output (by up to 13.3 times) and lowered the resonant frequency compared to the plain cantilever.

Making the holes a greater fraction of the substrate surface area could make the structure more pliable and increase the power-to-weight ratio. Chandrasekharan and Thompson [49] modelled a cantilever bimorph harvester with a regular honeycomb framework as their substrate, which provides a new means of tuning its resonant frequency. Their honeycomb cell walls were five times thinner than the substrate. Adeshara [42] modelled a similar auxetic honeycomb, but they dismiss it as less effective than the regular honeycomb, because the slightly greater mass required results in a lower power-to-weight ratio (though still an improvement over the plain). With such a weak structure, the auxetic substrate struggles to impose its lateral expansion upon the stiff PZT layers. This lack of authority shows that for auxetic components to improve energy harvesting, there must be a balance between making the structure flexible, and the authority it needs to stretch the piezoelectric material. When Umino *et al.* [43] modelled and tested a bimorph of a design similar to Adeshara's work, but with cell walls being only half the thickness of the substrate, they found that this auxetic design increased their PVDF's power output by 1.6 times and reduced the resonant frequency compared to a plain baseline. Grzybek *et al.* [50] also demonstrate the need for any auxetic component to be stiffer than the piezoelectric material being used. Their simulation of a square

Table 2.1 Summary of prior auxetic energy-harvesting work in chronological order

Author(s)	Device	Strengths	Weaknesses
Adeshara [42]	Re-entrant honeycomb bimorph harvester	Increased power to weight ratio, tuneable resonant frequency	Simulation only, thin cell walls limit output, more material than regular honeycomb
Li, Kuang and Zhu [40]	Rotating squares bimorph strain harvester	2.76 times power gain	Simulation only
Bellis and Bacigalupo [52]	Anti-tetrachiral strain gauge	More than six times sensitivity gain	Simulation only, difficult to manufacture
Umino *et al.* [43]	Re-entrant honeycomb bimorph harvester	1.6 times power output, tuneable resonant frequency	Small scale
Fang *et al.* [44]	Architected cuts in PVDF flag	Improved energy conversion efficiency, tuneable to lower wind speeds	Lower absolute energy output at peak (due to lower input windspeed)
Ferguson *et al.* [41]	Re-entrant strain harvester	14.5 times power gain, matched model to experiment	Small scale, prone to breaking
Farhangdoust [46]	Auxetic cantilever beam	2.51 times power gain	Simulation only, small scale
Zhang *et al.* [51]	Auxetic graphene electrodes	~2.7 times power gain over Ag, facilitates flexible harvesters	Required electrodes four times thicker than its piezoelectric layer
Eghbali *et al.* [48]	Circular auxetic buzzer enhancer.	Up to 13.3 times power gain, utilises existing PZT stock, matched model to experiment	Small scale
Grzybek *et al.* [50]	Auxetic cantilever beam	2.5 times power gain	Simulation only
Farhangdoust *et al.* [47]	Kirigami auxetic structure	Up to 19.1 times power gain	Simulation only, small scale

re-entrant honeycomb region (with a range of the number of cells within that area) within a longer cantilever compared two materials for the substrate: steel and polymer. With a steel substrate, the auxetic region was able to increase the power output in the attached PZT by up to 2.5 times, while with the soft polymer substrate, the gain was negligible.

Fang *et al.* [44] who modelled and tested a flag harvester with an auxetic design cut into the PVDF itself comparing an uncut sheet with unidirectional cuts, a square cut, and a fractal cut, resulting in improved energy conversion efficiency: from around 0.4% at best, up to 0.6%–0.7%. They found all these designs were able to increase the energy conversion efficiency from lower wind speeds than the uncut baseline. They found that these architected cuts could tune the resonant response of PVDF without diminishing its piezoelectric performance. Flexible harvesters such as this might benefit from the auxetic graphene film electrodes of Zhang *et al.* [51] , which increased the open-circuit voltage of their PVDF harvester by up to 1.7 times over an equivalent with silver electrode layers.

Forming a piezoceramic material into an auxetic shape was modelled by De Bellis and Bacigalupo [52], using an anti-tetrachiral lattice of PZT-5A sandwiched by a protective polymer with the same profile. This shape could make a strain gauge more than six times as sensitive, but the complicated shape would be difficult to manufacture from this brittle material, and due to the design's large empty spaces would greatly reduce the power output, meaning to be self-powered this design might have to cover a much larger area than a plain equivalent.

There is more available literature on the subject of piezoelectrically driven auxetic actuators [19,33,34,45,53], some of which could operate in reverse for harvesting. Fey *et al.* [45] made a re-entrant honeycomb structure from PZT, increasing the range of movement for actuation. While it could be used in reverse for harvesting, it would also be highly susceptible to damage under any excessive load it was exposed to.

Improvements in piezoelectric materials include flexible fibre composites [19,53,54]. Topolov and Bowen [53] discussed auxetic polyethylene composites incorporating piezoelectric blocks, serving as more flexible replacements for bulk ceramic piezoelectric materials; these allow higher strain (and lead-free) applications for actuation and harvesting. These piezocomposite devices are already used in some medical ultrasonic imaging sensors and sonar hydrophones, where their auxetic polymer matrix improves the coupling to the piezoelectric components [33,34,53]. If used for harvesting, the polymer might absorb some of the input energy, but these mechanical losses could be offset by being able to harvest from higher strains than a bulk ceramic could endure, resulting in a greater power output overall.

From all this literature, the combination of piezoelectric and auxetic materials has been found to increase the power output for a range of mechanically driven harvesters; much of these efforts thus far have been purely theoretical, however. The aim of the following research is to demonstrate large gains by focusing on optimising the substrate's auxetic design for power output; rather than for minimising weight or lowering resonant frequency, as much of the above work has.

Such designs could make low strain (<200 με) environments more suitable for energy harvesting; generating enough in situ for many structural health monitoring applications. As auxetic materials concentrate ambient stress, brittle piezoelectric materials could be easily damaged in turbulent or already high strain environments (where currently available harvesters would already be viable energy sources) [23]. The small oscillations caused by wind loading on a tower, pedestrians walking over a building's joists, or vehicles passing over the span of a bridge could all become sufficient energy sources for remote sensor nodes [46,55].

2.3 Auxetic region design and testing

To build on that prior research, the model from Ferguson *et al.* [41] was developed. The harvester was made larger, the piezoelectric material was changed from ceramic PZT to the more flexible MFC, and some alternative auxetic designs were trailed in the substrate. The optimised designs were made and tested under laboratory conditions. The aim was to increase their power output, to be a more viable source of energy for structural health monitoring sensors.

2.4 Model commonalities

COMSOL Multiphysics® 5.3b was used to model the prospective auxetic harvester designs, with the optimisation module to find the best parameters for each auxetic region. This model built on the previously validated proof-of-concept model [41], but with the optimisation module that process was able to explore more of the design space more quickly. The larger overall dimensions (given in Table 2.2) allowed a more complex auxetic geometry to be used in the substrate. Figure 2.1 shows an exploded layout of the model.

Table 2.2 Parameter values used in the COMSOL models

Part	Parameter		Value	Unit
Piezo-Layer	Length	L_p	85	mm
	Width	W_p	60	mm
	Thickness	t_p	0.3	mm
	Resistance	R	47.9	kΩ
Glue	Thickness	t_g	8	μm
	Thin elastic layers (TEL)	k_a	16.5	GN/m^3
Substrate	Length	L_s	300	mm
	Width	W_s	60	mm
	Thickness	t_s	0.9*	mm
	Clamping region	L_{clp}	25	mm
		W_{clp}	26	mm
	Auxetic region	L_{AR}	82	mm
	Auxetic buffer	L_{buf}	1.5	mm

*This thickness would later become 1.5 mm, the thinnest ultimately available in the required grade.

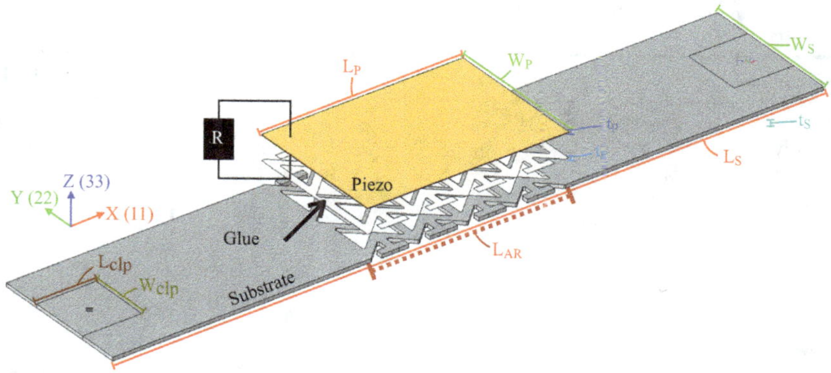

Figure 2.1 Exploded view of model with dimensional parameters; values in Table 2.2

For all the following models, there were many features in common, only the design within the auxetic region of the substrate was different. The model had three layers: an aluminium alloy (6082 T6) substrate, MFC as the piezoelectric layer, and epoxy as the bonding agent between them; all material parameters used in their modelling are listed in Table 2.3. The auxetic region was the central 82 mm (L_{AR}) of the substrate, which left a 1.5-mm buffer (L_{buf}) at either end to give the MFC a firm bond to the solid areas of the substrate.

The MFC's homogenised parameters came from Kuang and Zhu [26], which were selected due to them being derived specifically for energy harvesting purposes, and were experimentally validated using samples and an excitation method quite similar to the plain harvester used in this work. A virtual resistor was connected across the top and bottom surfaces of the MFC layer (emulating its encapsulated electrodes); its load set to maximise power transfer. The power output was calculated from the voltage and current in this resistor.

The glue between the constituents used TELs to model the imperfect adhesion at the interfaces. In the original work [41], these were found to be necessary to correctly emulate the experimental output. Throughout the following design process, the spring constant per unit area, k_a, determines the bonding strength of these TELs; its value was derived from preliminary experimental work on a similar plain sample. The effect of varying this parameter on the final models will be covered in Section 1.1.6.

The excitation throughout was the same 10 Hz, 50 μm extension (used as an upper strain limit of 200 με), analysed in the frequency domain (to save processing time). This excitation was applied using the clamping regions at either end of the substrate (indicated by black outlines in Figure 2.1) to emulate the experimental setup: the faces at one end were fixed, while those at the other end were pulled.

The stress limits were set to 200 MPa in the substrate (below the material's yield of ~250 MPa), and 17.5 MPa in MFC (this was set conservatively low, as the assumptions made to homogenise the composite material fail to account for any localised peak stress in its fibres).

Table 2.3 Material properties used in all models

Material	Property		Value	Unit
Macro-fibre composite (MFC) M8557-P2 [26,56] (Piezo-layer)	Density	ρ_p	5440	kg/m^3
	Compliance Matrix	s^E_{11}	31.89	μm^2/N
		s^E_{22}	50.84	μm^2/N
		s^E_{33}	112.3	μm^2/N
		s^E_{44}	463.0	μm^2/N
		s^E_{55}	413.2	μm^2/N
		s^E_{66}	188.3	μm^2/N
		s^E_{12}	−11.43	μm^2/N
		s^E_{13}	−13.45	μm^2/N
		s^E_{23}	−19.12	μm^2/N
	Coupling matrix	d_{31}	−283.4	pC/N
		d_{32}	−223.2	pC/N
		d_{33}	513.4	pC/N
		d_{15}	560	pC/N
	Relative permittivity	ε_{11}	1950	–
		ε_{33}	2544	–
Epoxy 3M DP 460 (Glue)	Density	ρ_g	1250	kg/m^3
	Poisson's ratio	ν_g	0.35	–
	Elastic modulus	E_g	1.00	GPa
Aluminium: 6082 T6 (substrate)	Density	ρ_s	2700	kg/m^3
	Poisson's ratio	ν_s	0.33	–
	Elastic modulus	E_s	70	GPa

Each auxetic design was built around parameters which could be varied. Using a combined parametric sweep, a general understanding of the effect each parameter had on the power output and peak stresses in each layer could be reached. The best results (i.e. highest power without exceeding stress limits) from this sweep became starting points for the optimiser.

The optimisation process for all designs used the Nelder–Mead [57] method, as it can have arbitrary, non-differentiable, constraints; such as a maximum stress limits in the aluminium or MFC. This takes the initial parameters and calculates its objective function (in this case, power) as a starting point, then walks around the solution space recalculating this function at various combinations of input parameters, attempting to find a maximum output. At each point, any breached constraint penalises its outcome to steer the solution away. All parameters and limits were scaled for the optimiser to prevent undue weighting. The disadvantage of any non-differentiable method is that it is impossible to know whether it has found the global optimum, or merely a local peak [57]. Repeating the optimisation from different starting points until a similar design is reached multiple times was a good indication of having found the best parameters for that auxetic region.

The baseline for all models was a plain harvester with only solid substrate under the MFC: its simulated output under the described excitation was 162 μW.

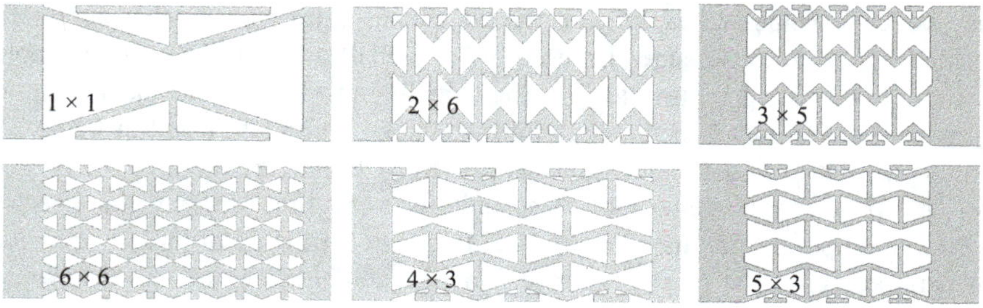

Figure 2.2 Some examples of M × N *re-entrant honeycombs*

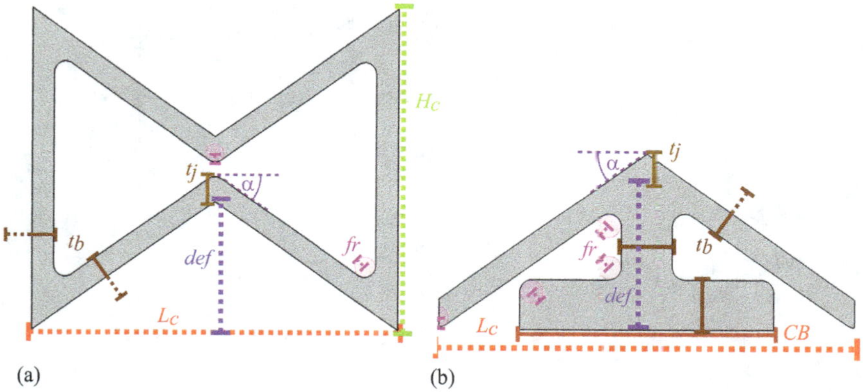

Figure 2.3 Parts for re-entrant honeycomb array: (a) hexagonal unit and (b) edge unit. Parameters are defined in Tables 2.4 and 2.5. An example full array is shown in Figure 2.4

2.5 Re-entrant honeycomb array

The re-entrant honeycomb is a widely discussed auxetic design [36,50,58]. Building on the prior proof-of-concept work [41], a single re-entrant hexagon unit can be repeated to form a honeycomb array fitted to the auxetic region; such as the examples in Figure 2.2.

An annotated re-entrant unit is shown in Figure 2.3(a). Partial units along the substrate's edges have a crossbeam to aid in lateral strain transfer, shown in Figure 2.3(b). An integer number (1 to 7) of these units are packed into the auxetic region in both axes: *M* full units across the substrate (the *y*-direction in Figure 2.1), and *N* full units along the substrate (its *x*-direction)*. Note the difference between

*It would be possible to build a re-entrant honeycomb framework that is not restricted to integer values (easiest to implement by subtracting a regular array of bowtie shapes) but this makes the crossbeams very difficult to implement and could lead to many ineffectual truncated beams at the edges.

odd and even values of M: if odd, the edge units mirror each other across the substrate; if even, these edges have an offset of half a unit length; the examples shown in Figure 2.2 demonstrate this. The units fitted together in the manner shown in Figure 2.4. Geometric parameters for the auxetic region are listed in Table 2.4, and its parametric sweep input values are listed in Table 2.5.

The $M \times N$ arrangement of these units was the first aspect to be selected. COMSOL's optimisation module struggles to handle a mixture of integer and floating parameters, so the parametric sweep of the combinations of values given in Table 2.5 was the first step. This encompassed all 49 available $M \times N$ arrays; each of which were built with every combination of the listed beam widths, t_b; deformation

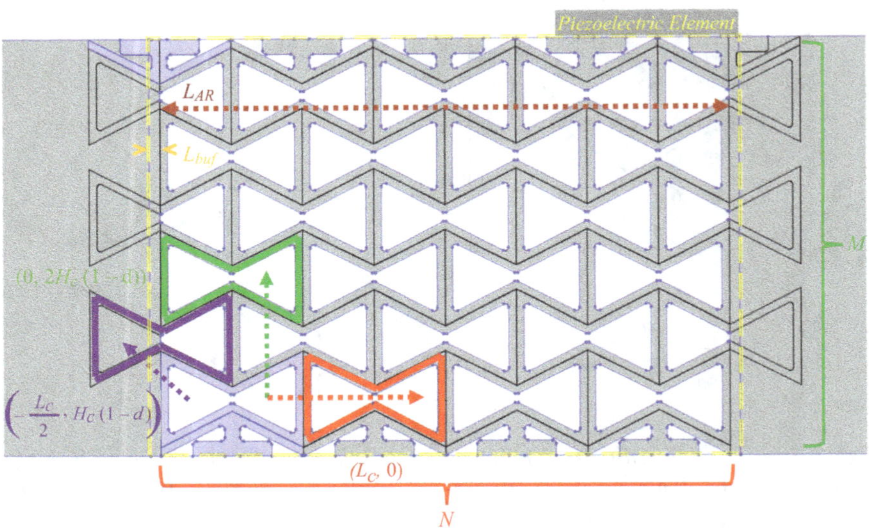

Figure 2.4 Example re-entrant honeycomb array before parts are united. Units placed manually are highlighted in blue with colour-coded duplication

Table 2.4 Auxetic region definitions with parametric variables in bold, see Table 2.5

Parameter		Definition	Unit
Cell length	L_c	$L_{AR}/\boldsymbol{N}$	mm
Cell angle	α	$tan^{-1}\frac{2\boldsymbol{d}(W_s-\boldsymbol{t_b})}{L_c(\boldsymbol{M}-\boldsymbol{d}(\boldsymbol{M}-1))}$	rad
t_b adjustment	t_j	$\boldsymbol{t_b}/(2cos\,\alpha)$	mm
Cell height	H_c	$\frac{W_s-2tj}{\boldsymbol{M}-\boldsymbol{d}(\boldsymbol{M}-1)}$	mm
Deformation	def	$\boldsymbol{d} \times H_c$	mm
Crossbeam length	CB	$\boldsymbol{t_b}+\boldsymbol{cb} \quad L_c(1-\boldsymbol{t_b}/2def)$	mm
Fillet radius	fr	0.5	mm

Table 2.5 Re-entrant honeycomb inputs for parametric sweep and optimisations

Parameter		**Swept values**	**Opt. range**	**Units**
No. of cells in W_s	M	1, 2, 3, 4, 5, 6, 7	Fixed	–
No. of cells in L_{AR}	N	1, 2, 3, 4, 5, 6,7	Fixed	–
Beam Width	t_b	0.7, 1, 2, 3	1–4.5	mm
Deformation	d	25, 33, 40	25–42	%
Crossbeam	cb	0, 50, 75	40–88	%

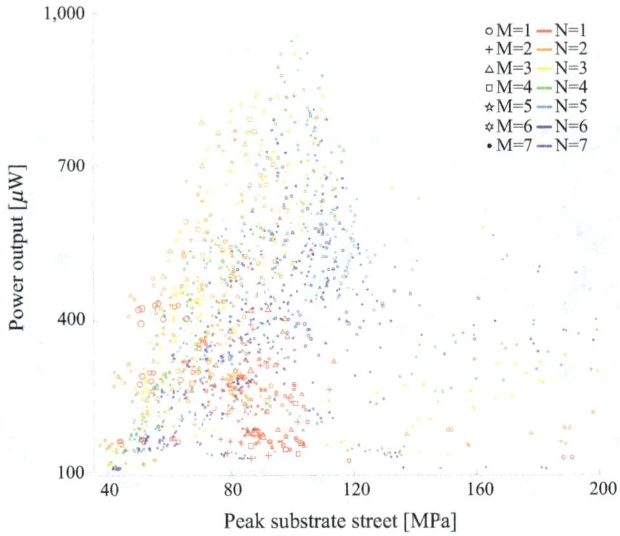

Figure 2.5 Distribution of re-entrant honeycomb parametric results, with overstressed and unviable models filtered out. Marker size indicates peak MFC stress

fractions, d, and crossbeam ratios, cb. This produces 1764 variations of the auxetic honeycomb. Some produce geometric errors (e.g. beam intersections), most of which were skipped using a parameter check (calculating the size of the gaps before building) to save processing time, but most were viable options. From the parametric sweep results, shown in Figure 2.5, the viable arrays found to produce the most power, within stress limits, were the starting points for the optimisation.

Some general tendencies can be extracted from these results. The chaotic appearance of Figure 2.5 demonstrates that the power output is too unpredictable to be calculated by analytic equations and that the finite-element models are necessary. All the best designs had M in the range of 4–7, while N was 3–5, indicating that short wide units were most effective at straining the MFC. Designs with fewer re-entrant units tended to fare poorly against designs with more, but thicker beams (with more authority) could overcome this disadvantage in some cases. All the best results for a given $M \times N$ array came from the most re-entrant available form with the longest crossbeam that would fit. The highest output was found from this sweep was 954 μW, from the 4 × 3 array where $t_b = 3$ mm, $d = 40\%$ and $cb = 75\%$. This design, and the six next highest $M \times N$ arrays, were the starting points for the optimisations.

The best optimised design was derived from this 4 × 3 array, and the stress distribution in the substrate and MFC are depicted in Figures 2.6 and 2.7, respectively. Its improved output of 986 μW is a gain of 6.08 times the plain baseline and is a small improvement over its parametric starting point. The optimised parameters were $t_b = 3.118$ mm, $d = 40.98\%$ and $cb = 81.69\%$. The increased output came by pushing the re-entrancy and crossbeam to their respective limits, with the beams almost touching; the tips are ~0.4 mm apart. Some of the other optimised designs improved their outputs by a greater margin through this process but still not enough

Figure 2.6 Selected optimal 4 × 3 re-entrant honeycomb design, shown as stress distribution in the auxetic region of the substrate

Figure 2.7 Selected optimal 4 × 3 re-entrant honeycomb design, shown as stress distribution in the MFC

to catch up on this selected design. In all cases, the re-entrancy and crossbeam length increased, while the beams occasionally became narrower to accommodate.

2.6 For rotating parallelograms

Rotating squares are a widely researched auxetic structure [35,36,38,39,59–63]. This design can easily be made by cutting elliptical slits in a sheet, leaving squares connected by hinges at their corners, as shown in Figure 2.8(a). These solid regions will preferentially rotate under tension; this pushes their corners outward and opens the slits up, leading to their auxetic behaviour [38,39,61]. By altering the relative lengths and angles between slits, arrays of rectangles, rhombuses or rhomboids can be created; these shapes are collectively called parallelograms and are all exemplified in Figure 2.8. As these designs share a similar construction, all of them could be processed in the same COMSOL model, and without any necessary integer parameters, the optimisation was also less constrained.

This rotating parallelogram model has six controllable parameters, depicted in Figure 2.9: h_A, the length of the baseline slits (the semimajor axis length of the baseline elliptical holes); h_b, the width of the slits (the semiminor axis length of all ellipses); r, the ratio of crossing slit length to the base slit length (for squares and rhombuses this is 1; less otherwise); R, the ratio of base slit length to grid spacing length (i.e. the distance between slit centres); θ, the angle between crossing gridlines (for squares and rectangles this is 90°; less otherwise); and φ, the angle between the entire grid (along the baseline slits) and the x-axis of the substrate (note, $\varphi = 0°$ for all examples in Figure 2.8). Values in the parametric sweep for all of these are listed in Table 2.6. This design was built outwards from the base slit centred in the middle of the auxetic region, which cut down the number of models through symmetry.

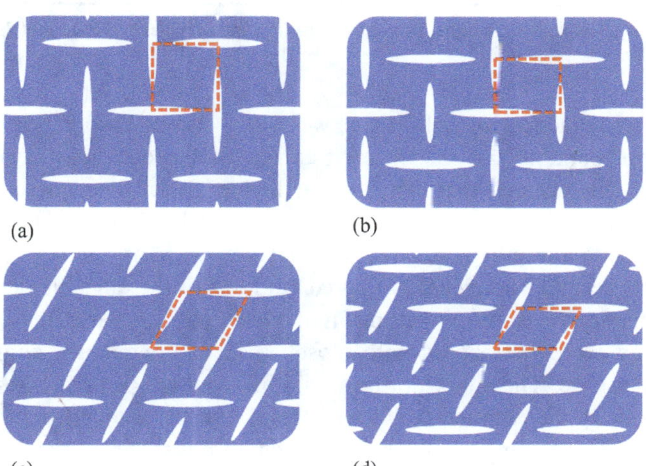

Figure 2.8 Diagrammatic rotating (a) square, (b) rectangle, (c) rhombus and (d) rhomboid arrangements

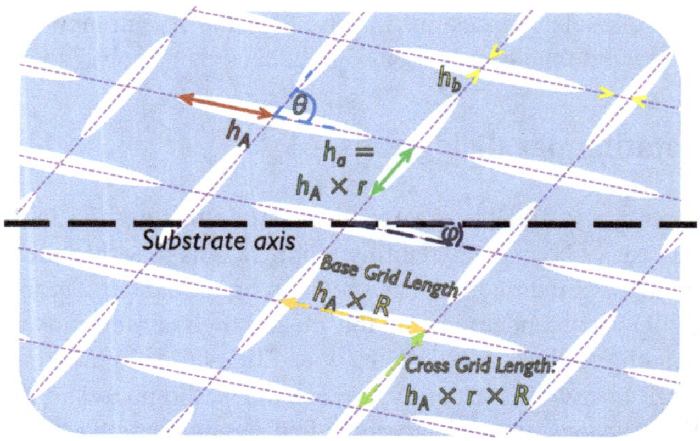

Figure 2.9 *Diagram showing parallelogram model parameters: h_A, base semimajor axis; r, crossing axis ratio; h_b, all semiminor axes; R grid spacing ratio; θ, angle between grids; φ, angle between grid and substrate*

Table 2.6 *Parameters swept in rotating parallelogram model*

Parameter		**Parametric values**	**Opt. range**	**Units**
Slit length	h_A	10, 7.5, 5	1.5–15	mm
Slit width	h_b	1, 0.5	0.5–1.5	mm
Slit ratio	r	100, 75, 50	45–100	%
Grid ratio	R	300, 200, 175	155–250	%
Slit angle	θ	90, 60, 45	35–90	°
Grid angle	φ	0, 40, 60,100, diag*	–45–100	°

*diag is an angle calculated to orient the parallelogram's long axis parallel to the substrate's[2].

To filter out impractical models, there were two checks before processing each design: before building the model, an estimate of the distance between nearest neighbouring slits ensured that there was enough material there to sustain a load; after building the geometry[†], there was a domain check to ensure the substrate had not been split into multiple pieces. If one failed, the sweep or optimiser skipped it to build the next design. These checks filtered out most of the unviable designs.

The parametric results, shown in Figure 2.10, indicated that the best designs for harvesting were rhomboids with closely packed long holes aligned almost across the substrate. The slit width primarily affected the peak stresses (by determining how sharp the elliptical holes were), and how close together the slits could be (i.e. how stiff the 'hinges' between the rotating units were). The grid angle, φ, could prime the rotating units to rotate outwards more easily. Rhomboids had the

[†]$diag = -\arctan(r\sin\theta/(1 + r\cos\theta))$

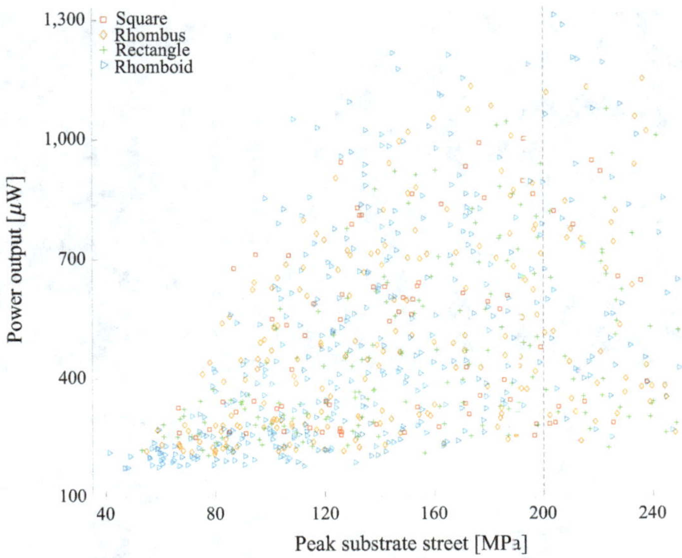

Figure 2.10 Parametric rotating parallelogram power vs. peak stress results, differentiated by rotating unit. Safety limit indicated by the dashed line

advantage of a greater lateral extension, due to their elongated sides giving greater leverage than a square, and pointed corners granting more outward sweeping expansion than a rectangle; increasing the power output from the attached MFC. During preliminary investigations (often without any constraining piezoelectric layer) having the long axis of the rotating region run parallel to the applied force tended to induce the largest auxetic response; hence the *diag* angle included in Table 2.6. This was no longer the case with MFC in place, as no models at this angle feature in the top results. Most of the highest power results were slightly over the substrate stress safety limit, but within the material's actual limit; as this could be amended through the optimiser they were also used as seeds for the optimiser. The best design found in this parametric sweep was among these, with an output of 1313 μW, where h_A =10 mm, $h_b = 0.5$ mm, $R = 175\%$, $r = 75\%$, $\theta = 45°$, and $\varphi = 100°$.

The best results from the parametric sweep were all similar in appearance so, while they formed the basis for most optimisations, some runs started from very different initial conditions to ensure a broader foundation. The optimised results rarely ventured far from their starting points, unless that had a very low output. The best optimised result had an output of 1488 μW, where $h_A = 10.54$ mm, $h_b = 0.518$ mm, $R = 153.6\%$, $r = 61.61\%$, $\theta = 62.45°$ and $\varphi = 76.86°$ and is shown in Figure 2.11. This features long slits cutting across the substrate, like many of the best parametric results. Its power output is 9.19 times greater than the plain baseline, much more than the prior honeycomb but with almost double the peak stress (197 MPa) to contend with. Where the maximum strain the harvester is likely to be subjected to in use is known well in advance, this could be a good trade-off.

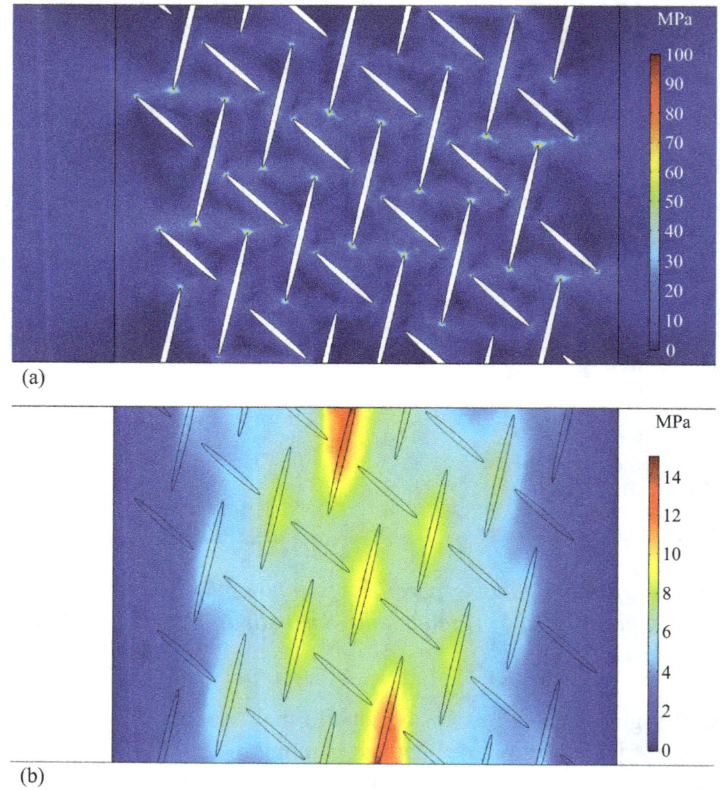

Figure 2.11 Selected rotating rhomboid auxetic region design, with slice through the (a) substrate and (b) MFC to show stress concentrations

2.7 Rotating hexagrams

The previous parallelogram designs were built on two crossing grids of parallel slits. An arrangement of three such intersecting grids results in regular triangles and hexagons, arranged into hexagrams (also known as the kagome-cut) [35,62], as depicted in Figure 2.12. It responds similar to the rotating parallelograms under tension; every unit rotating counter to its neighbours, leading to the auxetic behaviour.

This design has four controllable parameters: h_A, the semimajor length of the slits; h_b, the semiminor width of the slits; R, the grid length ratio (controlling the slit spacing); and φ, the angle between the grid and the substrate. These are all depicted in Figure 2.13, with parametric sweep values listed in Table 7. The grid angle is fixed at 120° to keep the tessellation regular. This design was built outwards from a

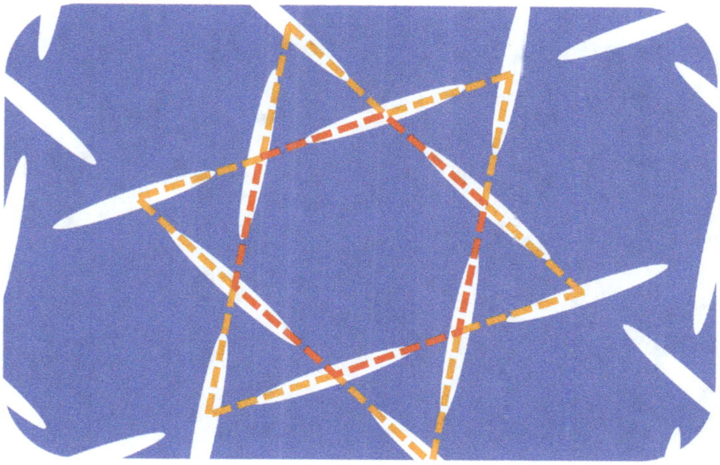

Figure 2.12 Rotating hexagram array example, with a hexagon unit highlighted in red and its associated triangle units in orange

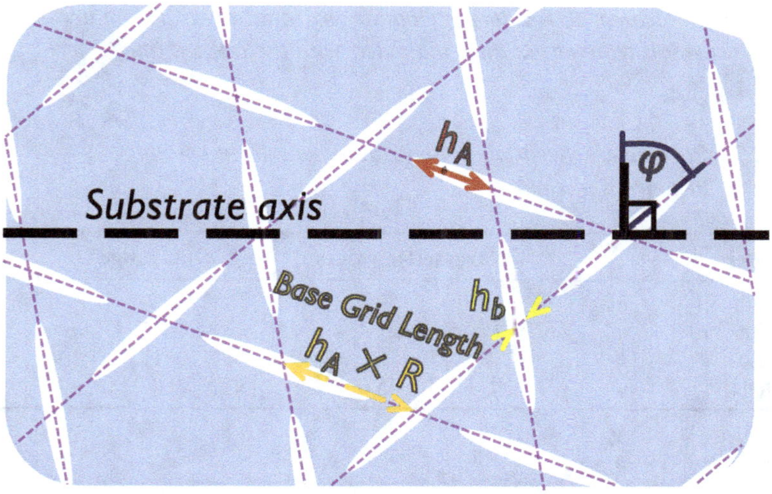

Figure 2.13 Rotating hexagram array parameters: h_A, ellipse semimajor length; h_b, semiminor length; R, grid spacing ratio; φ, angle between grid and substrate

central cluster of six slits arranged around a hexagon unit. The parametric sweep results are shown in Figure 2.14, where the highest output model produced 2139 μW but had a peak stress of 330 MPa ($h_A = 15$ mm, $h_b = 0.5$ mm, $R = 125\%$ and $\varphi = 45°$).

All the highest power models from the parametric sweep had long holes, with the tightest spacing available. They were also significant over the peak stress limit,

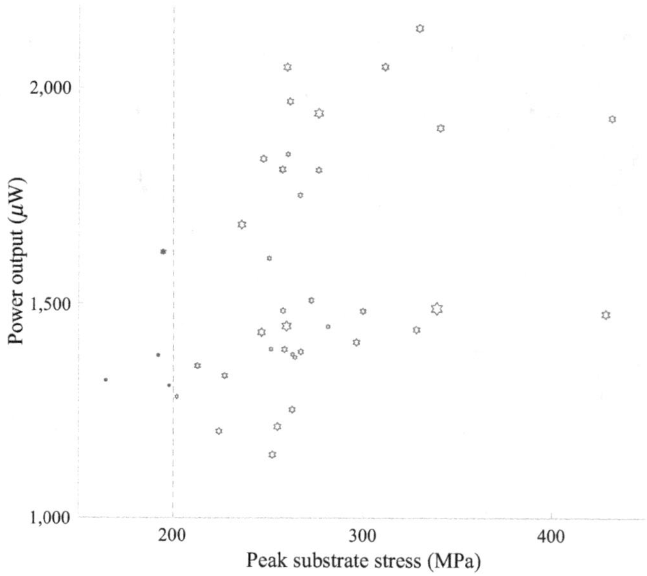

Figure 2.14 Parametric results distribution for rotating hexagram arrangement. Substrate stress limit indicated with the dashed line

Table 2.7 Parameters swept and optimisation ranges in rotating hexagram models

Parameter		Parametric values	Opt. range	Units
Slit length	h_A	15, 10, 5	3.5–17	mm
Slit width	h_b	1, 0.5	0.3–1.5	mm
Grid ratio	R	125, 150	110–175	%
Grid angle	φ	0, 15, 30, 45	−20–50	°

as were almost all the parametric results. The hinge regions were under a great deal of tensile and shear stress when acted upon. The width of the slits can influence the flexibility of these hinges, while also directing more stress there. The angle to the substrate, φ, can prime the units to rotate.

As before, the best parametric results were taken as starting points for the optimiser. For this model, the optimum power output had to go down compared to their starting point, in order to reach a design with a tolerable peak stress. The final design, shown in Figure 2.15, had a peak substrate stress of 189 MPa and an output of 1733 μW, a gain of 10.7 times the plain baseline ($h_A = 8.87$ mm, $h_b = 0.85$ mm, $R = 127.5\%$ and $\varphi = 41.86°$). This design spreads the stress out over more small rotating regions than the top parametric designs.

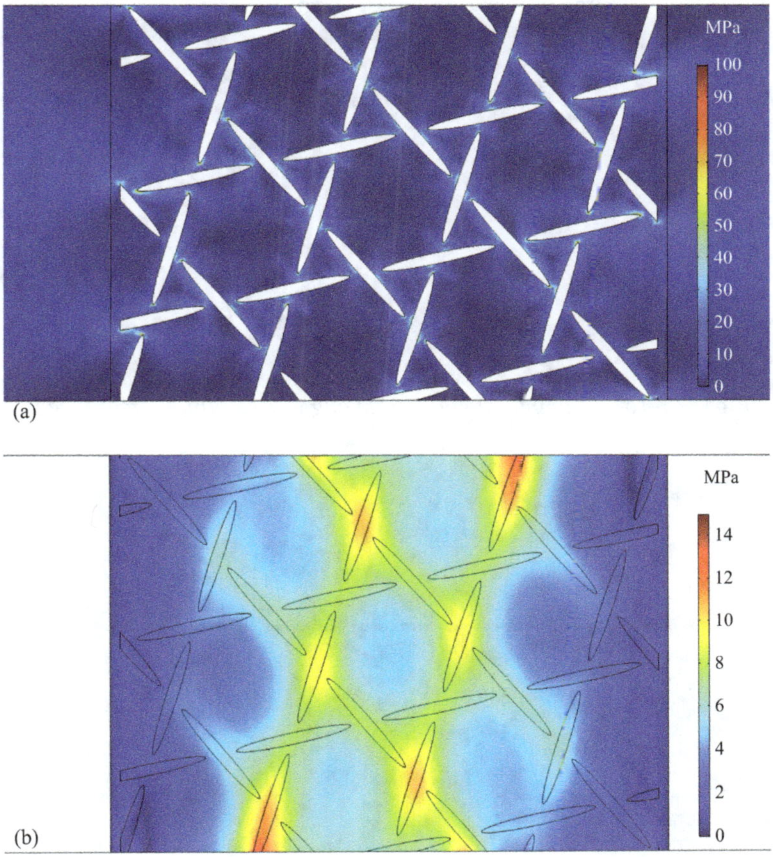

Figure 2.15 Selected optimal rotating hexagram arrangement, shown with stress map of the auxetic region of the (a) substrate and (b) MFC

The rotating hexagram design would be ideal for increasing the energy output of a harvester from a very low strain amplitude environment ($\ll$200 με), but only if these excitations could be well known in advance. If an amplitude much less than this were the expected maximum, this design could have delivered the highest gain of any discussed in this chapter.

2.8 Rotating triangles

A three-pointed slit (made from hemi-ellipses regularly oriented at 120°) can make the equilateral rotating triangle design[‡], depicted in Figure 2.16. Under tension, the

[‡]Irregular triangles are possible, even though tessellation becomes difficult to build in COMSOL.

Figure 2.16 Example of rotating triangle array (indicated in red)

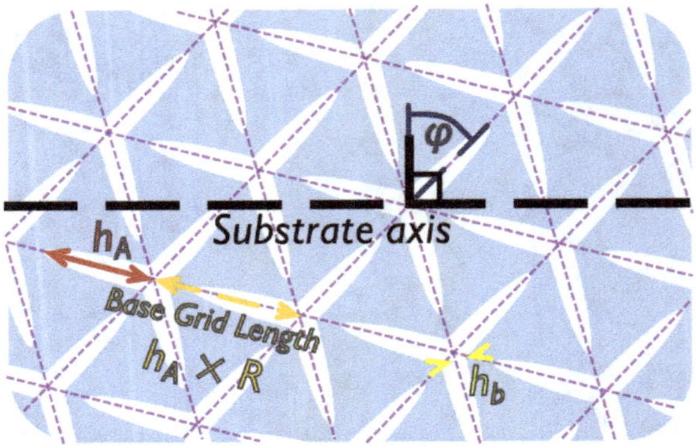

Figure 2.17 Rotating triangle parameters: h_A, semimajor length; h_b, semiminor width; R, grid ratio; φ, angle between the grid and the substrate

triangles are primed to rotate outwards, expanding these holes and giving the auxetic behaviour [35,38].

The parameters for the rotating triangle model were similar to the rotating hexagrams: h_A, the semimajor hole length (i.e. the length from the centre of the hole to its tips); h_b, the semiminor width of the hemi-elliptical slits; R, the grid ratio (dictates hole spacing); and φ, the angle between the hole grid and the substrate. How these parameters are used to construct each model are shown in Figure 2.17, with their parametric sweep values listed in Table 2.8. Designs were built outwards from the central hole. Before processing, the model checks the substrate is intact and deletes any unattached areas caught between the substrate edge and the corner of a hole.

The spread of parametric results is shown in Figure 2.18. The rotating triangle designs generally have greater outputs when the holes are long and closely packed. This structure was much better able to distribute the stress than the triangles in the prior hexagram designs, in part because all the rotating

Table 2.8 Parameters used in rotating triangle models

Parameter		**Parametric values**	**Optimisation range**	**Units**
Slit length	h_A	10, 5, 3	2.5–15	mm
Slit width	h_b	1, 0.5	0.45–1.5	mm
Grid ratio	R	66, 75, 90, 100, 125	60–175	%
Grid angle	φ	0, 15, 30	−33–33	°

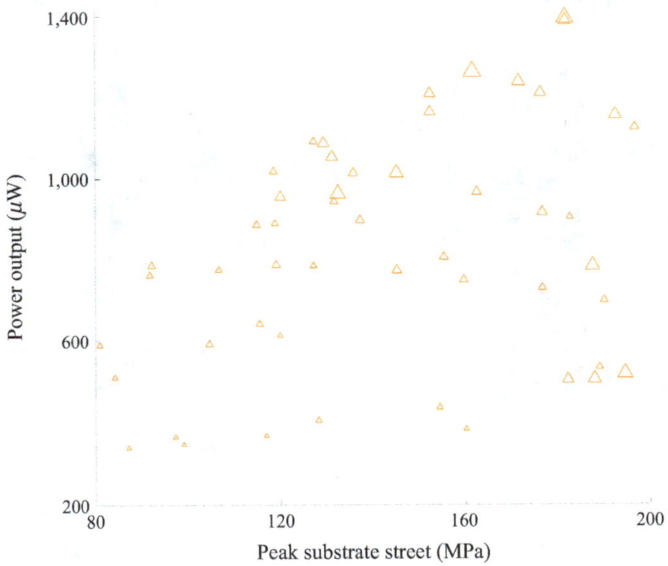

Figure 2.18 Simulated results for the rotating triangle designs

sections here were of the same size. The best design was found to have an output of 1398 μW (with parameters $h_A = 10$ mm, $h_b = 0.5$ mm, $R = 66\%$ and $\varphi = 15°$). This and nearby designs were the seeds for the optimisation, supplemented by some new initial designs with longer holes than covered in the parametric sweep.

The most optimal rotating triangle arrangement produced an output of 1656 μW, a gain of 10.22 times the plain baseline ($h_A = 12.28$ mm, $h_b = 0.494$ mm, $R = 67.61\%$ and $\varphi = 2.97°$), and is shown in Figure 2.19. A small wedge of material was removed from this design (seen as the rightmost gap at the lower edge), as it was only attached to the rest of the substrate by <0.3 mm of material and its absence had no significant loss in the power output.

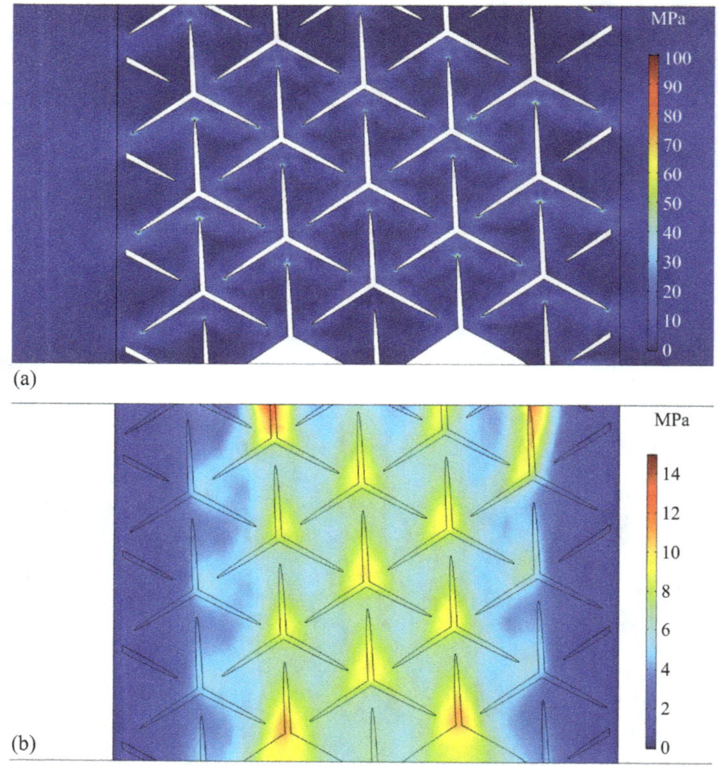

Figure 2.19 Stress concentrations in optimised rotating triangle design: (a) substrate-only view and (b) MFC

2.9 Modelling conclusion

These four auxetic substrate designs, shown in Figure 2.20, and a plain baseline were manufactured to compare their practical performance to this theoretical one. The adhesion between the substrate, epoxy, and MFC has up to this point been assumed to have a value offspring constant per unit area, $k_a = 16.5$ GN/m^3 in the TELs at these interfaces; this was based on some preliminary experimental work on a plain substrate. The effect of varying this parameter on each model at its optimal load is shown in Figure 2.21. This can be used to find the most appropriate value to properly model the sample by comparison to the experiment.

2.10 Experimental method

The substrates were first milled from a sheet of aluminium 6082 T6 1.5 mm thick to the plain's outer dimensions. Cutting the auxetic designs required electric discharge machining (EDM) because of the narrow features (less than 0.4 mm) [64–66]. The

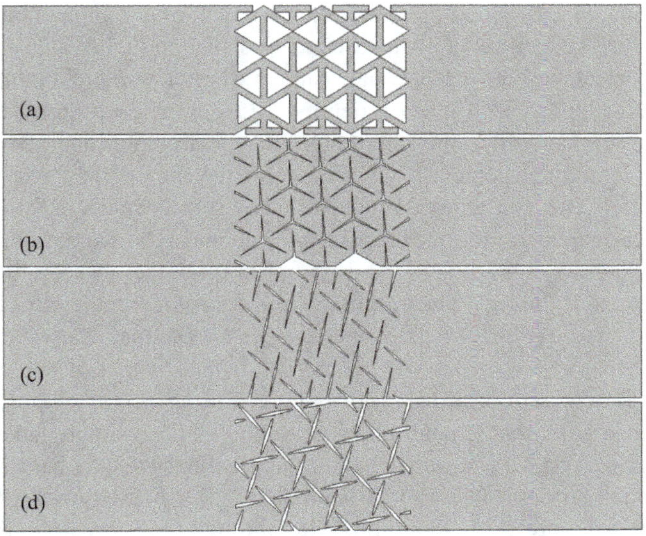

Figure 2.20 Final auxetic substrate designs: (a) re-entrant honeycomb; (b) rotating triangles; (c) rotating parallelograms and (d) rotating hexagrams

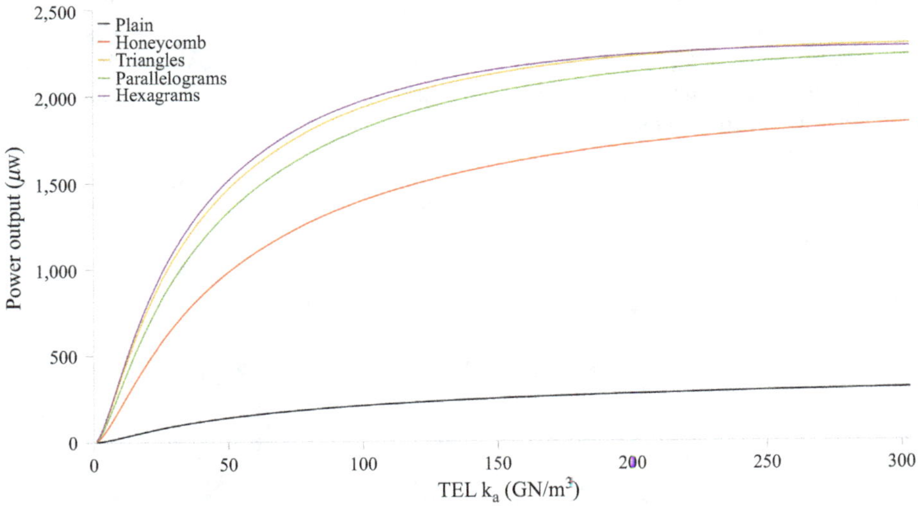

Figure 2.21 Thin elastic layer spring constant per unit area, k_a, *for all designs*

MFC was bonded on top by applying epoxy (Scotch-Weld™ 460 [67]) to the middle of the substrate then clamping the sample between boards with a card wrapped in resin-proof tape [68] in a vice for over 24 h at room temperature. Wires were soldered to the MFC's exposed terminals to their encapsulated electrodes and taped over to protect the joint. Due to the cost of MFC, only five harvester samples could be made: one of each final design, and a plain baseline, as shown in Figure 2.22.

The experimental procedure has each harvester sample clamped in the Instron ElectroPlus™ E10000 to a depth of 25 mm at each end. The machine applied a harmonic strain to the sample for 200 s at an amplitude of 100 $\mu\varepsilon$ peak-to-peak (i.e. total displacement of 25 μm), at a frequency of 10 Hz to compare to the prior model. The excitation was controlled by displacement and was preceded by a 3 s linear extension to the midpoint of the sinusoidal excitation, positioned 1 μm more than the amplitude (so to 13.5 μm in this case) to keep the sample in tension throughout the test. The amplitude and frequency could later be changed and the test was redone to find the sample's response across a range of excitations.

The performance of each harvester is judged by its power output at the optimum resistance load; this is determined from the MFC's voltage generated under each excitation. The MFC's output wires were connected to a bank of resistors where the load was controlled by the Agilent U2751A Switch Matrix mechanical relay, shown in Figure 2.23, which ratcheted through these resistors. It connected one of the four row resistors (R1–R4 in Table 2.9) to one of the eight column ones (C1–C8) in turn, giving 32 steadily increasing resistance values incrementing every 5 s; diagrammatic circuit shown in Figure 2.24. The voltage across the attached resistor was recorded by a data log (National Instrument cDAQ-9174 with NI 9229 card) at 2000 samples per second. The total load resistance was thus the current resistor in parallel with the 1 MΩ input impedance of this data log. These load increases and voltage measurements were controlled with LabVIEW and began only after the excitation had achieved a steady amplitude (this took at least 1–5 s, depending on frequency).

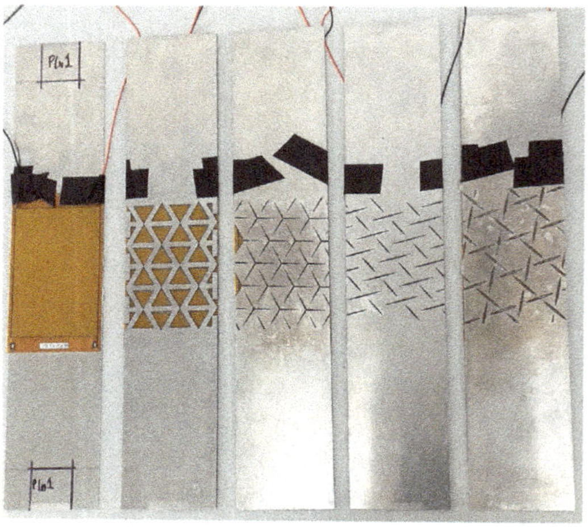

Figure 2.22 Photograph of all samples. Auxetic samples seen from the underside to better show the designs

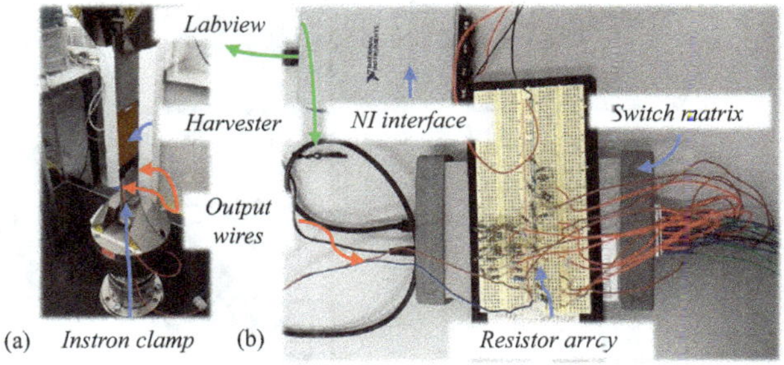

Figure 2.23 Annotated photographs of (a) sample harvester in Instron, and (b) switch matrix and resistor relay, with NI LabView interface

Table 2.9 Resistance values for row and column nodes, shown in Figure 2.24. Total resistance is R(i) + C(j)

Node	Resistance (kΩ)
R1	0
R2	100.1
R3	411.5
R4	845
C1	1.000
C2	10.00
C3	21.88
C4	32.89
C5	50.94
C6	67.1
C7	79.1
C8	99.9

The data was subsequently processed to calculate the cycle power output from the voltage at each load. Figure 2.25 shows an example of an (a) input excitation and (b)–(d) the resulting electrical outputs of the honeycomb sample excited at 10 Hz, 25 μm (note that the time axis for these electrical measurements is offset to the excitation). The switch matrix initial activation and switching between nodes every 5 s could cause spikes in the measured voltage (where the Switch Matrix is briefly open, leaving the NI Interface's internal impedance as the only load), which can be seen in Figure 2.25(b). All voltage data within 0.25 s either side of these switches were thus filtered out from further analysis to remove these transient electrical effects (leaving the white gaps in part [c]). This gave 4.5 s of steady voltage signal at each attached load resistance from which to calculate the cycle power output.

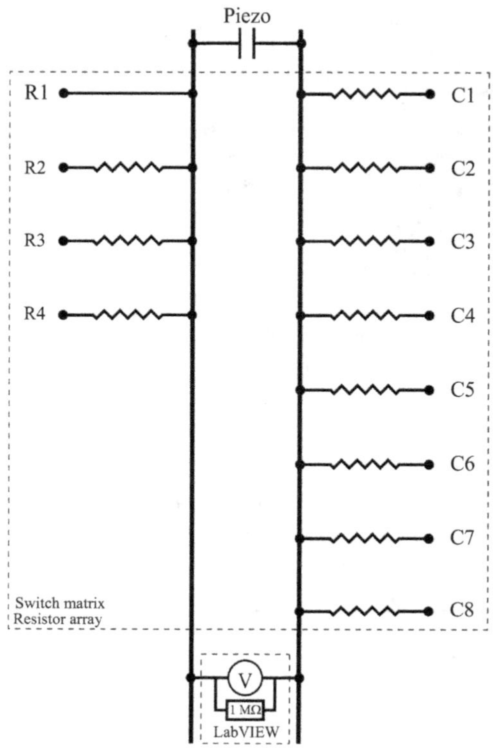

Figure 2.24 Switch matrix circuit diagram, where an R node was joined to a C node to provide the load on the MFC during voltage measurements

An average of the instantaneous power in this timespan would usually give a satisfactory estimate of the cycle power, but if the output were near the noise floor (most common for the lower amplitudes and frequencies, and especially for the plain sample) then this could bias that average. The gradient of the accumulated energy (found by cumulatively integrating the instantaneous power over time) at each resistance was therefore used instead, shown in Figure 2.25(d). At most excitations the difference was negligible. This set of gradients then supplies each point in the power versus load results shown below.

From these points, the optimum power output and load resistance could be found using the following equations. The first gives the piezoelectric power output at a given resistance (when excited far below the material's resonant frequency) and assumes the stresses, T_1, T_2 and T_3[§], on the piezoelectric material are of constant and uniform sinusoidal amplitudes, though this is rarely the case. Fortunately, for fitting purposes their equivalent values need not be calculated: for any given

[§] T_3 is negligible beside T_1 and T_2 for surface mounted piezoelectric harvesters.

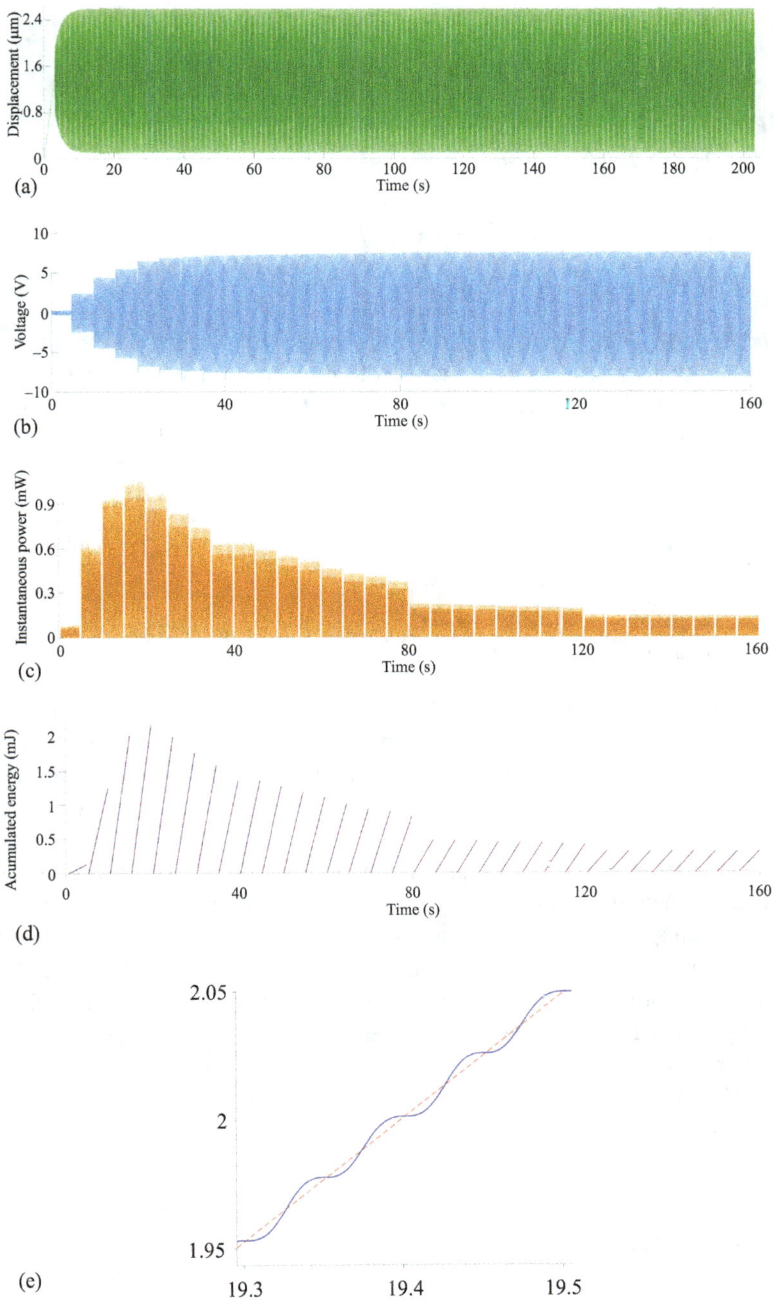

Figure 2.25 *Example of a single 100 με, 10 Hz excitation from the Honeycomb sample, showing: (a) input displacement against time; with the resulting (b) voltage output, (c) instantaneous power output, (d) accumulated energy at each resistance, and (e) close up of part of (d)*

excitation of a particular sample, fitting factors A and B are constant across any load resistance.

$$\boldsymbol{P} = \frac{\boldsymbol{R}(2\pi f)^2}{2\left(1 + \boldsymbol{R}^2\left(2\pi f \quad \varepsilon_0\varepsilon_{33} \quad \frac{L_p \quad W_p}{t_p}\right)^2\right)}(d_{31}T_1 + d_{32}T_2 + d_{33}T_3)^2$$

$$\mathrm{A} = (2\pi f)^2(d_{31}T_1 + d_{32}T_2 + d_{33}T_3)^2/2$$

$$\mathrm{B} = \left(2\pi f \quad \varepsilon_0\varepsilon_{33} \quad L_p \quad W_p/t_p\right)^2$$

The experimental fitting used the following form to find the optimum load, R_{Opt}, and output, P_{Opt}, of each excitation directly from these fitting parameters. If this load fell outside the tested resistance range, then it could only be considered a guideline because the fitting becomes unstable without some datapoints either side of its peak.

$$y = \frac{A \quad x}{1 + B \quad x^2}$$

$$R_{Opt} = 1/\sqrt{B}$$

$$P_{Opt} = A/2\sqrt{B}$$

Error bars were derived from the noise floor and accuracy of all the measuring instruments, the quality of data fitting at each step, and a factor to account for variations between repeated experiments (arising largely from misaligning the sample in the clamps, i.e. deviations from vertical or intended clamping length).

To test a more realistic use case, a sample of arbitrary data was generated to emulate the typical noisy strain experienced by a bridge in use [14]. This excitation was used on all samples, across a total fixed load of 448 kΩ, to provide a consistent input for comparison. The excitation displacement is shown in Figure 2.26. The procedure was otherwise similar to the above method. Gain was calculated by comparing each sample's accumulated energy output during the excitation to that of the plain sample.

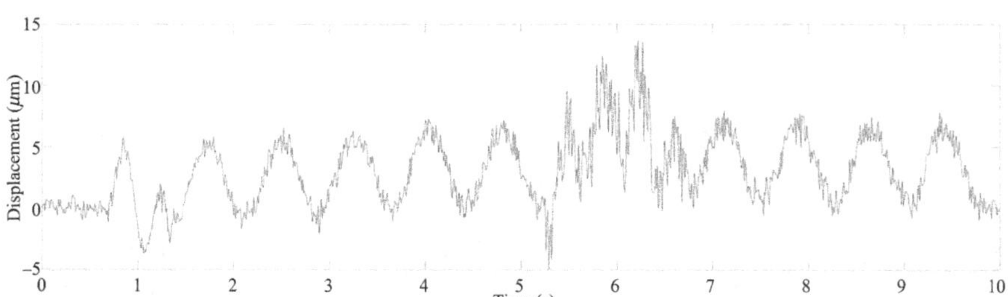

Figure 2.26 The input excitation to emulate the strain on a real bridge [14]

2.11 Experimental results and discussion

The highest output results from each of the six samples at the same excitation (10 Hz, 100 $\mu\varepsilon$ peak-to-peak) are shown in Figure 2.27. The corresponding models are also shown here. The k_a values used in each model's TELs and the experimental optimum values they emulate are listed in Table 2.10. These values were derived by comparison to Figure 2.21.

Recall that during the design stage, the models' gain was predicted to be (in order from least to highest output): re-entrant honeycomb (a gain of 6.09×); rotating parallelograms (9.19×); rotating triangles (10.23×); and rotating hexagrams (10.70). The baseline plain sample performed far better than the model anticipated (with a k_a value almost double that found in the preliminary work), which lowered all the auxetic samples' experimental gains relative to it. The lower than anticipated gain factors of the rotating unit designs in Table 2.10 are due to excess epoxy leaking into the narrow holes (which would reduce the sample's

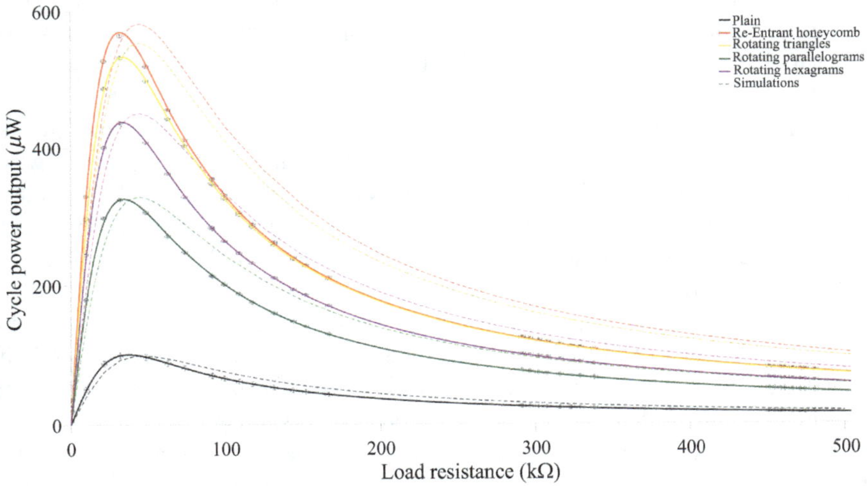

Figure 2.27 Best resistance sweep results from all samples at 10 Hz, 100 $\mu\varepsilon$ (25 μm) excitations, with equivalent models as dashed lines in the same colours

Table 2.10 Optimum experimental results from Figure 2.27

Harvester	R_{Opt} (kΩ)	P_{Opt} (μW)	Gain	k_a (GN/m^3)
Plain	37.1	100.7	–	32.8
Honeycomb	31.9	569.9	5.66	25.5
Triangles	34.0	533.1	5.29	14.4
Parallelograms	34.4	326.1	3.24	10.5
Hexagrams	33.7	438.5	4.35	11.5

flexibility), which caused the experimental gains not to replicate the order found in the models.

Only the plain and honeycomb design exceeded the preliminary k_a value, resulting in a higher power output than anticipated. This resulted in the honeycomb design having the highest experimental gain, despite being the lowest model gain. The auxetic model k_a values were found to be quite similar to each other and within the expected range. This indicates that the TELs are an appropriate way to emulate the bonding adhesion losses in various designs. More precise bonding than is possible by hand, with a smaller amount of epoxy more evenly applied to the substrate, would reduce the overflow and improve the power output dramatically.

The increased thickness of the aluminium substrate leads to an increased power output from the expectations found during modelling (from 0.9 mm thick, rather than the available 1.5 mm), for all but the rotating parallelogram sample (due to its low k_a value which counteracted the greater authority of the thicker substrate). The greatest increase from the modelled expectations was on the plain and honeycomb samples, due to their higher than modelled k_a values.

The same k_a value can be used to emulate a particular sample at any input excitation, as shown in Figure 2.28. This holds for all the different auxetic designs. The gain factors above remain approximately constant at any given input.

These results match the expected linear and parabola relation to their respective input. The amplitude sweep models slightly underestimate the experimental outputs at higher strains. This could indicate that assuming the materials are always in the linearly proportional regime is not as valid at these higher excitations. The frequency sweeps are generally well matched to the experimental data. The hexagram has the largest deviation, due to some low datapoints bringing their fitting down more significantly than other samples.

The concentrating effect of the auxetic region is also applied to the dynamic excitations. The results from the bridge excitation on the plain and re-entrant honeycomb samples are shown in Figure 2.29, with the relative gain of all samples listed in Table 2.11. The voltage tracks of all samples had a similar shape, with the only significant difference being their amplitude. These gain factors were within the experimental variation of those in Table 2.10, indicating the auxetic designs operate under arbitrary excitations in a similar way to the controlled sinusoidal ones.

These results demonstrate that any of these auxetic designs would be an improvement on the baseline. The cost to manufacture each sample was very high due to the optimised designs having such narrow features that they could only be cut by EDM. Tighter controls on these features at the modelling stage would have allowed much cheaper laser cutting to have been used, but these designs would have had lesser outputs. Depending on the intended application and lifespan, this may be an acceptable trade-off. Using the auxetic substrate is an easy-to-implement method of increasing the power output from piezoelectric material. By concentrating the ambient strain energy into this material the output magnitude is increased but otherwise does not alter the output signal's timing or waveform; making it simple to integrate with existing control circuitry for self-powering sensor nodes. This is ideal for use in structural health monitoring.

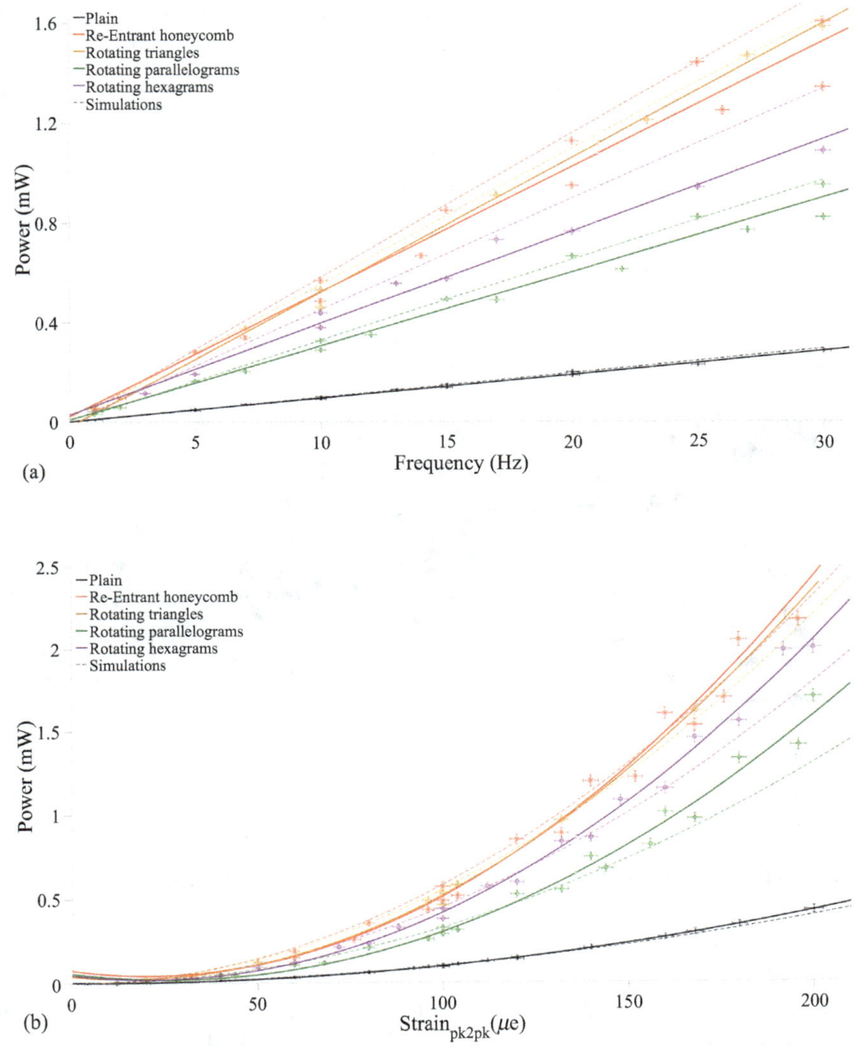

Figure 2.28 Results from (a) frequency and (b) amplitude sweeps on all samples. Experimental fits in solid lines, with equivalent models in the same colours with dashed lines

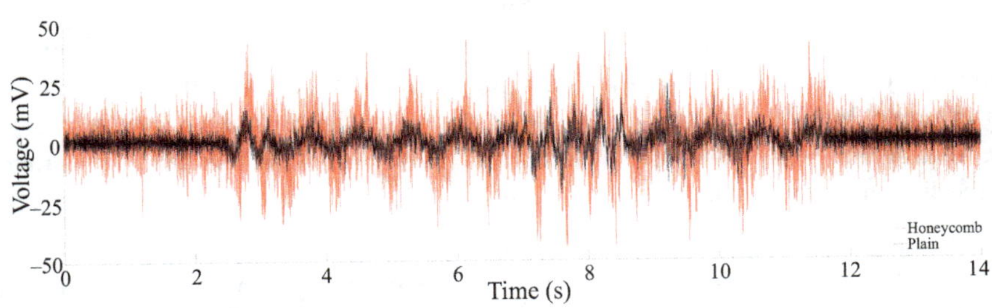

Figure 2.29 Voltage output of plain and honeycomb samples under the bridge emulating excitation

Table 2.11 Gain results from all samples under the bridge emulating excitation

Harvester	E_{total} (nJ)	Gain
Plain	0.470	–
Honeycomb	3.006	6.39
Triangles	1.862	3.96
Parallelograms	1.526	3.25
Hexagrams	2.083	4.43

Figure 2.30 Potential application of surface-mounted auxetic vibration energy harvester in situ under a bridge

2.12 Future challenges and research opportunities

This work above could be used to improve the viability of vibration energy harvesters in low strain environments. There are numerous ways this research could be further developed. For example, some longer term field tests could find that the above stress limits were too stringent or too lax for the intended real-world applications. Where the harvester is surface mounted on some spacers, as in Figure 2.30, simply placing piezoelectric material on both faces of the auxetic region would be an improvement; this was prohibitively expensive here, but models suggest that this would add an extra ~66% to power output[||].

Many other flat auxetic designs have been researched [34–36,38,39,58,63,69–71], and some of these could prove even more effective than those tested here. There are, for example, hierarchical auxetic designs (e.g. smaller slits within the solid regions of the rotating units) that could also be utilised in future [44]. Most auxetic designs found in the literature, and those tested above, use regular patterns to make them easier to

[||]It is not a straightforward doubling of the output because the second layer constrains the substrate more than the one alone; though this could be countered by making the substrate thicker.

model, but there is promise in non-uniform auxetic designs here. For example, beams or hinges in the middle of the auxetic region could be narrower than those at the periphery, using a radial position dependence in some geometric parameters. Considering the stress concentrations in the above auxetic designs, such as Figure 2.7, the greatest substrate stress is often found where the auxetic region joins the rest of the substrate, yet the stress in the piezoelectric material tends to be the lowest over these regions. These regions experience the most direct strain, while being least able to flex. Lightning bolt kinks in the re-entrant honeycomb beams could induce greater flexibility and invite more auxetic behaviour to the substrate. This would give an auxetic region which flexes outwards evenly and thus strain the piezoelectric material more uniformly; delivering more power without excessively stressing it.

There is scope for building slits into MFC, between fibres, to encourage their motion; reminiscent of the work putting slits into PVDF sheets [44]. If these lined up in some way with the auxetic substrate (not necessarily matching its shape exactly), this could allow freer motion, thus inducing more power from the strained piezoelectric fibres. Long-term field testing would be needed to ensure these cuts would not cause severe degradation to the MFC.

Further developments on the piezo-composite materials could also support these auxetic harvesters. MFC is made from parallel fibres, and thus the auxetic nature of the substrate is only partially utilised. Some lateral strain is inevitably lost to pulling apart the polymer encapsulation between fibres. If the fibres were arranged in a spiral or zig-zag pattern or had an interlaced arrangement of fibres, then more of the lateral motion could be harnessed. By designing this fibre arrangement in tandem with the substrate, the benefits from the auxetic concentration could be further improved upon. Manufacturing these advanced composites would be difficult however; sintering or laser cutting thicker fibres into the desired shape would be possible [5,45].

There is also the prospect to develop this work into three-dimensional auxetic structures. Many 3D auxetic designs have been demonstrated where the framework collapses in on itself in all directions under compression [35]. Placing piezoelectric blocks within these frameworks would allow a stress to be applied from all sides, greatly increasing the power output of compressionbased harvesting methods (as most ceramic piezoelectric materials are stronger in compression than tension [72,73] their stress concentration is more tolerable). Integrating many such blocks into an auxetic network could be used to make a highly flexible piezoelectric metamaterial (though aligning the asynchronous outputs of the many blocks would likely require complicated rectification, similar to ZnO fibre devices [74]). Such harvesters could be incorporated into the feet of vibrating machinery to draw power while isolating them from the surroundings. The technique could also make harvesting from impacts, such as footsteps on pavements or ocean waves hitting harbour walls, more viable.

2.13 Conclusion

Overall, this work has shown how using auxetic structures can dramatically increase the power output of vibration energy harvesters. This could easily be applied to infrastructural health monitoring, where the relatively low strains would otherwise be too small to harvest useful amounts of power from.

In this chapter, many auxetic harvester designs have been simulated for comparison. Four optimised designs, and a plain baseline, were selected to be manufactured: the re-entrant honeycomb (experimental gain of 5.66×); the rotating triangles (5.29×); rotating parallelograms, specifically rhomboids (3.24×); and the rotating hexagrams (4.35×). These gains remained consistent at a range of input excitations. The values for spring constant per unit area, k_a, in the models' TELs at the glue interfaces were of a similar range to each other and were found capable of emulating all samples, demonstrating their effectiveness as a fitting parameter for any future auxetic energy harvester designs.

List of acronyms

Acronym	Meaning	
EDM	Electrical discharge machining	Method of cutting metal
MFC	Macro-fibre composite	Composite piezoelectric material
PVDF	Polyvinylidene difluoride	Polymer piezoelectric material
PZT	Lead zirconate titanate	Ceramic piezoelectric material
TEL	Thin elastic layer(s)	COMSOL boundary condition

References

[1] Ni, Y.-Q. and Xia, Y.-X.: 'Strain-Based Condition Assessment of a Suspension Bridge Instrumented with Structural Health Monitoring System', *Int. J. Struct. Stab. Dyn.*, 2016, **16**(04), p. 23.

[2] Capella, J. V., Bonastre, A., Ors, R., and Peris, M.: 'A step forward in the in-line river monitoring of nitrate by means of a wireless sensor network', *Sensors Actuators, B Chem.*, 2014, **195**, pp. 396–403.

[3] Kong, L.B., Li, T., Hng, H.H., Boey, F., Zhang, T., and Li, S.: 'Waste Energy Harvesting: Mechanical and Thermal *Energies'* (Springer, 2014)

[4] Perpetuum: 'Perpetuum: Rail Applications', https://perpetuum.com/rail-applications/, accessed October 2017.

[5] Mescia, L., Losito, O., and Prudenzano, F.: 'Innovative Materials and Systems for Energy Harvesting Applications' (2015).

[6] Chen, J., Zhu, G., Yang, J., *et al.*: 'Personalized Keystroke Dynamics for Self-Powered Human - Machine Interfacing', *ACS Nano*, 2015, **9**(1), pp. 105–116.

[7] US Navy, Roberts, M.J., Baca, A.J., Lai, W.W., Baldwin, L.C., and Owens, M.T.: 'Energy Harvesting Device using auxetic materials – US 9048761 B1'. United States Patent Office, US 9048761 B1, 2015.

[8] Elvin, N. and Erturk, A.: 'Advances in Energy Harvesting *Methods'* (Springer, 2013).

[9] Kuang, Y., Daniels, A., and Zhu, M.: 'A sandwiched piezoelectric transducer with flex end- caps for energy harvesting in large force environments', *J. Phys. D. Appl. Phys.*, 2017, **50**(34), pp. 1–14.

[10] Karimi, M., Karimi, A.H., Tikani, R., and Ziaei-Rad, S.: 'Experimental and theoretical investigations on piezoelectric-based energy harvesting from bridge vibrations under travelling vehicles', *Int. J. Mech. Sci.*, 2016, **119**, pp. 1–11.

[11] Arafa, M., Akl, W., Majeed, M., Al-Hussain, K., and Baz, A.: 'Energy harvesting of gas pipeline vibration', *Act. Passiv. Smart Struct. Integr. Syst.*, 2010, **7643**, p. 76430L.

[12] Chiu, M.C., Karkoub, M., and Her, M.G.: 'Energy harvesting devices for subsea sensors', *Renew. Energy*, 2017, **101**, pp. 1334–1347.

[13] Le, M.Q., Capsal, J.-F., Lallart, M., *et al.*: 'Review on energy harvesting for structural health monitoring in aeronautical applications', *Prog. Aerosp. Sci.*, 2015, **79**, pp. 1–11.

[14] Yoon, H.J., Song, K.Y., Choi, C., Na, H.S., and Kim, J.S.: 'Real-Time Distributed Strain Monitoring of a Railway Bridge during Train Passage by Using a Distributed Optical Fiber Sensor Based on Brillouin Optical Correlation Domain Analysis', *J. Sensors*, 2016, **2016**.

[15] Pozzi, M., Canziani, A., Durazo-Cardenas, I., and Zhu, M.: 'Experimental characterisation of macro fibre composites and monolithic piezoelectric transducers for strain energy harvesting', in 'Smart Structures (NDE)' (2012), p. 834832.

[16] Chew, Z.J., Ruan, T., and Zhu, M.: 'Strain Energy Harvesting Powered Wireless Sensor System Using Adaptive and Energy - Aware Interface for Enhanced Performance', *IEEE Trans. Ind. Informatics*, 2017, **13**(6), pp. 3006–3016.

[17] Ikeda, T.: *'Fundamentals of Piezoelectricity'* (Oxford Science Publications, 1990, 1st edn.).

[18] Qin, Q.H.: *'Advanced mechanics of piezoelectricity'* (2013).

[19] Bowen, C.R., Topolov, V.Y., and Kim, H.A.: 'Modern Piezoelectric Harvesting *Materials'* (Springer, 2016).

[20] Jaffe, B., Cook Jr., W.R., and Jaffe, H.: *'Piezoelectric Ceramics'* (Academic Press Inc., 1971).

[21] Marin, A., Priya, S., Srinath, C., *et al.*: 'Mechanical Energy Harvesting for Powering Distributed Sensors and Recharging Storage Systems'. Virginia Tech, 2013.

[22] Upadrashta, D. and Yang, Y.: 'Experimental investigation of performance reliability of macro fiber composite for piezoelectric energy harvesting applications', *Sensors Actuators, A Phys.*, 2016, **244**, pp. 223–232.

[23] Bhullar, S.K.: 'Three decades of auxetic polymers: A review', *E-Polymers*, 2015, **15**(4), pp. 205–215.

[24] Smart Material: 'Macro Fiber Composite (MFC)' (2014), p. 8.

[25] Moheimani, S.O.R. and Fleming, A.J.: 'Piezoelectric Transducers for Vibration Control and Damping' (Springer, 2006).

[26] Kuang, Y. and Zhu, M.: 'Evaluation and validation of equivalent properties of macro fibre composites for piezoelectric transducer modelling', *Compos. Part B Eng.*, 2018.

[27] Priya, S. and Inman, D.J.: 'Energy harvesting technologies' (Springer, 2009).

[28] Ruan, T., Chew, Z.J., and Zhu, M.: 'Energy-Aware Approaches for Energy Harvesting Powered Wireless Sensor Nodes', *IEEE Sens. J.*, 2017, **17**(7), pp. 2165–2173.

[29] Abdelkefi, A.: 'Aeroelastic energy harvesting: A review', *Int. J. Eng. Sci.*, 2016, **100**, pp. 112–135.

[30] Shi, Y., Hallett, S.R., and Zhu, M.: 'Energy Harvesting behaviour for Aircraft Composites Structures using Macro-Fibre Composite: Part I – Integration and Experiment', *Compos. Struct.*, 2016.

[31] Bowen, C.R., Kim, H. a., Weaver, P.M., and Dunn, S.: 'Piezoelectric and ferroelectric materials and structures for energy harvesting applications', *Energy Environ. Sci.*, 2014, **7**(1), pp. 25–44.

[32] Evans, K.E., Nkansah, M.A., Hutchinson, I.J., and Rogers, S.C.: 'Molecular network design' (1991), **353**, pp. 124–124.

[33] Alderson, A.: 'A triumph of lateral thought', *Chem. Ind.*, 1999, **10**, pp. 384–391.

[34] Mir, M., Ali, M.N., Sami, J., and Ansari, U.: 'Review of mechanics and applications of auxetic structures', *Adv. Mater. Sci. Eng.*, 2014, **2014**, pp. 1–18.

[35] Saxena, K.K., Das, R., and Calius, E.P.: 'Three Decades of Auxetics Research – Materials with Negative Poisson's Ratio: A Review', *Adv. Eng. Mater.*, 2016, **18**(11), pp. 1847–1870.

[36] Lim, T.C.: 'Analogies across auxetic models based on deformation mechanism', *Phys. Status Solidi - Rapid Res. Lett.*, 2017, **11**(6).

[37] Alderson, K., Alderson, A., Ravirala, N., Simkins, V., and Davies, P.: 'Manufacture and characterisation of thin flat and curved auxetic foam sheets', *Phys. Status Solidi Basic Res.*, 2012, **249**(7), pp. 1315–1321.

[38] Grima, J.N. and Gatt, R.: 'Perforated sheets exhibiting negative Poisson's ratios', *Adv. Eng. Mater.*, 2010, **12**(6), pp. 460–464.

[39] Pozniak, A.A., Wojciechowski, K.W., Grima, J.N., and Mizzi, L.: 'Planar auxeticity from elliptic inclusions Artur', *Malays. J. Microbiol.*, 2016, **12**(2), pp. 177–181.

[40] Li, Q., Kuang, Y., and Zhu, M.: 'Auxetic piezoelectric energy harvesters for increased electric power output', *AIP Adv.*, 2017, **015104**.

[41] Ferguson, W.J.G., Kuang, Y., Evans, K.E., Smith, C.W., and Zhu, M.: 'Auxetic structure for increased power output of strain vibration energy harvester', *Sensors Actuators A Phys.*, 2018, **282**, pp. 90–96.

[42] Adeshara, J.: 'Design optimization of geometrical parameters and material properties of vibrating bimorph cantilever beams with solid and honeycomb substrates for maximum', *All Theses. Pap. 1717.*, 2013.

[43] Umino, Y., Tsukamoto, T., Shiomi, S., Yamada, K., and Suzuki, T.: 'Development of vibration energy harvester with 2D mechanical metamaterial structure', in Suzuki, F. (Ed.): 'Power MEMS 2017' (Power MEMS, 2017), pp. 1–4.

[44] Fang, L., Li, J., Zhu, Z., Orrego, S., and Kang, S.H.: 'Piezoelectric polymer thin films with architected cuts', *J. Mater. Res.*, 2018, **33**(03), pp. 330–342.

[45] Fey, T., Eichhorn, F., Han, G., *et al.*: 'Mechanical and electrical strain response of a piezoelectric auxetic PZT lattice structure', *Smart Mater. Struct.*, 2016, **25**(1), p. 015017.

[46] Farhangdoust, S.: 'Auxetic cantilever beam energy harvester', in 'SPIE Smart Structures and Nondestructive Evaluation for Industry 4.0, Smart Cities, and Energy Systems' (SPIE, 2020), p. 29.

[47] Farhangdoust, S., Georgeson, G., Ihn, J.B., and Chang, F.-K.: 'Kirigami auxetic structure for high efficiency power harvesting in self-powered and wireless structural health monitoring systems', *Smart Mater. Struct.*, 2020.

[48] Eghbali, P., Younesian, D., Moayedizadeh, A., and Ranjbar, M.: 'Study in circular auxetic structures for efficiency enhancement in piezoelectric vibration energy harvesting', *Sci. Rep.*, 2020, **10**(1), pp. 1–11.

[49] Chandrasekharan, N. and Thompson, L.L.: 'Increased power to weight ratio of piezoelectric energy harvesters through integration of cellular honeycomb structures', *Smart Mater. Struct.*, 2016, **25**(4), p. 045019.

[50] Grzybek, D., Sikora, W., Kata, D., and Micek, P.: 'Comparative Numerical Analysis of a Piezoelectric Harvester Based on Non-auxetic and Auxetic Material', in '21th International Carpathian Control Conference' (2020), pp. 1–5.

[51] Zhang, H., Wen, P., Li, P., *et al.*: 'Enhanced output performance of flexible piezoelectric energy harvester by using auxetic graphene films as electrodes', *Appl. Phys. Lett.*, 2020, **117**(10).

[52] De Bellis, M.L. and Bacigalupo, A.: 'Auxetic behavior and acoustic properties of microstructured piezoelectric strain sensors.', *Smart Mater. Struct.*, 2017, **26**, p. 085037.

[53] Topolov, V.Y. and Bowen, C.R.: 'High-performance 1-3-type lead-free piezo-composites with auxetic polyethylene matrices', *Mater. Lett.*, 2015, **142**, pp. 265–268.

[54] Bowen, C.R., Betts, D.N., Kim, H.A., and Topolov, V.Y.: '2-2 composites based on [011]-poled relaxor-ferroelectric single crystals: Analysis of the piezoelectric anisotropy and squared figures of merit for energy harvesting applications', *Microsyst. Technol.*, 2014, **20**(4–5), pp. 709–717.

[55] Brownjohn, J.M.W., Stefano, A. De, Helmut, Y.X., and Emin, W.A.: 'Vibration-based monitoring of civil infrastructure: challenges and successes', 2011, pp. 79–95.

[56] Smart Material: 'MFC', https://www.smart-material.com/MFC-product-main.html, accessed October 2017.

[57] Nelder, J.A. and Mead, R.: 'A Simplex Method for Function Minimization', *Comput. J.*, 1965, **7**(4), pp. 308–313.

[58] Masters, I.G. and Evans, K.E.: 'Models for the elastic deformation of honeycombs', *Compos. Struct.*, 1996, **35**(4), pp. 403–422.

[59] Grima, J.N. and Evans, K.E.: 'Auxetic behavior from rotating squares', *J. Mater. Sci. Lett.*, 2000, **19**(17), pp. 1563–1565.

[60] Grima, J.N., Zammit, V., Gatt, R., Alderson, A., and Evans, K.E.: 'Auxetic behaviour from rotating semi-rigid units', *Phys. Status Solidi Basic Res.*, 2007, **244**(3), pp. 866–882.

[61] Grima, J.N., Gatt, R., Alderson, A., and Evans, K.E.: 'On the auxetic properties of "rotating rectangles" with different connectivity', *J. Phys. Soc. Japan*, 2005, **74**(10), pp. 2866–2867.

[62] Shan, S., Kang, S.H., Zhao, Z., Fang, L., and Bertoldi, K.: 'Design of planar isotropic negative Poisson's ratio structures', *Extrem. Mech. Lett.*, 2015, **4**, pp. 96–102.

[63] Mizzi, L., Azzopardi, K.M., Attard, D., Grima, J.N., and Gatt, R.: 'Auxetic metamaterials exhibiting giant negative Poisson's ratios', *Phys. Status Solidi - Rapid Res. Lett.*, 2015, **9**(7), pp. 425–430.

[64] Gostimirović, M., Kovač, P., Sekulić, M., Rodić, D., and Pucovsky, V.: 'Machining Characteristics of Electrical Discharge Machining – a Review', *J. Prod. Eng.*, 2016, 19.

[65] Kar, C., Surekha, B., Jena, H., and Choudhury, S.D.: 'Study of Influence of Process Parameters in Electric Discharge Machining of Aluminum - Red Mud Metal Matrix Composite', in 'Procedia Manufacturing' (Elsevier B.V., 2018), pp. 392–399.

[66] Patel, J.D. and Maniya, K.D.: 'A Review on: Wire cut electrical discharge machining process for metal matrix composite', *Procedia Manuf.*, 2018, **20**, pp. 253–258.

[67] 3M: 'Adhesives DP460 Off-White and DP460NS' (3M, 2017), pp. 1–8.

[68] CFS Fibreglass: 'Resin Proof Tape 50mm wide', https://www.cfsnet.co.uk/acatalog/Resin-Proof-Tape-50mm-wide-TTACEL_050.html, accessed May 2019.

[69] De Bellis, M.L. and Bacigalupo, A.: 'Auxetic behavior and acoustic properties of microstructured piezoelectric strain sensors', *Smart Mater. Struct.*, 2017, **26**(8).

[70] Cabras, L. and Brun, M.: 'Auxetic two-dimensional lattices with Poisson's ratio arbitrarily close to - 1', *Proc. R. Soc. A Math. Phys. Eng. Sci.*, 2014, **470**(2172), pp. 1–26.

[71] Smith, C.W., Grima, J.N., and Evans, K.E.: 'Novel mechanism for generating auxetic behaviour in reticulated foams: Missing rib foam model', *Acta Mater.*, 2000, **48**(17), pp. 4349–4356.

[72] Zheng, Y.H., Li, Y.F., Wang, S.W., and Zhao, Y.: 'Experiment on yield strength of PZT-4 piezoelectric generating column', *Appl. Mech. Mater.*, 2014, **441**, pp. 62–65.

[73] Fett, T., Munz, D., and Thun, G.: 'Tensile and bending strength of piezoelectric ceramics', *J. Mater. Sci. Lett.*, 1999, **18**(23), pp. 1899–1902.

[74] Wang, Z.L., Chen, J., and Lin, L.: 'Progress in triboelectric nanogenerators as a new energy technology and self-powered sensors', *Energy Environ. Sci.*, 2015, **8**(8), pp. 2250–2282.

Chapter 3
Energy harvesting from human body

Mostafa G. Abdelmageed[1] and Ahmed M.R. Fath El-Bab[2]

Abstract

This chapter reviews the concepts used to harvest energy inside the human body and compare the harvesters' output in each concept with each other. Also, the chapter introduces a solution that avoids drawbacks in previous work done in this area. Implantable energy harvesters target to harvest energy from internal human body energy (mostly) or external energy sources. Examples of internal body energy are expanding and contracting arteries, blood flow and pressure in veins and arteries, vibrating organs, body heat, and chemical reactions. The chapter starts with current trends and challenges in the field of harvesting energy inside the human body. It categorizes the sources of energy inside the human body. Section 3.2 reviewed the previous work done in this field and classified it according to the harvesting concept (electromagnetic, piezoelectric, fuel cell, and electrostatic). Section 3.3 introduces the proposed solution, which utilizes the blood flow and pressure energies inside a large vein to deform piezoelectric beams and generate electricity. Future challenges and research opportunities are discussed in Section 3.4, and finally, Section 3.5 ends with the conclusion.

3.1 Current trends and challenges in the area

The human body contains many internal sources of energy that could be harvested to be used in another form for powering devices for a specific application, as shown in Figure 3.1. Usually, this energy is turned to electrical energy to directly power an impeded medical device or being stored in a capacitor or battery to be used later.

The first source is the expanding and contracting of large arteries. The expansion and contraction forces could be used to deform the piezoelectric material ribbon [1] (piezoelectric material generates electric charges when subjected to mechanical stress and discussed in detail in Section 3.2). The expanding and contracting of large

[1]Mechanical Design and Production Engineering Department, Faculty of Engineering, Cairo University, Cairo, Egypt
[2]Mechatronics and Robotics Department, School of Innovative Design Engineering, E-JUST, Alexandria, Egypt

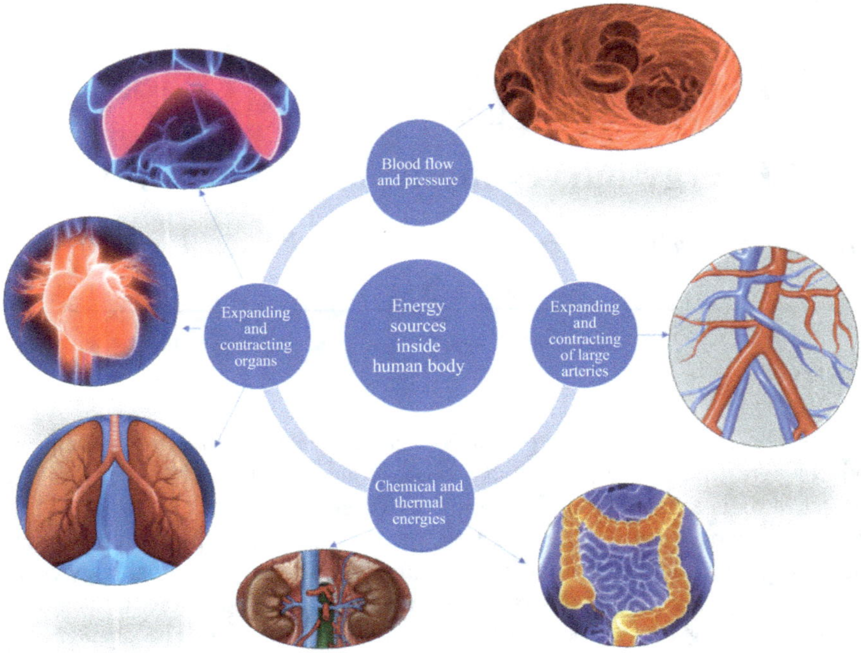

Figure 3.1 Classification of human's internal energies

arteries can also be used to move a coil inside the magnetic field [2] wrapped up around the artery to generate pulsations of electrical charges. Major drawbacks associated with this concept are the need for special surgery to implant this type of harvesters and the restriction-induced by the harvester on the artery walls.

The second source is the power of blood flow and pressure inside veins, arteries, and heart cavity. In this case, the blood flow could be used directly to rotate a miniature turbine attached to the blood vessel [3,4]. Also, blood flow drag force can be used to deform micro-beams fabricated from a piezoelectric material such as lead zirconate titanate (PZT) ceramics, implanted in the bloodstream [5]. Other research uses pressure variation inside the blood vessel to deform a diaphragm covering a cavity and attached to the PZT material. The pressure variation deforms the diaphragm, and hence, the PZT layer generates pulsations of electricity [6]. Using the blood pressure inside the heart cavity, a researcher designed a micro-harvester consisting of flexible bellows connected to a spiral PZT material. When pressure varies inside the cavity, the bellows deforms, causing the PZT spiral to deform and generate electricity [7]. The limitation of this method is the negative effect of partially blocking the blood path. This blocking may result in body pressure variation and a high probability of blood clots formation due to the blood cells' sticking on the harvester surface.

The third and powerful source is expanding and contracting (vibrating) organs. Research studies were done depending on the motion of organs. Heart beating could be used to operate a harvesting system like that one used automatic clockwork of a wristwatch and be glued to the outer heart surface [8]. Another research used a flexible PZT material that could be stuck to the heart surface, lungs, and diaphragm. This material uses the contraction and expansion force, which generate cyclic forces to the PZT to generate pulsatile electrical signals [9,10]. A problem associated with this technique is the load and restriction exerted on the vital organ besides an extra surgery to fix these harvesters will be needed.

The ultimate source is the chemical and thermal energies. Here, in body chemical reaction, a common body fluid (like glucose) is used as fuel for the biofuel cell. These reactions result in electricity production [11]. For thermal energy harvesting, the harvester uses the temperature gradient through the human body to generate electric potential using the Seebeck effect [12]. Simultaneously, others combined both sources to gain more power [13].

3.2 Review of related works and possible solutions

This part addresses the previous trials for harvesting energy from internal body energy such as organs motion, blood flow/pressure, and arteries deformation. The review is divided into four parts according to the concept used in harvesting energy such as electromagnetic, piezoelectric, biofuel cell, and electrostatic.

3.2.1 Electromagnetic concept

In the electromagnetic concept, the authors used the human internal energy to move coils inside a magnetic field or vice to generate electrical potential. Pfenniger *et al.* [2] developed a harvesting system by winding a flexible conductive coil made of shape memory alloy-SMA (for its elasticity) around the targeted artery (iliac artery). Two rings of strong magnets that create a magnetic field parallel with the artery axis are fixed, as shown in Figure 3.2. When the pressure changes inside the artery, its wall expands and contracts, causing the coil to deform and cut the magnetic field and generate electrical charges. The average power harvested was about 42 nW at a resisting load of 3.3 Ω.

In another research, Pfenniger *et al.* [3] studied a miniaturized turbine generator's performance attached to an internal thoracic artery, which utilizes only 1% of the cardiac output flow. The miniaturized turbine of the type tesla converts the blood pressure inside the artery into mechanical torque on the shaft through friction between blood and rotating discs mounted on the turbine shaft, as shown in Figure 3.3. The generated power by this system was about 1 mW for a pressure difference of 35 mmHg.

In the same manner of using turbine generators, Kim *et al.* [4] developed a tiny turbine consisting of four fixed coils with ferrite cores and a rotor carrying six pairs of permanent magnets, as shown in Figure 3.4. With the advantage of a wide working range (blood flow from 1.4 to 2.7 L/min and blood pressure from 6.75 to 54.75 mmHg). The targeted arteries are the aorta, main pulmonary artery, and right and left pulmonary arteries. The harvester was tested in vitro and in vivo; the maximum output power was about 3.4 mW at a pressure drop of 54.75 mmHg and a flow rate of 2.68 L/min.

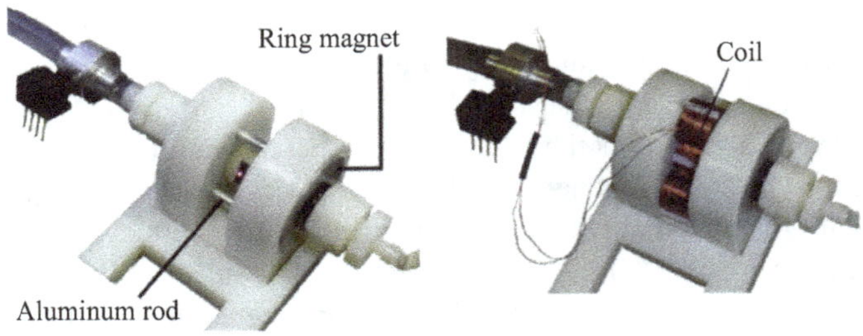

Figure 3.2 Flexible coil around an artificial artery and ring magnets arrangement [2]

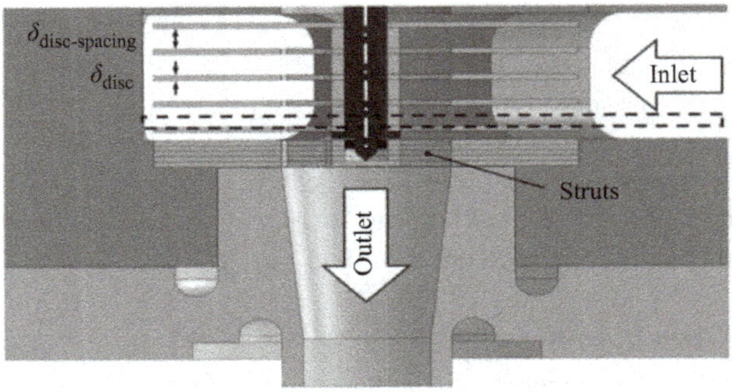

Figure 3.3 Rotor discs and stator arrangement of the tesla turbine generator [3]

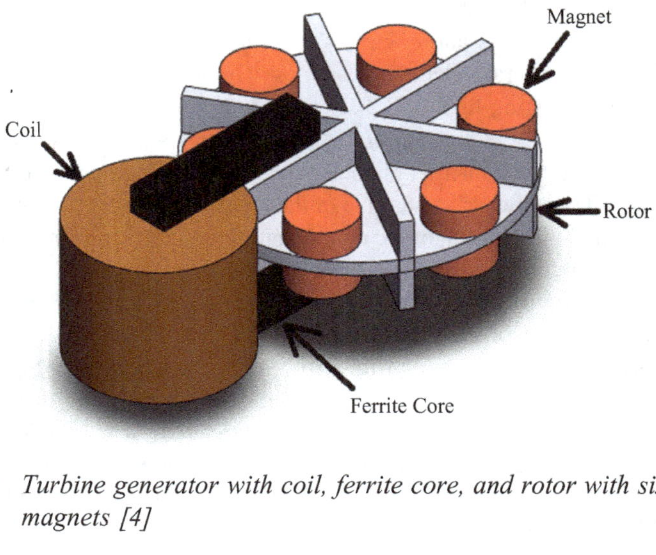

Figure 3.4 Turbine generator with coil, ferrite core, and rotor with six magnets [4]

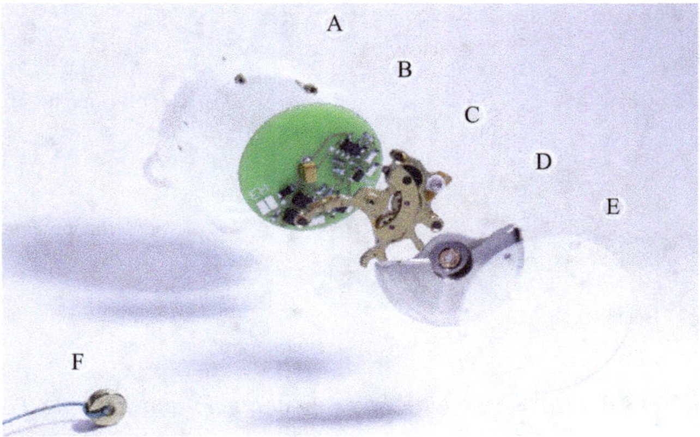

Figure 3.5 Assembly of the pacemaker with the harvesting system. (a and e) casing, (b) electronic circuit, (c) gearing system, (d) oscillation weight, and (f) micro-generator [8]

Zurbuchen *et al.* [8] developed a pacemaker (PM) with a harvesting system depending on the automatic Swiss clockwork wrest watch. In this design, the harvester is fixed on the outer surface of the heart. It translates its motion into electricity through oscillating weight connected to a micro-generator by transmission gears, as shown in Figure 3.5. The harvesting system was tested in vitro (on an artificial heart that simulates the real human heart motion) with an output harvested mean power of 85 μW and in vivo on a domestic pig with a successful scenario.

Tholl *et al.* [14] developed an energy harvester with a lever mechanism deflected by the blood flow inside the right ventricular outflow tract and utilizes the resulting torque to trigger an internal coil attached to the flexible beam. The coil is between a pair of magnets, as shown in Figure 3.6. When the inner coil is triggered, it vibrates between the magnets and generates electricity for each systole and diastole. The harvester weight is 6.4 g with a volume of 2 cm^3. It was tested in vitro. It gave an output power of 14 $\pm$ 8 μW at 60 bpm and 82 $\pm$ 17 μW at 200 bpm.

Zurbuchen *et al.* [15] introduced an energy harvesting system based on electromagnetic induction, in which 19 iron discs separate a set of 20 neodymium permanent magnets discs. These discs are aligned and suspended inside an array of copper coils with a wire diameter of 60 μm using two spiral springs, as shown in Figure 3.7. The device is targeted to be implanted inside the heart cavity. It utilizes the heart motion (acceleration) to vibrate the magnets' stack inside the coil and generate electricity by electromagnetic induction. The harvester diameter was 5 mm with a length of 26 mm, and it was tested in vivo using 60-kg domestic pigs. The output power was about 0.78 and 1.7 μW for a heart rate of 84 and 160 bpm, respectively.

Pancharoen *et al.* [16] developed an electromagnetic energy harvester that can be embedded in the hip. The harvester utilized the patient's hip movement to vibrate a magnet surrounded by a copper coil to generate electricity. The harvester consists of a middle mass [two samarium cobalt (SmCo) cylindrical magnets spaced by ferrite

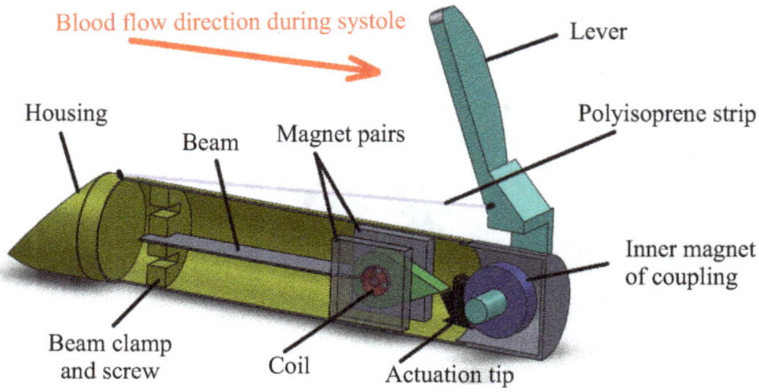

Figure 3.6 Harvester with lever mechanism and housing [14]

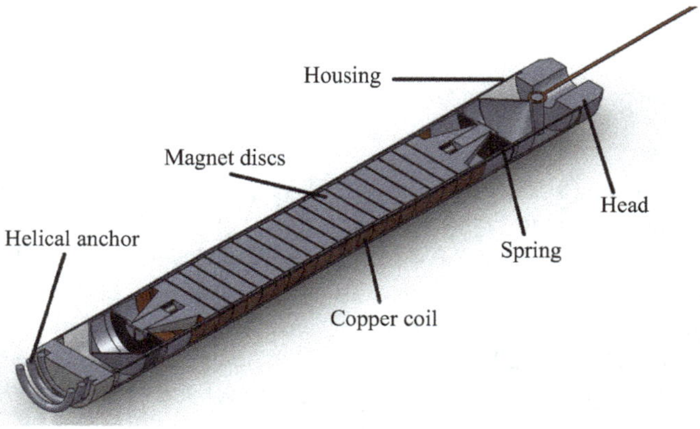

Figure 3.7 Magnet discs arrangement connected by two springs and surrounded by the copper coil [15]

spacer and top and bottom tungsten cylinders]. Secondly, top and bottom SmCo magnets were used to form a magnetic spring with the middle mass and enclosed inside a Teflon tube. Finally, a copper coil with 1,100 turns of 50 μm copper wire was used, as shown in Figure 3.8. The device was tested in vitro at a frequency of 1 Hz. The generated power was about 0.96 μW for a resistance load of 250 Ω.

Beyaz *et al.* [17] presented a kinetic energy harvester that could be implanted inside the human body and utilize the body movement to induce a motion of magnet between coils to generate electricity. The device consists of a bottom and top metal stator, each containing three copper coils, a moving metal housing carrying two magnets, and separated from the stators using a ball bearing. Finally, the system is enclosed in a polydimethylsiloxane (PDMS) frame for biocompatibility, as shown in Figure 3.9. The system was tested using a test bench. The output harvested voltage was about 1.05 mV at a coil's resistance of 2.2 Ω.

Sahara *et al.* [18] proposed an energy harvester that utilizes the expansion and contraction of muscles to generate electricity throughout a transmission system to

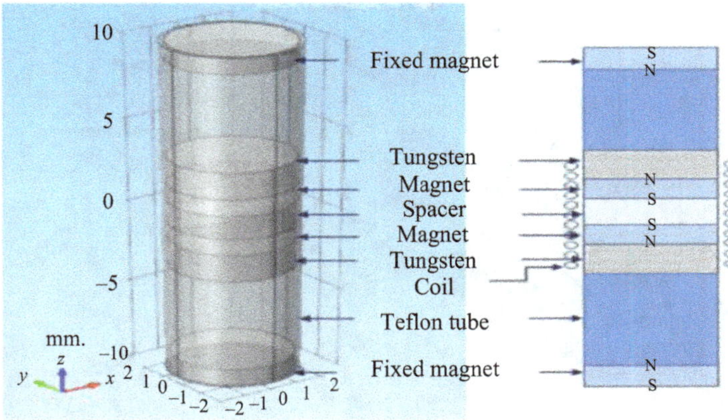

Figure 3.8 Assembly of the electromagnetic harvester with the middle mass composition and Teflon casing [16]

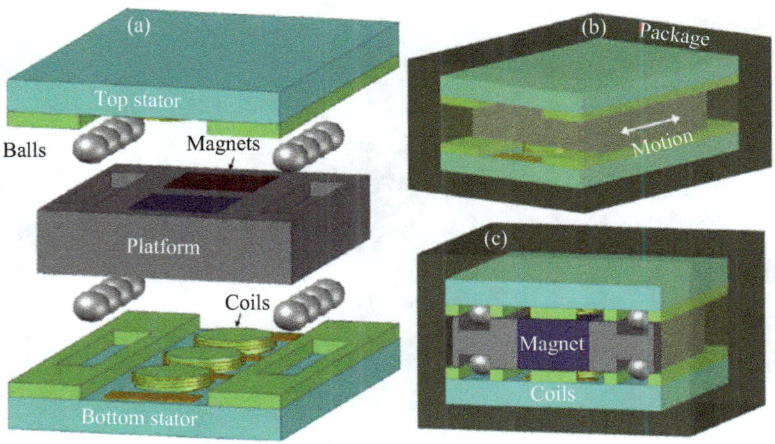

Figure 3.9 (a) Components of the electromagnetic harvester, (b) motion direction of the moving platform, and (c) assembly of the harvester [17]

translate the muscle's linear motion to a rotational motion. This rotational motion operates a small electromagnetic induction generator, as shown in Figure 3.10. A prototype of length 28 mm and weight of 1.3 g was tested in vitro and in vivo using the gastrocnemius muscle of a toad to determine the harvester's efficiency. The output was found to be 19.4% average efficiency with 111-μW harvested power.

Haeberlin *et al.* [19] developed an electromagnetic energy harvester that utilizes the intracardiac blood flow to activate a small turbine in the bloodstream to generate electricity. The harvester is about 0.34 cm^3, weighs 1.3 g, and consists of a closed housing with a micro-generator whose rotating shaft is coupled with a two-bladed rotor through a magnetic coupling, as shown in Figure 3.11. When this device is placed in the bloodstream, the blood exerts drag forces on blades and is converted into torque. This

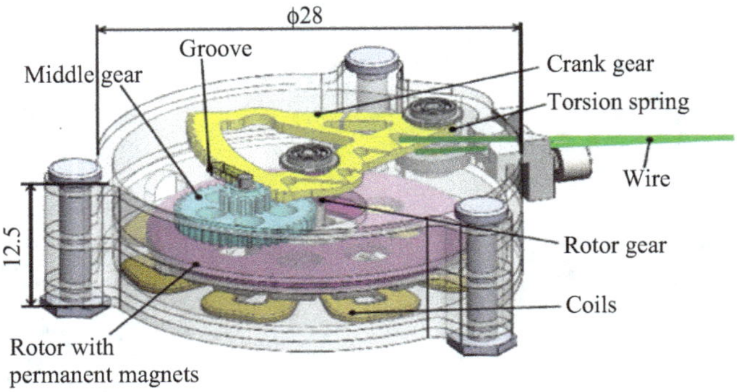

Figure 3.10 Electromagnetic harvester with transmission system and generator [18]

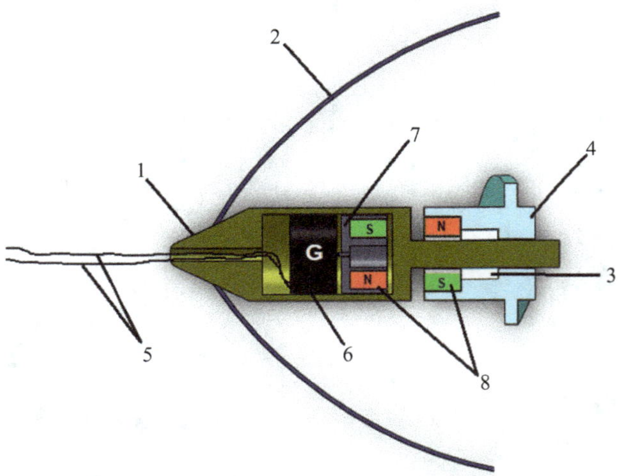

Figure 3.11 Turbine generator components: (1) housing, (2) anchoring system, (3) bearing, (4) turbine blades, (6) microgenerator, (7) back-irons, and (8) coupling magnets [19]

torque rotates the generator shaft through the magnetic coupling. The device was tested in vivo on a domestic big, and the resulted output power was about $10.2 \pm 4.8\ \mu W$.

Pfenniger *et al.* [20] used the magneto-hydrodynamics' effect to generate electricity from artery wall deformation. The harvester size is about 3 cm^3, and consists of the main chamber (MC) connected to a buffer chamber (BC) through four channels and surrounded by four neodymium (NdFeB) magnets to create a strong magnetic field perpendicular to the channel's direction, as shown in Figure 3.12. A highly electrically conductive fluid fills the main chamber. When the artery expands, the MC is squeezed, causing the fluid to pass to the BC through the channels, cuts the magnetic field, and generates electricity. During artery contraction, the pressure decreases in the BC, causing the fluid to back again to the MC, cutting the magnetic field again and generate electricity. The maximum output power harvested from this device was about 135 μW.

Zurbuchen *et al.* [21] used the concept of mass imbalance in a commercial clockwork automatic wristwatch to harvest energy from cardiac wall motion. The working principle depends on utilizing the heart wall's motion (where the harvester is fixed) to oscillate an imbalanced mass, which in turn wind a spring until it reaches a specific threshold and then released to rotate a micro-generator that generates electricity. The harvester was tested in vitro using a robot arm that simulates the target point's motion on the cardiac wall, and the generated mean power ranges from 23 to 30 μW. The harvester was also tested in vivo using 65-kg female Swiss alpine sheep with an average heart rate of 90 bpm, and the harvested power was about 16.7 μW, as shown in Figure 3.13.

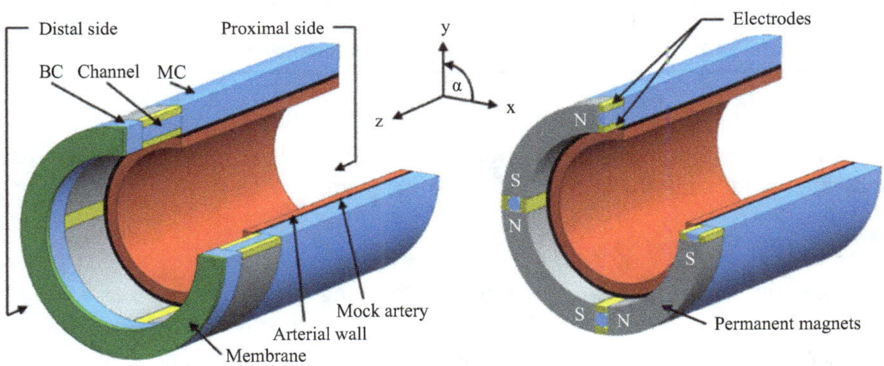

Figure 3.12 Magneto-hydrodynamic generator assembly [20]

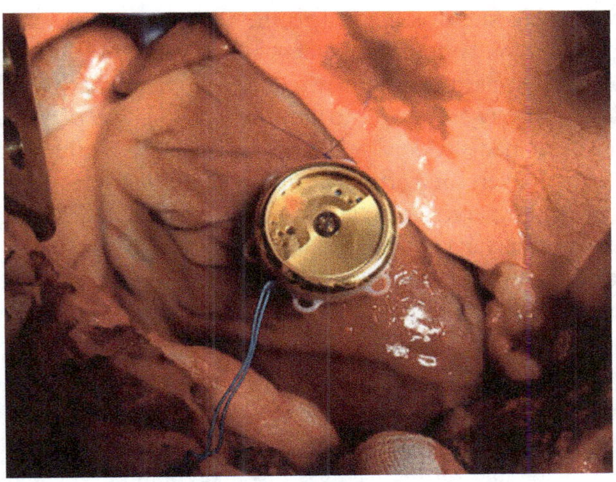

Figure 3.13 In vivo testing of the harvester [21]

Table 3.1 Electromagnetic concept of previous work

Author	Generated power	Region of implantation	Special surgery
Pfenniger *et al.* [2]	42 nW	Iliac artery	y
Pfenniger *et al.* [3]	1 mW	Internal thoracic artery	y
Kim *et al.* [4]	3.4 mW	Arteries	y
Zurbuchen *et al.* [8]	85 μW	Outer surface of the heart	y
Tholl *et al.* [14]	14±8 μW at 60 bpm 82±17 μW at 200 bpm.	Inside right ventricular	y
Zurbuchen *et al.* [15]	0.78 & 1.7 μW for 84 & 160 bpm	Inside the heart cavity	y
Pancharoen *et al.* [16]	0.96 μW	Hip	y
Beyaz *et al.* [17]	1.05 mV	Inside the human body	y
Sahara *et al.* [18]	111 μW	Muscles	y
Haeberlin *et al.* [19]	10.2±4.8 μW	Intracardiac blood flow	y
Pfenniger *et al.* [20]	135 μW	Arteries	y
Zurbuchen *et al.* [21]	16.7 μW	Cardiac wall	y

Table 3.1 summarizes the contribution of previous work in harvesting internal body energy utilizing the concept of electromagnetic.

3.2.2 Piezoelectric concept

Piezoelectric materials have a unique transducing character that can convert mechanical energy into electrical energy and vice versa. It could be used as an actuator to convert electrical signals into mechanical motion by applying an electrical charge on both surfaces of the piezo-crystal material, which induces strain in the material causing extension or contraction. On the contrary, piezoelectric materials could be used as a sensor by converting mechanical stress into electrical signals. So, when any stress is applied to the PZT material's surface, electrical charges are generated on both the piezo-crystal surfaces.

Many studies used piezoelectric materials in the energy harvesting process utilizing internal human body energy due to this property. Zhang *et al.* [1] used a flexible piezoelectric material, namely, polyvinylidene fluoride (PVDF) films, coated with an aluminum layer to convert the aorta internal blood pulsatile pressure into electrical charges by wrapping the harvester around the artery. Testing of the harvester was done in vitro by replacing the aorta with a latex tube, as shown in Figure 3.14. This tube is connected to an intra-aortic balloon pump that simulates the pulsation of blood inside the aorta. The peak harvested power was about 681 nW at a beating rate of 70 bpm and a sampling resistance of 33 MΩ. Also, in vivo experiment was done on a domestic porcine through surgery. At which, the harvester was wrapped around the ascending aorta. The output power was 30 nW for the same conditions.

Deterre *et al.* [7] introduced the design, fabrication, and testing of an energy harvesting system suitable for the leadless PM and could be implanted directly in the cardiac cavity. The harvester consists of a casing with micro-bellows. This

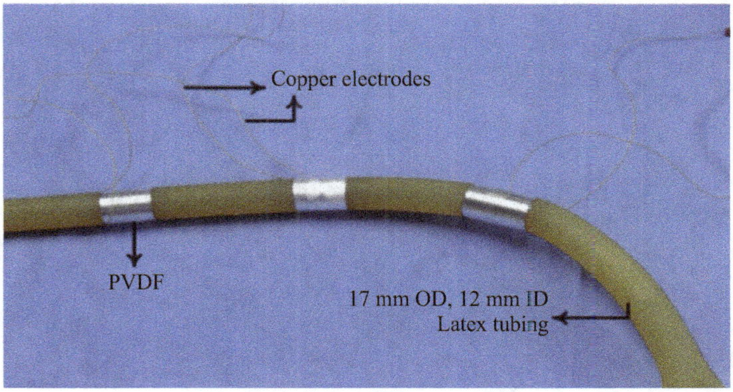

Figure 3.14 Flexible PVDF harvesters wrapped around latex tube (in vitro testing) [1]

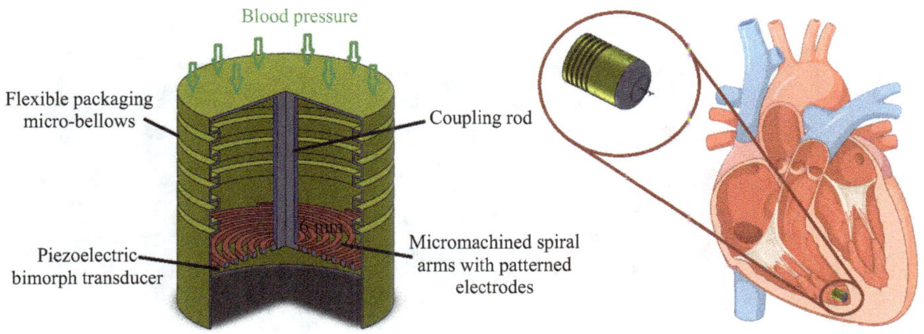

Figure 3.15 Place of harvester inside the cardiac cavity and its composition [7]

micro-bellows is connected internally by a rod with a piezoelectric spiral made of PZT, as shown in Figure 3.15. When the pressure inside the cardiac cavity changes, the micro-bellows compressed and expanded, causing the PZT spiral to deform up and down and generate electricity. The harvester components fabricated using MEMS techniques and three PZT spiral designs were tested to reach the maximum efficiency. The maximum output power was for the third spiral design (having ten arms with one turn per arm), which gave three $\mu J/cm^3$/cycle.

Dagdeviren *et al.* [10] reported an advanced piezoelectric material and harvester design with a high mechanical to electrical conversion efficiency with flexible and biocompatible properties. The harvester targets the organs with expansion and contraction motion such as the heart, lungs, and diaphragm to induce stresses in the piezoelectric harvester, generating electrical charges for each compression and extension. The device was tested in vivo using bovine and ovine hearts, lungs, and diaphragm in open surgery, as shown in Figure 3.16, and the generated maximum power was 1.2 $\mu W/cm^2$.

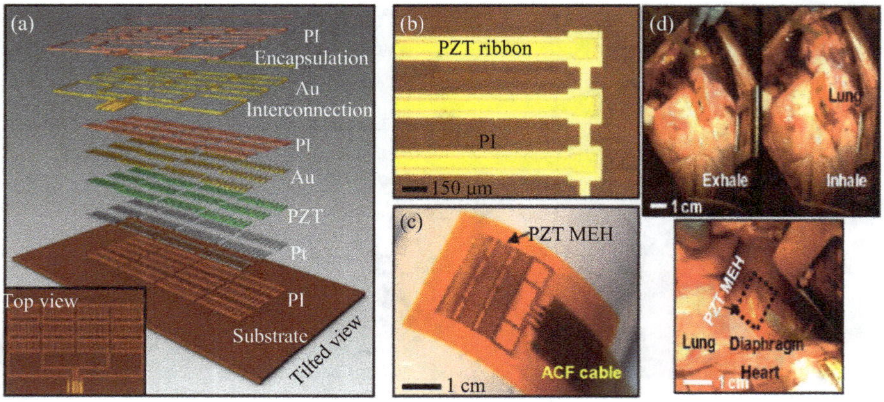

Figure 3.16 (a) Composition of the flexible PZT material, (b) PZT under a microscope, (c) harvester patch, and (d) in vivo testing of the harvester [10]

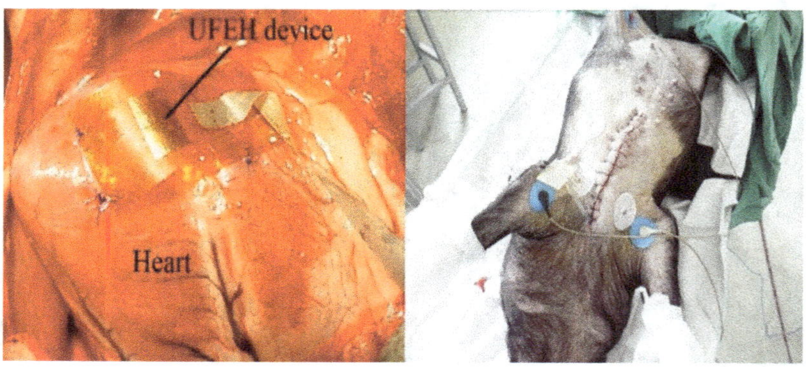

Figure 3.17 In vivo testing of the flexible harvester [9]

Lu *et al.* [9] developed an energy harvester based on an ultra-flexible piezoelectric material using MEMS fabrication techniques to overcome the brittleness and fragility of PZT ceramic properties. The harvester is designed to be fixed on the heart, utilizing its expansion and contraction to deform the harvester and generate electricity. The harvester was tested in vivo on a swine, as shown in Figure 3.17. The harvested peak-to-peak voltage was about 3 V at a blood pressure of 76 mmHg and heart beating rate of 125 bpm.

Jiang *et al.* [22] developed an energy harvester composed of high performance and thin 7 × 7 array of PZT/epoxy 1–3 composite with 120 × 120 μm^2 each. They connected the array elements in parallel with a double layer of 100-μm diameter twisted copper wires to allow flexible deformation of the PZT array. The system was encapsulated in the PDMS layer, as shown in Figure 3.18. The harvester was

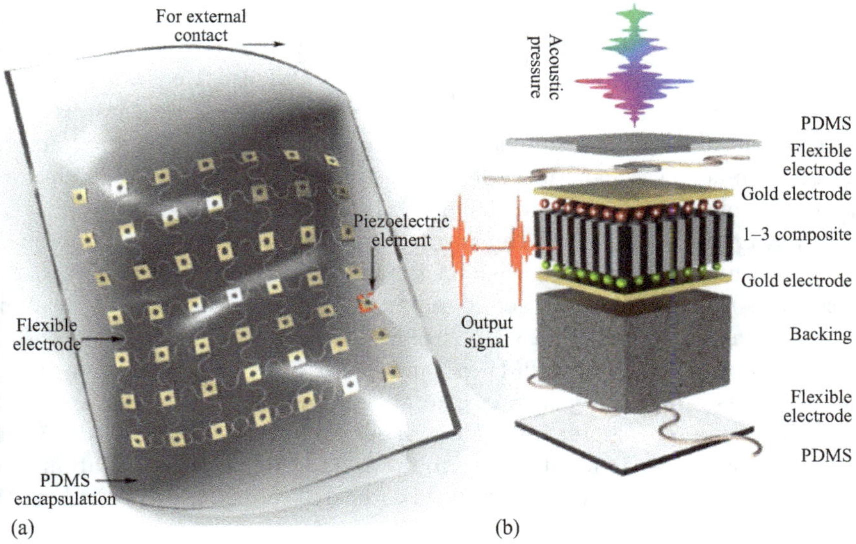

Figure 3.18 Array of the PZT elements connected with wavy copper electrodes and the PZT element composition [22]

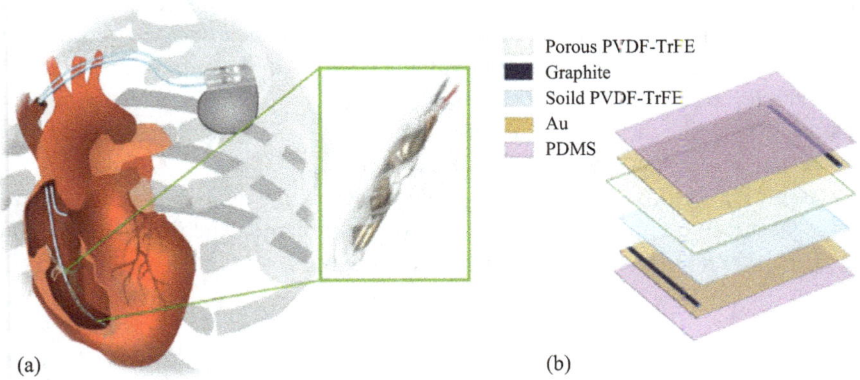

Figure 3.19 (a) PZT element wrapped around pacemaker lead and (b) PZT layer composition [23]

tested in vitro using an ultrasonic probe and different pork tissue thicknesses inserted between the ultrasonic transmitter and the harvester. The output harvested voltage was 0.91 Vpp and a current of 2.5 μA at a pork tissue of 14 mm and 1.07 Vpp and 3 μA without pork tissue.

Dong *et al.* [23] developed a piezoelectric harvester that consists of a 10×10 helical PZT elements wrapped around PM lead and covered by the PDMS layer, as shown in Figure 3.19. This design means no extra implantation surgery is needed as the PM lead is implanted during the PM implantation operation. The working principle depends on the shaking during heart beating. The harvester utilized those movements to induce deformation in the PM lead. Hence, it stresses the PZT helical elements to generate electrical charges. The harvester was tested in vitro with a shaker that simulates the shakes inside the heart cavity, and the harvested voltage was about 0.6 Vpp. It is also tested in vivo on two porcine, and the output voltage was about 0.65 Vpp.

Dong *et al.* [24] introduced an energy harvesting system that depends on the concept of piezoelectricity in which a flexible porous PVDF–TrFE was fabricated as a strip of dimensions 5×20 mm^2 and wrapped around the PM lead in a spiral way, as shown in Figure 3.20. The harvesting idea is based on utilizing the heart shaking to deform the PM lead and hence twist the PVDF strip, which induces stresses in it and generates electrical charges. The harvester was tested in vivo on a dog. The output voltage was about 0.5 V and a current of 43 nA. The authors claimed that adding tip mass to the end of the harvester could improve the output harvested energy by 1.8 times compared to the harvester without the tip masses.

Hwang *et al.* [25] reported a flexible single crystalline piezoelectric lead magnesium niobate-lead titanate (PMN-PT) harvester that could be used to directly stimulate the heart without an external energy source, which could be considered an artificial PM. The device was tested in vivo on a rat heart through stimulation electrodes in open surgery to stimulate the rat heart muscle, as shown in Figure 3.21. The harvester was also tested on a test bench and showed the capability to generate a maximum output voltage of 8.2 V and a current of 145 μA for the harvester area of 1.7×1.7 cm^2 at a frequency of 0.3 Hz. The future work includes fixing the harvester on porcine organs to activate the harvester from contraction and expansion of the organ's muscle.

Han *et al.* [26] reported a three-dimensional piezoelectric polymer used in energy harvesting applications and biomedical implants. The three-dimensional shape was created through a controlled and nonlinear buckling process of

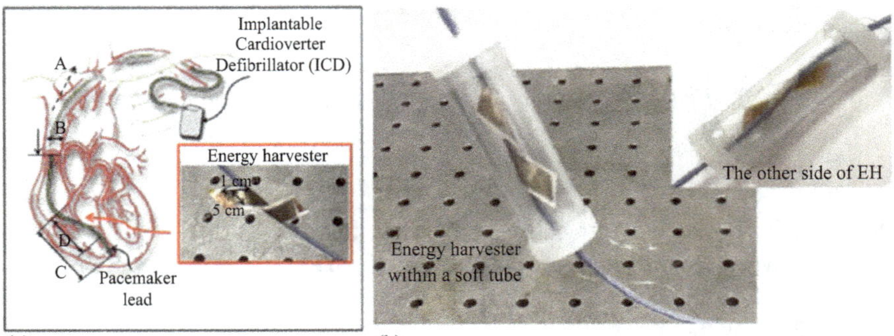

(a) (b)

Figure 3.20 (a) PVDF harvester wrapped around pacemaker lead and (b) the harvester inside a soft tube [24]

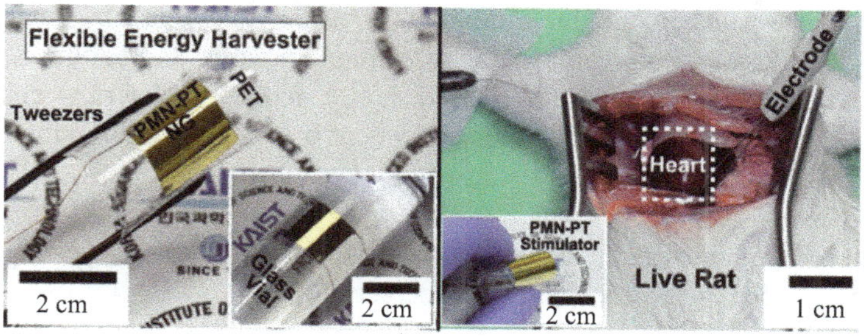

Figure 3.21 Flexible PZT harvester, and it is in vivo testing on rats [25]

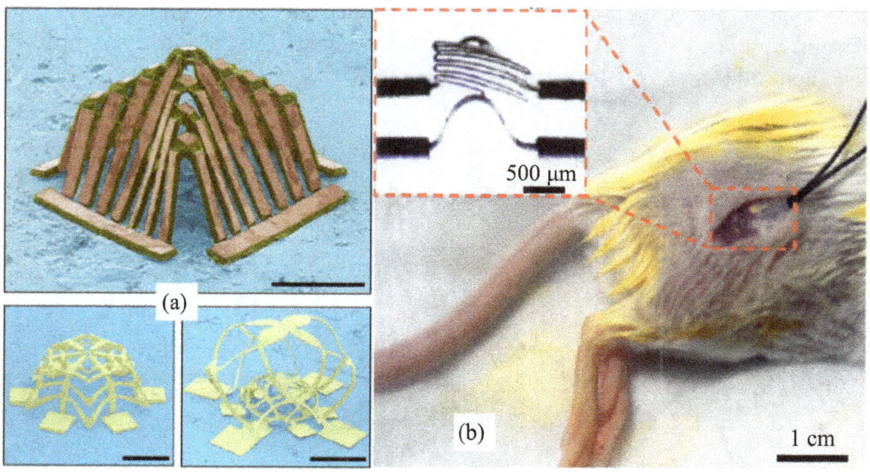

Figure 3.22 (a) Some of the 3D created PZT shapes and (b) in vivo testing of the harvester [26]

two-dimensional electrode patterns with thin-film piezoelectric into complicated 3D shapes, improving their operation modes. More than 20 different shapes were created, as shown in Figure 3.22(a). Different shapes were tested to show the versatility of the 3D approach, and the resultant mean square voltage ranged from 2 to 790 mV. One of the 3D harvesters' shapes was tested in vivo by implanting the harvester in the subcutaneous region near the mouse leg's right flank. The harvested voltage was more than one mV with the mouse's movement, as shown in Figure 3.22(b).

Li *et al.* [27] introduced a piezoelectric energy harvester based on a single-wire generator (SWG) technique. In this technique, a single nanowire of PZT is fabricated by the vapor deposition technique then fixed tightly to a flexible polyimide substrate using a silver paste. Then, the harvester is enclosed by a layer of flexible polymer to keep out of body fluids. The harvester was tested in vivo in two positions using a Sprague Dawley adult rate of 220 g. The first experiment was done by gluing the harvester on the rat's diaphragm using synthetic absorbable tissue adhesive. During the rate breathing, the diaphragm expands and contracts, causing the PZT wire to be stretched and released, hence generates electrical charge pulsations of 2 mV and 4 pA. The second experiment was done by gluing the harvester on the rat's heart. The harvested voltage was about 3 mV and 30 pA, as shown in Figure 3.23.

Table 3.2 summarizes the contribution of previous work in harvesting internal body energy utilizing the concept of piezoelectric.

3.2.3 Fuel cell concept

Two concepts of the fuel cell can be implanted inside the human body. The first one is the biofuel cell (BFC), in which a common body fluid (like glucose) is used as fuel for the biofuel cell, and reactions result in electricity production. Second is the microbial fuel cell (MFC), in which the cell is implanted in the intestinal, utilizing its content and microorganisms to generate electricity.

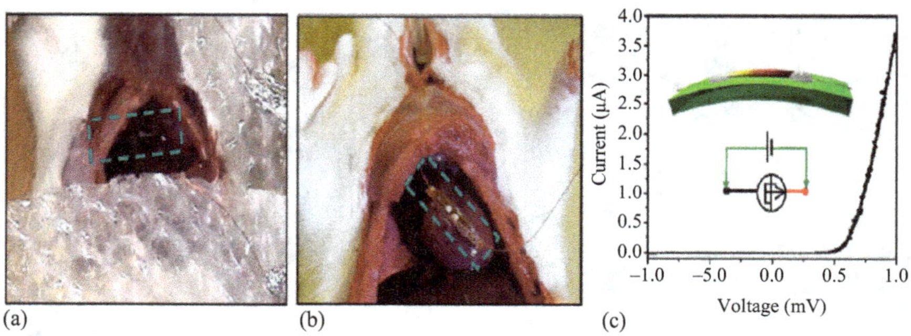

Figure 3.23 In vivo testing of the harvester on rat diaphragm and heart [27]

Table 3.2 Piezoelectric concept of previous work

Author	Generated power	Region of implantation	Special surgery
Zhang *et al.* [1]	30 nW	Ascending aorta	y
Deterre *et al.* [7]	3 μJ/cm^3/cycle	Cardiac cavity	n
Dagdeviren *et al.* [10]	1.2 μW/cm^2	Heart, lungs, and diaphragm	y
Lu *et al.* [9]	3 V	Heart	y
Jiang *et al.* [22]	0.91 Vpp, 2.5 μA	Skin	y
Dong *et al.* [23]	0.65 Vpp	Inside the heart cavity	n
Dong *et al.* [24]	0.5 V, 43 nA	Pacemaker lead	n
Hwang *et al.* [25]	8.2 V, 145 μA	Heart surface	y
Han *et al.* [26]	1 mV	Subcutaneous region near The right flank	y
Li *et al.* [27]	3 mV and 30 pA 2 mV and 4 pA	Heart and diaphragm	y

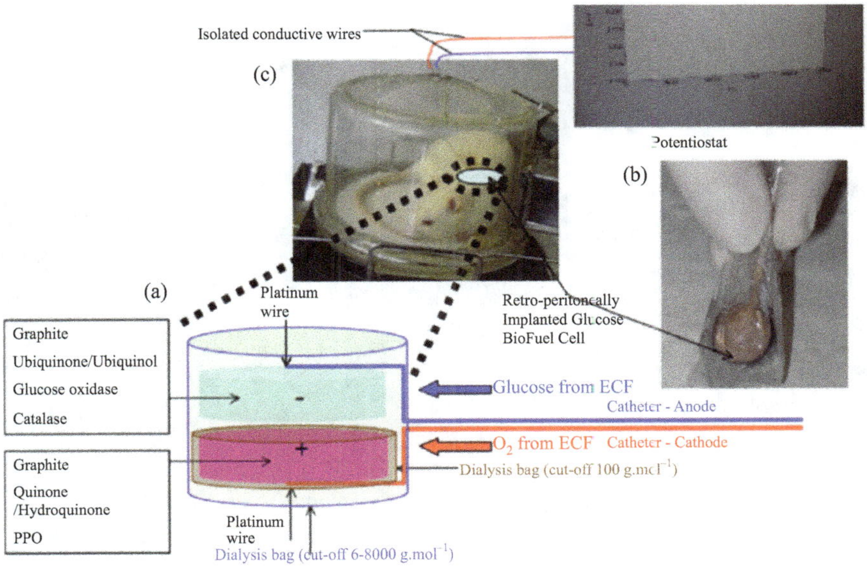

Figure 3.24 Composition of the biofuel cell and testing it in vitro [11]

For biofuel cells, Zebda *et al.* [12] reported that the most common BFC is the glucose biological fuel cell that uses glucose as fuel due to its availability in the body fluids. Cinquin *et al.* [11] presented the first working implantable glucose biofuel cell. The cell comprises the anode that consists of a compacted graphite disc having glucose oxidase, ubiquinone, and catalase. The cathode also consists of a compacted graphite disc having polyphenol oxidase and quinhydrone, as shown in Figure 3.24. Both electrodes are enclosed in dialysis bags that permit glucose and dioxygen to flow inside the cell. When the ubiquinol is oxide inside the cell, the electrical charges are generated. The cell was tested in vitro using glucose solution, and the resulted maximum harvested power was about 1.65 μW. The cell was also tested in vivo with a male Wistar rat of 514 g weight by implanting the cell in its retroperitoneal space. Results showed the cell capability to harvest around 2 μW for several hours.

Han *et al.* [28] developed a MFC consisting of $10 \times 10 \times 2.5$ cm^3 and made of Plexiglas. Each chamber has an electrode, the first one is an anode made of active carbon fiber cloth with a dimension of 6×6 cm^2, and the second electrode is the cathode made of carbon paper electrode, as shown in Figure 3.25. A carbon paper was placed between the two chambers to reduce the cathode's oxygen diffusion in order not to affect the electricity generation. The cell was tested experimentally, and a maximum power density of 73.3 mW/m^2 could be obtained at 0.55 V OCV.

Katic *et al.* [13] developed an energy harvester that combines both the glucose fuel cell and the thermoelectric generator (with an area of 2.3 mm^2) and supported with a converter and control circuit to extract maximum power with the highest efficiency. The harvester could harvest about 66 μW at an efficiency of 89.5%.

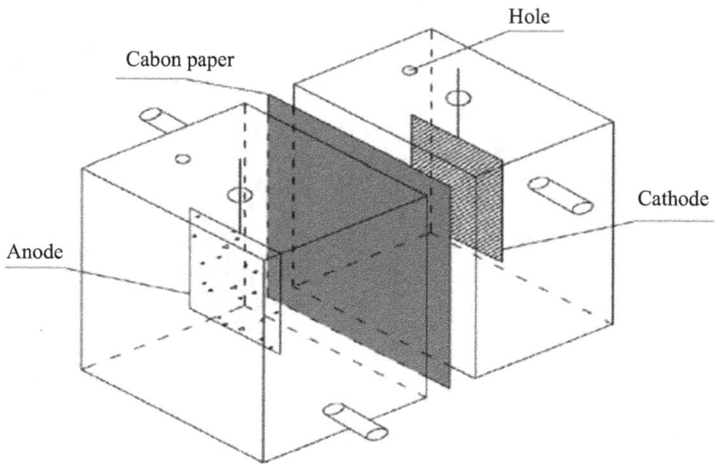

Figure 3.25 Composition of the microbial fuel cell [28]

Table 3.3 Fuel cell concept previous work

Author	Generated power	Region of implantation	Special surgery
Cinquin *et al.* [11]	2 μW	Retroperitoneal space	y
Han *et al.* [28]	73.3 mW/m^2	Large intestine	y
Katic *et al.* [13]	66 μW	Retroperitoneal space	y

Table 3.3 summarizes previous work's contribution to harvesting internal body energy utilizing the fuel cell concept.

3.2.4 Electrostatic concept

Deterre *et al.* [29] used the concept of electrostatic combs to harvest energy from pressure variation inside the heart chamber, which is a step toward a leadless PM that could be implanted in right or left ventricles. The harvester working principle utilizes the pressure variation inside heart chambers to deform micro bellows, as shown in Figure 3.26. This micro-bellows is connected by a rod with the moving combs (red). During bellows deformation, those moving combs have a relative displacement with the static combs (blue), which leads to energy generation. The harvester was designed, optimized, fabricated, and tested with achieving harvested energy of a few tens of micro-joules.

Based on the review, the concept of piezoelectric is chosen for building a new harvesting device. It is compatible with the MEMS fabrication technology. It can be miniaturized to fit inside the human body besides its advantages over the other concepts such as having the superiority in endurance, free maintenance, and high cyclic life. Using these materials in applications where reachability is approximately

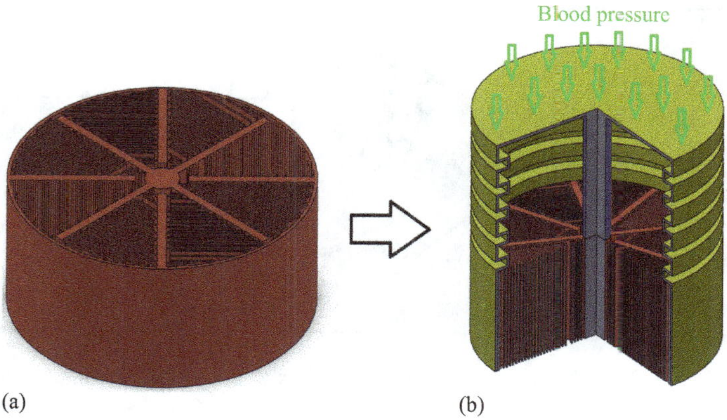

Figure 3.26 (a) Electrostatic comp-drive and (b) assembly of the electrostatic harvester and the concept [29]

Table 3.4 Electrostatic concept previous work

Author	Generated power	Region of implantation	Special surgery
Deterre *et al.* [29]	Few tens of μJ	cardiac cavity	y

impossible, like implanted harvesters inside the human body, is more suitable. Another advantage is that the working principle of the piezoelectric is not restricted by light like in solar cells, or motion between components like in the electromagnetic concept, or the need for a catalyst like glucose in biofuel cells.

Table 3.4 summarizes previous work's contribution to harvesting internal body energy utilizing the electrostatic concept.

3.3 Proposed solutions

One of the proposed and new solutions is the blood flow and pressure inside the large veins, where drawing some of the blood flow and pressure energy would not have a severe effect on the circulatory system. Most research avoided dealing with the concepts of flow and pressure inside veins because of their low values, unlike arteries with high pressure that could supply high power to the harvesting device. The problem associated with arteries is that they carry the oxygenated blood to vital organs and has only one way to the organs, so blocking it with a harvester may seriously affect the amount of oxygen delivered to organs and cause problems. In contrast, veins carry the deoxygenated blood back to the lungs and mostly bypass partially or wholly blockage cases.

This solution was initiated by Abdelmageed *et al.* [5] to overcome the problem of making an extra surgery to implant the harvester inside the human body by integrating the harvester in the biomedical device wires that pass through the

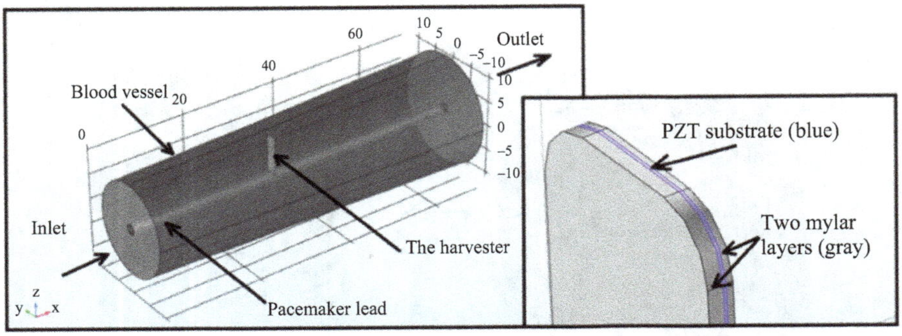

Figure 3.27 Harvester configuration based on the flow concept [5]

superior vena cava (SVC) vein. The study started with extensive research about the safe limit of SVC blockage and its effects on the circulatory system to determine the harvester's dimensions. This study introduces two possible scenarios.

The first one utilizes the pulsatile blood flow nature inside the vein to deform micro PZT beams integrated into the PM lead and generate electricity. The optimum number of harvesters per lead is eight harvesters (based on the allowable obstruction limit induced by 30% blockage of SVC). The harvester cross-section area is 2 mm by 8 mm (based on Hemodialysis Catheters' cross-sectional area) and arranged at 120° spacing between, as shown in Figure 3.27. The simulation and modelling results showed enough power (sum of about 9 μW with suitable synchronization circuit) at a loading resistance of 35 MΩ that could be used to recharge the PM battery.

The second one [6] depends on covering the PM lead (which is a wire inserted through SVC during PM implantation surgery) with a PDMS layer and creates some cavities in this layer. These cavities would be covered with a thin diaphragm of PDMS that will be in contact with the pulsatile blood pressure inside SVC. Under this diaphragm, there will be a PZT layer with an area of 2.5×6 mm^2 and thickness of 0.028 mm working as a cantilever fixed in the cavity wall from one side and free from the other side, as shown in Figure 3.28. When the pressure harmonically varies inside SVC, the PDMS diaphragm deflects up and down, causing the PZT layer to deform and hence generate pulsations of electricity for each heartbeat. The harvesting process was simulated and showed that power of 52 nW could be harvested for each cavity, which could be used to recharge the PM battery and avoid replacing it.

This study lakes the micro-fabrications and vital testing in vivo to ensure the compatibility and sufficient power generation to power the implanted device.

3.4 Future challenges and research opportunities

As discussed in Section 3.1, most of the disadvantages of previous techniques are inducing some restriction on moving vital organs, blocking blood or air flow stream, or needing special surgery to implant the harvester in its targeted position

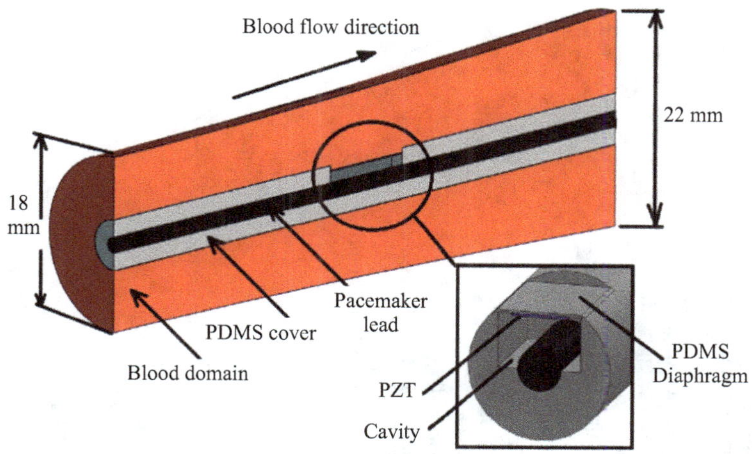

Figure 3.28 Harvester configuration based on the pressure concept [6]

inside the human body. But why is a harvester necessary to be implanted inside the human body? Of course, to power an impeded biomedical assisting device like a PM. So, suppose that the harvester is designed (taking into consideration the biomedical effects) to be integrated into the implanted devise like [29] or attached to a part of it like [5]; in that case, it will be an excellent research point.

This research discipline requires many conditions to be successful. First, to have a strong background in the field and previous work done to see if the new idea is applicable or not and start where the others ended. Second, to have a specialized simulation program to test the new design before the fabrication, this step is necessary to save time and money and predict the resultant output power and determine if the design is reliable or not. Third, fabricating the system—which needs MEMS fabrication techniques—and tests it in vitro then in vivo to approve the device compatibility to be tested on a human's body. The last step is to test the device on the human body by implanting the device inside patients and making statistical analysis and medical study to determine the device's biocompatibility and its side effects. Those steps and conditions are the real challenges that must be overcome to successfully use the internal energy inside the human body and utilize it to power a vital biomedical implantable device.

Regarding the proposed solution in Section 3.3, there are many investigations and research that could be done by conducting more experimental work to address the effect of the harvester's implantation inside the SVC and partially blocks the bloodstream. The harvesting system should also be fabricated in the MEMS scale and tested in vivo for reliability and effectiveness to see its capability for powering cardiac PM for both concepts flow and pressure.

More development could be done on the experimental setup used in Section 3.3 to be as a guide for researchers working in the field of biomedical engineering, such as using elastic piping with conical shape to simulate the SVC shape and mechanical properties physically.

Table 3.5 Output power summary from previous work

Concept	Author	Generated power	Region of implantation	Special surgery
Electromagnetic	Pfenniger *et al.* [2]	42 nW	Iliac artery	y
	Pfenniger *et al.* [3]	1 mW	Internal thoracic artery	y
	Kim *et al.* [4]	3.4 mW	Arteries	y
	Zurbuchen *et al.* [8]	85 μW	Outer surface of the heart	y
	Tholl *et al.* [14]	14±8 μW at 60 bpm82±17 μW at 200 bpm	Inside right ventricular	y
	Zurbuchen *et al.* [15]	0.78 and 1.7 μW for 84 and 160 bpm	Inside the heart cavity	y
	Pancharoen *et al.* [16]	0.96 μW	Hip	y
	Beyaz *et al.* [17]	1.05 mV	Inside the human body	y
	Sahara *et al.* [18]	111 μW	Muscles	y
	Haeberlin *et al.* [19]	10.2±4.8 μW	Intracardiac blood flow	y
	Pfenniger *et al.* [20]	135 μW	Arteries	y
	Zurbuchen *et al.* [21]	16.7 μW	Cardiac wall	y
Piezoelectricity	Zhang *et al.* [1]	30 nW	Ascending aorta	y
	Deterre *et al.* [7]	3 μJ/cm^3/cycle	Cardiac cavity	n
	Dagdeviren *et al.* [10]	1.2 μW/cm^2	Heart, lungs, and diaphragm	y
	Lu *et al.* [9]	3 V	Heart	y
	Jiang *et al.* [22]	0.91 Vpp, 2.5 μA	Skin	y
	Dong *et al.* [23]	0.65 Vpp	Inside the heart cavity	n
	Dong *et al.* [24]	0.5 V, 43 nA	Pacemaker lead	n
	Hwang *et al.* [25]	8.2 V, 145 μA	Heart surface	y
	Han *et al.* [26]	1 mV	Subcutaneous region near The right flank	y
	Li *et al.* [27]	3 mV and 30 pA2 mV and 4 pA	Heart and diaphragm	y
	Mostafa *et al.* [5]	9 μW	Superior vena cava (SVC)	n
	Mostafa *et al.* [6]	52 nW	Superior vena cava (SVC)	n
Biofuel cell	Cinquin *et al.* [11]	2 μW	retroperitoneal space	y
	Han *et al.* [28]	73.3 mW/m^2	Large intestine	y
	Katic *et al.* [13]	66 μW	Retroperitoneal space	y
Electrostatic	Deterre *et al.* [29]	Few tens of μJ	Cardiac cavity	y

3.5 Conclusion

The human body could be considered an energy factory for various types of energies such as kinetic, chemical, and flow power. These types of energies can contribute to power implanted devices inside the human body through the energy harvesting process. The energy harvesting process should have some specifications related to the physical harvester itself, like the harvester's biocompatibility to be implanted inside the human body, the small dimensions of the implanted harvester, and the applicability of implanting the harvester in its position without an extra or big surgery. While the other part is associated with the harvesting process, which could be clarified through answering questions like; Will the harvesting process affect human health in the long run? Is the harvested energy enough to supply the implanted devices with the needed power? Does the harvesting process affect the human's activity or his health and fitness state? Could the harvesting process be affected by external energy sources like electromagnetic waves? And so on.

The previous trials to harvest power from internal human energy are concluded in Table 3.5:

List of acronyms

BC	Buffer chamber
BFC	Biofuel cell
MC	Main chamber
MFC	Microbial fuel cell
NdFeB	Neodymium
PDMS	Polydimethylsiloxane
PM	Pacemaker
PMN-PT	Lead magnesium niobate-lead titanate
PVDF	Polyvinylidene fluoride
PZT	Lead zirconate titanate
SmCo	Samarium cobalt
SVC	Superior vena cava
SWG	Single-wire generator

References

[1] H. Zhang, X.-S. Zhang, X. Cheng, *et al.*, "A flexible and implantable piezoelectric generator harvesting energy from the pulsation of ascending aorta: In vitro and in vivo studies," *Nano Energy*, vol. 12, pp. 296–304, 2015, doi: 10.1016/j.nanoen.2014.12.038.

[2] A. Pfenniger, L. N. Wickramarathna, R. Vogel, and V. M. Koch, "Design and realization of an energy harvester using pulsating arterial pressure,"

Med. Eng. Phys., vol. 35, no. 9, pp. 1256–1265, 2013, doi: 10.1016/j.medengphy.2013.01.001.

[3] A. Pfenniger, R. Vogel, V. M. Koch, and M. Jonsson, "Performance analysis of a miniature turbine generator for intracorporeal energy harvesting," *Artif. Organs*, vol. 38, no. 5, 2014, doi: 10.1111/aor.12279.

[4] S. H. Kim, C. H. Yu, and K. Ishiyama, "Rotary-type electromagnetic power generator using a cardiovascular system as a power source for medical implants," *IEEE/ASME Trans. Mechatron.*, vol. 21, no. 1, pp. 122–129, 2016, doi: 10.1109/TMECH.2015.2436910.

[5] M. G. Abdelmageed, A. M. R. Fath El-Bab, and A. A. Abouelsoud, "Design and simulation of pulsatile blood flow energy harvester for powering medical devices," *Microelectron. J.*, vol. 86, pp. 105–113, 2019, doi: 10.1016/j.mejo.2019.02.021.

[6] M. G. Abdelmageed, A. M. R. Fathelbab, and A. A. Abouelsoud, "Design and simulation of an energy harvester utilizes blood pressure variation inside superior vena cava," Mechatronics and Robotics Department , E-JUST, Alexandria 21934, Egypt," in *SICE Annual Conference 2019*, 2019, pp. 1107–1112.

[7] M. Deterre, E. Lefeuvre, Y. Zhu, M. Woytasik, B. Boutaud, and R. D. Molin, "Micro blood pressure energy harvester for intracardiac pacemaker," *J. Microelectromech. Syst.*, vol. 23, no. 3, pp. 651–660, 2014, doi: 10.1109/JMEMS.2013.2282623.

[8] A. Zurbuchen, A. Haeberlin, L. Bereuter, *et al.*, "The Swiss approach for a heartbeat-driven lead- and batteryless pacemaker," *Hear. Rhythm*, vol. 14, no. 2, pp. 294–299, 2017, doi: 10.1016/j.hrthm.2016.10.016.

[9] B. Lu, Y. Chen, D. Ou, *et al.*, "Ultra-flexible piezoelectric devices integrated with heart to harvest the biomechanical energy," *Sci. Rep.*, vol. 5, no. 1, p. 16065, 2015, doi: 10.1038/srep16065.

[10] C. Dagdeviren, B. Yang, Y. Su, *et al.*, "Conformal piezoelectric energy harvesting and storage from motions of the heart, lung, and diaphragm.," *Proc. Natl. Acad. Sci. USA*, vol. 111, no. 5, pp. 1927–1932, 2014, doi: 10.1073/pnas.1317233111.

[11] P. Cinquin, C. Gondran, F. Giroud, *et al.*, "A glucose biofuel cell implanted in rats," *PLoS One*, vol. 5, no. 5, pp. 1–7, 2010, doi: 10.1371/journal.pone.0010476.

[12] A. Zebda, J. Alcaraz, P. Vadgama, *et al.*, "Challenges for successful implantation of biofuel cells," *Bioelectrochemistry*, vol. 124, pp. 57–72, 2018, doi: 10.1016/j.bioelechem.2018.05.011.

[13] J. Katic, S. Rodriguez, and A. Rusu, "A high-efficiency energy harvesting interface for implanted biofuel cell and thermal harvesters," *IEEE Trans. Power Electron.*, vol. 33, no. 5, pp. 4125–4134, 2018, doi: 10.1109/TPEL.2017.2712668.

[14] M. Tholl, A. Haeberlin, B. Meier, *et al.*, "An intracardiac flow based electromagnetic energy harvesting mechanism for cardiac pacing," *IEEE Trans. Biomed. Eng.*, vol. 66, no. 2, pp. 530–538, 2019, doi: 10.1109/TBME.2018.2849868.

[15] A. Zurbuchen, A. Haeberlin, L. Bereuter, *et al.*, "Endocardial energy harvesting by electromagnetic induction," *IEEE Trans Biomed. Eng.*, vol. 65, no. 2, pp. 424–430, 2018, doi: 10.1109/TBME.2017.2773568.
[16] K. Pancharoen, D. Zhu, and S. P. Beeby, "A Hip implant energy harvester," *J. Phys. Conf. Ser.* 557, 2014, doi: 10.1088/1742-6596/557/1/012038.
[17] M. T. Beyaz, H. C. Baelhadj, S. Habibiabad, S. S. Adhikari, H. Davoodi, and V. Badilita, "A non-resonant kinetic energy harvester for bioimplantable applications," *Micromachines*, p. 12, 2018, doi: 10.3390/mi9050217.
[18] G. Sahara, W. Hijikata, K. Tomioka, and T. Shinshi, "Implantable power generation system utilizing muscle contractions excited by electrical stimulation," *Eng. Med.*, 2016, doi: 10.1177/0954411916638889.
[19] A. Haeberlin, Y. Rösch, M. Tholl, *et al.*, "Intracardiac turbines suitable for catheter-based implantation – an approach to power battery- and leadless cardiac pacemakers?," *IEEE Trans. Biomed. Eng.*, vol. 9294, no. c, 2019, doi: 10.1109/TBME.2019.2932028.
[20] A. Pfenniger, D. Obrist, A. Stahel, V. M. Koch, and R. Vogel, "Energy harvesting through arterial wall deformation: design considerations for a magneto-hydrodynamic generator," *Med. Biol. Eng. Comput.*, vol. 51, no. 7, pp. 741–755, 2013, doi: 10.1007/s11517-012-0989-2.
[21] A. Zurbuchen, A. Pfenniger, A. Stahel, *et al.*, "Energy harvesting from the beating heart by a mass imbalance oscillation generator," *Ann. Biomed. Eng.*, vol. 41, no. 1, pp. 131–141, 2013, doi: 10.1007/s10439-012-0623-3.
[22] L. Jiang,Y. Yang, R. Chen, *et al.*, "Flexible piezoelectric ultrasonic energy harvester array for bio-implantable wireless generator," *Nano Energy*, vol. 56, pp. 216–224, 2019, doi: 10.1016/j.nanoen.2018.11.052.
[23] L. Dong, A. Closson, M. Oglesby, *et al.*, "In vivo cardiac power generation enabled by an integrated helical piezoelectric pacemaker lead," *Nano Energy*, vol. 66, p. 104085, 2019, doi: 10.1016/j.nanoen.2019.104085.
[24] L. Dong, X. Han, Z. Xu, *et al.*, "Flexible porous piezoelectric cantilever on a pacemaker lead for compact energy harvesting," *Adv. Mater. Technol.*, vol. 4, no. 1, pp. 1–9, 2019, doi: 10.1002/admt.201800148.
[25] G. Hwang, H. Park, J. Lee, *et al.*, "Self-powered cardiac pacemaker enabled by flexible single crystalline PMN-PT piezoelectric energy harvester," *Adv. Mater.*, vol. 26, no. 28, pp. 4880–4887, 2014, doi: 10.1002/adma.201400562.
[26] M. Han, H. Wang, Y. Yang, *et al.*, "Three-dimensional piezoelectric polymer microsystems for vibrational energy harvesting, robotic interfaces and biomedical implants," *Nat. Electron.*, vol. 2, no. 1, pp. 26–35, 2019, doi: 10.1038/s41928-018-0189-7.
[27] Z. Li, G. Zhu, R. Yang, A. C. Wang, and Z. L. Wang, "Muscle-driven in vivo nanogenerator," *Adv. Mater.*, vol. 22, no. 23, pp. 2534–2537, 2010, doi: 10.1002/adma.200904355.
[28] Y. Han, C. Yu, and H. Liu, "A microbial fuel cell as power supply for implantable medical devices," *Biosens. Bioelectron.*, vol. 25, no. 9, pp. 2156–2160, 2010, doi: 10.1016/j.bios.2010.02.014.
[29] M. Deterre, "Toward an Energy Harvester for Leadless Pacemakers," Universit´e Paris Sud, 2013.

Part II

RF energy harvesting

Chapter 4

Cognitive and energy harvesting-based D2D communication in wireless multimedia sensor networks underlying multi-tier cellular networks

Fereidoun H. Panahi[1] *and Farzad H. Panahi*[1]

Abstract

In this chapter, we use a statistical approach based on stochastic geometry to model and evaluate the performance of a cognitive D2D-enabled wireless multimedia sensor network (WMSN) in the presence of a multi-tier multi-channel cellular network. It will be shown that energy harvesting can efficiently power D2D communications in the WMSN. We will also observe that cognitive channel access can help improve the quality of service of D2D connections in the WMSN.

4.1 Motivations

As the capacities of devices to capture and play video continue to develop, demands for dealing with mobile multimedia data traffic are also growing. The demands have affected not only the cellular networks but also sensor networks. Wireless multimedia sensor networks (WMSNs), i.e., wireless sensor networks (WSNs) for targeting the multimedia data, have been developed. In the WMSN, a sensor that can gather multimedia information is called a multimedia sensor; it sends the collected data to a central aggregator (CA) for reporting. The multimedia sensor can be used in various fields, such as multimedia transmission, traffic control, nature, people, and machinery monitoring. An application in WMSNs may require several months of operating lifetime. Since WMSN consists of sensors, spectral and energy-efficient technologies are needed to ensure the quality of service (QoS) and to overcome limited computing power and energy of sensors. The WMSN may be applied to future smart cities, surveillance networks, or industries when these problems of limitation are minimized. Moreover, a centralized structure

[1]Department of Electrical Engineering, University of Kurdistan, Sanandaj, Kurdistan, Iran

is one of the fundamental deficiencies that cause more problems. Unlike the typical WSN system, the amount of traffic is very large in the WMSN. It is a significant burden for the CA, and it can degrade the overall performance of WMSN. Thus, technologies such as device-to-device (D2D) communication should be considered in the WMSN to reduce the dense overhead to the central system by distributing the traffic as much as possible.

4.2 Introduction

Harvesting energy from non-traditional sources such as ambient interference emerges as an attractive approach to power low-energy wireless communication devices and sensor nodes. To improve spectrum utilization and mitigate the scarcity of spectrum, innovative solutions such as cognitive radio (CR) and D2D technology have been recently proposed to underlay current cellular and WSNs. On the other hand, WMSNs have been improved with the increase of multimedia files and data. A sensor that can collect multimedia data is called a multimedia sensor in WMSN. In WMSNs, a centralization problem can occur because of large-size multimedia data. Therefore, it is necessary to deploy D2D communication into WMSNs. In this chapter, we focus on a cognitive D2D-enabled WMSN based on cellular networks. More specifically, we consider a cognitive D2D-enabled WMSN based on a multi-tier and multi-channel cellular network where the sensors are able to use only the harvested radio frequency (RF) energy from the ambient interference that results from the concurrent downlink cellular transmissions. After harvesting sufficient energy, each cognitive sensor performs spectrum sensing to opportunistically access a predefined nonexclusive D2D channel. In the WMSN, a sensor can be connected to a CA via a designated communication channel or connected to another sensor via a D2D link (on a predefined nonexclusive D2D channel). One fundamental question is how to share the spectrum resources between cellular and D2D communications. The D2D can be divided into two groups depending on the form of spectrum sharing: in-band and out-of-band. The in-band corresponds to D2D using the cellular spectrum, and the out-of-band refers to D2D using bands other than the cellular band. The equivalent interpretation of out-of-band D2D is that a proportion η of the orthogonal frequency-division multiple access (OFDMA) resource blocks is assigned to D2D while the remaining resource blocks are used by cellular. In-band D2D can be divided into two more categories: overlay and underlay. Overlay means the usage of orthogonal time/frequency resources by cellular and D2D, while underlay means that cognitive D2D sensors have opportunistic access to the time/frequency resources of cellular users. With this approach, our analysis is conducted for two D2D spectrum sharing scenarios: overlay and underlay in-band D2D. Also, in underlay in-band D2D, cognitive D2D communication is considered. We use the term "cognitive D2D communication" in the sense that spectrum sensing is performed at each cognitive sensor prior to transmission to ensure that the received interference from any neighboring cellular base station (BS) on the channel designated for D2D transmission is smaller than a predetermined sensing threshold γ; otherwise, the channel is not available to be used by the cognitive D2D sensor. In this chapter, we use a statistical approach based on stochastic geometry to model and evaluate the performance of the cognitive D2D-enabled WMSN in the presence of a multi-tier multi-channel cellular network. It will be shown that energy harvesting (EH) can efficiently power D2D communications. We will also observe that cognitive channel access can help improving the QoS of D2D connections in the WMSN.

4.3 Green wireless sensor network

Wireless sensors form an essential component of IoT-based smart applications. They are miniaturized, low-cost devices capable of detecting parameters of interest and regularly transmitting the findings to a collection point. They are mainly battery powered. A collection of integrated wireless sensors known as WSN is being deployed. The WSN architecture consists of wireless sensor nodes connected to a BS operating as a sink node. The nodes supply energy and the computing resources are of considerable significance in formulating WSN requirements. For a WSN, there are two topologies, which can be either infrastructure based or ad hoc based. In infrastructure-based topology, the BS performs the main network core functions of scheduling, interference management, allocation of resources, routing, and acting as a link to the local area network. On the other hand, there is no centralized BS in ad-hoc-based system architecture, because each sensor will take on the task of coordinating the system [1]. An overview of low-power wireless networking sensor standards, including Bluetooth, LR-WPAN, Long Range (LoRa), UWB, WiMAX, Mobile Communication, and Wi-Fi, is outlined in [2–8].

Wireless sensors are configured to exchange data continuously and independently over a long period of time. The important examples of WSNs that continuously transmit data are healthcare, weather forecasts, traffic updates, water quality, and embedded system applications. As mentioned, the wireless sensors in a WSN are operated by batteries. Nevertheless, the replacement and servicing costs of depleted batteries are prohibitive and are difficult to replace, particularly in remote areas due to geographical limitations. Developing energy-efficient solutions to improve battery life and reduce replacement and servicing costs has thus been crucial for WSN's sustainability.

Energy-efficient approaches have thus become critical. The overview of the current major established energy-saving strategies is as follows: (a) radio optimization techniques [9–14], (b) sleep/wakeup schemes [15–19], (c) green energy and wireless charging strategies [20–29], (d) D2D communication [30,31], and (e) CR [32–34]. Figure 4.1 provides the simplified taxonomy of the green WSNs.

A detailed overview of the major energy-saving strategies is discussed in this section. In this chapter, we focus on radio optimization, sleep/wakeup, green energy and wireless charging, D2D communication, and CR techniques. For more

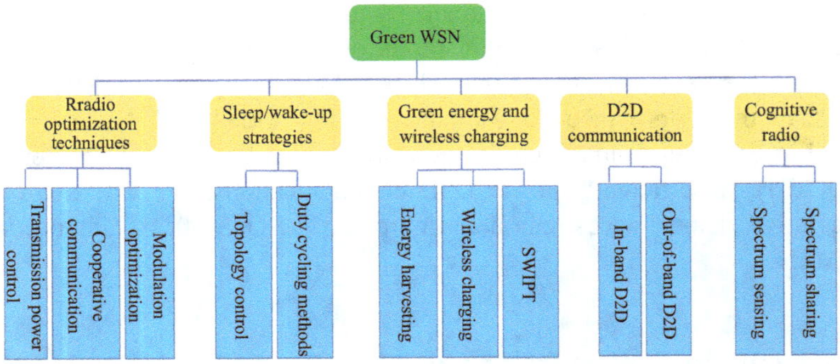

Figure 4.1 Taxonomy of energy-saving strategies for green WSNs

information on the other energy-saving strategies such as energy-efficient routing and WSN's architecture, and data reduction mechanisms please refer to [35,36].

4.3.1 Radio optimization techniques

The radio unit is the most dominant energy-consuming unit in the WSN. The radio unit energy exhaustion is triggered by two parts: (a) the circuit power and (b) the broadcasting signal power. In short-range communications, more energy is used for powering the circuit, while the transmitted signal utilizes more power in long-range communications. Several studies [10,37] have looked at techniques to improve energy efficiency (EE) by dynamically adjusting the transmission power level. Also in [9], the authors suggested an adaptive energy-saving cooperative topology, in which sensor nodes with higher residual energy are allowed to raise transmitting power leading to other sensor nodes to reduce their own transmitting power. The suggested topology further decreases interference and increases connectivity due to a decrease in transmission power. However, there is supposed to be an increase in delay, as more hops will be required for packet forwarding. Nonetheless, the issue of delay can be solved by cooperative communication between adjacent sensor nodes, which produces a virtual multi-antenna (spatial diversity) environment. Reference [12] evaluated the energy consumption of both single-input single-output (SISO) and virtual multi-antenna [multiple-input and multiple-output (MIMO)] systems and found that virtual multi-antenna systems would achieve higher energy savings and reduce end-to-end (E2E) delays over some distances of propagation. The authors in [13], on the other hand, investigated the relationship between energy usage, transmission time, and bit error rate. The study revealed that optimizing the transmission time may have reduced the energy consumption needed to reach a given rate of bit error as well as the need for delay. In addition, the authors in [14] provided a comparative analysis on the EE of three modulation schemes in order to select the optimal modulation scheme which results in the lowest energy consumption at different distances between nodes. Several studies [38–40] have considered transmission power and rate allocation in the uplink of a cellular wireless network to tackle the issue of joint transmission power and rate allocation.

4.3.2 Sleep/wake up techniques

Switching off non-active transceivers (i.e., putting them into sleep mode) has become one of the main solutions to the realization of EE in information and communication technologies, as it can save a significant amount of resources. The idea behind the proposed solution is to exploit the dense and redundant deployment of sensor nodes, contributing to a limited region of coverage. The on/off switching strategy of the sensor nodes is more suitable for enhancing the WSN EE and prolonging the battery life of the wireless sensors. That being said, the issue of coverage should be considered, and the remaining active nodes should ensure it. The authors in [15] suggested a subset mechanism capable of maintaining the network coverage while minimizing the energy consumption of the network by activating only a subset of nodes, with the minimum overlap area. In the meanwhile, the authors in [16] modeled the EE issue as a subset selection problem of active connected sensors for the correlated data payload gathering. The sensor information of non-active sensor nodes can be acquired from those of active nodes using spatial correlation, thus making it a good strategy.

Authors in [41] have developed an adaptive staggered sleep protocol (ASLEEP) for efficient power management in WSNs targeted to periodic data acquisition. The proposed protocol dynamically adjusts nodes' sleep schedules to match the network demands, even in time-varying operating conditions. Moreover, it does not require any a priori knowledge of the network topology or traffic pattern. On the contrary, to have a sensor node switched on/off dependent on the network traffic activity, one can exploit the duty cycling schemes. Duty cycling schemes can be grouped into three types: asynchronous, on-demand, and scheduled rendezvous. Meanwhile, duty cycling-based protocols may be the most important of all the energy-efficient techniques [18,42]. This should be taken into consideration, however, that the low duty cycle has the potential to retain a significant amount of energy but can result in high delays in communication. The protocol parameters may be adjusted for ease before deployment to reduce the delay, but this can result in inflexibility, or adjustable configurations can be deployed to indicate instant traffic conditions. In addition, as discussed in [43], the active node duration for optimizing power consumption is a function of traffic load, delay requirements, buffer overflows, or harvested energy.

4.3.3 Energy harvesting and wireless charging techniques

The main characteristics of wireless sensors' energy sources such as sustainability and reliability, as well as greenhouse gas emission cuts, can be achieved by advancements in renewable energy technologies (RETs) [44]. The RET is one of the innovative ways of handling the WSN EE problem in rural and remote areas. Batteries are difficult to replace in this terrain profile due to geographical limitations (challenging terrain) which make access to these sites difficult [20]. The solar cells have low maintenance requirements and good efficiency, with a life expectancy of 20–30 years. In addition, new sensor technologies have arisen that can harness power from their immediate surroundings, such as wind and kinetic energy [21]. The harvested energy is then transferred to electrical signals, either immediately consumed or saved for later use. For instance, using solar panels to charge a rechargeable battery throughout the day. Nodes turn to conservative mode at night, pulling energy from the stored electricity.

The development of wireless power charging technologies has allowed energy-constraint devices to achieve more controllable functionality, thus increasing the sustainability and efficiency of WSNs [35]. We can see today that the idea of wireless power charging has already been implemented in various applications, such as power medical sensors and implantable devices [45], restocking sensors embedded in concrete walls [46], and powering a ground sensor from an unmanned aerial vehicle [47]. Generally speaking, energy transfer strategies can be categorized into (a) non-radiative coupling-based charging, which can also be grouped into three types: magnetic inductive coupling [48], magnetic resonance coupling [49], and capacitive coupling [50]; and (b) radiative RF-based charging, which can also be divided into two approaches: RF power beamforming directive and RF power transfer non-directive. However, the attainable amount of coupling capacitance in capacitive coupling depends on the device's available area [51]. Nonetheless, generating enough power density for charging is difficult for a typical-sized portable electronic device, which imposes a challenging design limitation. As for the directive RF power beamforming, the restriction is simply that the charger needs to know the precise position of the energy receiver [52]. Because of the apparent disadvantage and limitations of two aforementioned methods, wireless charging is typically realized using three other methods: magnetic resonance coupling, magnetic inductive coupling, and non-directive RF radiation [1]. Electric energy is emitted in non-directive RF radiation schemes as electromagnetic radiation with a frequency range from 300 GHz to as low as 3 kHz RF spectrum [53]. The RF energy transfer is appropriate for far-field communications. Several studies have investigated the wireless energy delivery to deployed sensor nodes [24,25]. Wireless energy delivery is a new area that needs to be explored by wireless charging technologies, as it provides the environment where sensor nodes are able to exchange energy between neighbours. Wireless network nodes are thus expected in the foreseeable future to incorporate energy sharing schemes by harvesting energy from the environment and transferring some of this energy to other sensor nodes, creating a self-sustaining network. Multi-hop EH techniques have been researched to realize this paradigm [54], thereby opening up a new dimension in the development of wireless charging protocols and energy cooperative networks, as well as energy-efficient routing.

Unlike EH from other sources, such as solar, wind and vibrations, RF EH has these characteristics: (a) For RF energy harvesters, RF sources can provide controllable and constant energy transfer over distances; (b) In a fixed RF EH network, due to the fixed distance the harvested energy is predictable and relatively stable over time; and (c) Since the amount of RF energy harvested depends on the distance from the RF source, there may be significant differences in the amount of RF energy harvested by the network nodes at different locations. Ambient RF EH techniques [55,56] provide the ability to convert the RF signals received to electricity. This has therefore recently emerged as an alternative approach for running low-power devices [57], including wireless sensors. The fact that ambient RF EH helps to capture and recycle environmentally sustainable resources such as broadcast TV, radio, and cellular signals, which are practically free and universally available, makes these strategies much more appealing. In this context, one can install battery-free wireless devices powered by ambient RF energy, and they can run indefinitely. EH is a technological breakthrough that can be used to resolve hurdles preventing the real-world deployment of WSNs. As mentioned, solar EH in particular has been typically used in overcoming this hurdle. It should be remembered, though, that WSNs running on solar power suffer from energy scarcity during nighttime hours. Hence, to tackle this problem, the TV broadcasting airwaves or interfering signals in a cellular network can be used as energy sources for powering wireless sensor nodes. The authors in [58] have measured the output of a rectenna continuously for 7 days; from the results of this measurement, they showed that RF energy can always be harvested. Note that a rectenna is a rectifying antenna that consists of an antenna and a rectifier circuit. Based on this, they developed an RF EH WSN prototype to show the effectiveness of RF EH for the usage of a WSN.

Simultaneous wireless information and power transfer (SWIPT) is a strategy capable of reducing overall recharging costs to increase WSN 's lifetime. SWIPT is a wireless communication technique through which it is possible to receive information and harvest energy from a received signal, and the harvested energy can be utilized for relaying information or processing purposes. Recently, SWIPT has drawn interest in the field of wireless communication networks [53,59]. In the era of 5G mobile communications, SWIPT technology within various forms of modern communication networks may be of crucial significance for energy and information transmission. Power consumption in WSNs is split into three parts: sensing, data processing, and communication. Among the three, communication is the most expensive function of a typical sensor node [60]. Local data processing is therefore critical to reduce the power consumption of WSNs since the failure of any node may re-route the network structure. Furthermore, robust communication protocols are crucial for the efficient use of energy sources. Also, clustered WSNs with SWIPT that recharge low-energy relay nodes in cooperative clustered WSNs are discussed in [27], wherein the authors consider a WSN consisting of several sensor clusters and a sink node that gathers data from the sensors in the clusters. In order to enable SWIPT, each cluster's head node operates as an information and power transferrer. The relay node receiver harvests the RF energy from the cluster's head node and recharges the batteries.

Thanks to the characteristics of WSNs, the mobile sink/reader has been proved and selected as a robust and energy-efficient data collection technique for gathering data from WSNs' sensors [61]. On the other hand, the SWIPT technology has been considered to be a practical and feasible method of energy charging for WSNs

[27–29], particularly in hardly accessible (extreme) locations. It is then possible for the mobile sink/reader to adopt SWIPT to charge the sensors and gather data simultaneously [62]. The authors in [63] proposed an optimization framework for mobile data collection for EH WSN systems. Therefore, it is meaningful to investigate, evaluate and optimize the SWIPT in WSN scenarios.

4.3.4 D2D communication

As improvements in the capabilities of devices for capturing and playing video continue to grow, demands to deal with mobile multimedia data traffic also increase [64]. The demands not only have affected the cellular networks but also the sensor networks. WMSNs, i.e., WSNs for targeting multimedia data, have been developed [65]. A sensor capable of collecting multimedia data is called a multimedia sensor in WMSN. The multimedia sensor can be used in a number of areas, such as multimedia communication, traffic control, and monitoring environment, people, and machines.

Many problems may arise in WMSN such as cost of deployment, energy-efficient routing, network coverage, and connectivity [66]. A centralized structure is one of the fundamental shortcomings that cause these problems. In previous studies, all data generated were finally sent to a CA. Unlike the typical WSN system, the traffic volume in WMSN is very large. This is a major burden on the CA, and can reduce WMSN's overall performance. D2D communication should therefore be considered in WMSN. In fact, in D2D communications, devices communicate with each other autonomously without any centralized control and collaborate to gather, share, and forward information in short-range networks such as WSNs, LTE Direct, Wi-Fi Direct, or Bluetooth Low Energy (BLE) [67,68]. D2D communication/networking is a promising concept to improve resource utilization, the network coverage and facilitate new types of wireless peer-to-peer services, while at the same time, increasing EE of communications. The D2D communication in WMSN can be used for different services, such as video streaming platforms and mobile aggregation systems. In the case of the streaming service, local streaming broadcast to personal devices in a stadium may be considered. The data can be delivered to local networks. The traffic should be distributed as much as possible in order to decrease the dense overhead to the central system. D2D connections of WMSN can be considered in the services. Also, mobile aggregation is one of the candidates to use D2D connections for WMSN, which implies that sensor networks utilize a mobile device as an aggregator [69]. Multi-hop communication, where a D2D connection is adopted, should be assumed to support the mobile aggregator.

4.3.5 Cognitive radio techniques

For two reasons, CR can be seen as an important solution to higher spectral efficiency (SE) and EE in wireless communication systems. First, the EE-related functionalities can be incorporated into the cognitive operating cycle. Second, from a green viewpoint, spectrum is a natural resource that should not be wasted on unused licensed channels but shared efficiently [30,34]. Typically sensor nodes in a WSN use a fixed spectrum assignment scheme for data communication, and the performance is limited because of restricted processing and communication power. More and more applications like WMSNs have been raised in recent years to change the classical data-only, delay-tolerant assumptions in the design of these networks. Furthermore, with the growing number of commercial applications, sensors are installed in heavily populated areas and can experience greater interference levels in the unlicensed frequency bands. Thus, the need to implement a frequency-agile approach leads to the fast development of CR in WSN, which can flexibly select its transmission spectrum, control its transmitting power, and thus serve high-bandwidth applications with enhanced network life. CR networks (CRNs) allow for the opportunistic usage of licensed spectrum bands under the condition that the activity of licensed band users is not affected [56]. Specifically for WSNs based on CR [70,71], nodes in a sensor network function independently of each other due to their correlated measurements. They may sense the same spectrum to be vacant. So, a key component of a WSN's CR-based framework is how to share the available spectrum fairly with limited communications between the sensor nodes.

4.4 EH-aided cognitive D2D-enabled WMSN based on HetNets

CR compatible with D2D technique can be an aid to enhance system SE and EE. CR emerges as a striking technology that aims to overcome the spectrum scarcity by allowing unlicensed devices to access the licensed frequency bands. With D2D communication, sensors in proximity can establish a direct link, thereby offloading the WSN infrastructure and providing improvement of EE. In this chapter, the objective is to model and analyze a cognitive D2D-enabled WMSN based on a multi-tier and multi-channel cellular network where the D2D sensors are able to harvest RF energy from ambient interference that results from the concurrent downlink cellular transmissions. Each cognitive D2D sensor can harvest energy from ambient interference and uses one of the channels allocated to cellular users to communicate with the corresponding sensor node. Furthermore, to consider D2D requirements under the practical application scenarios, we assume a framework for wireless multimedia sharing, where sensors can store popular multimedia files and share files via D2D communication. Stochastic geometry is used to evaluate the performance of the cognitive D2D-enabled WMSN in the presence of the multi-tier cellular network. It will be shown that EH can efficiently power D2D communications. We will also observe that cognitive channel access can help improve the QoS of D2D connections in the WMSN.

4.4.1 Network topology

We consider a cognitive D2D-enabled WMSN based on a K-tier HetNet composed of a *macrocell tier overlaid* with an extremely dense *tier* of small cells (*SCs*) [72–76]. The spatial distribution of cellular BSs of all tiers is assumed to be captured using collocated and independent homogeneous Poisson point processes (HPPPs), i.e., Φ_j with intensity λ_j ($j \in \mathcal{K} = \{1, \ldots, K\}$). Cellular users are scattered according to another independent HPPP Φ_U with density λ_U. Similarly, sensors in the WMSN are placed according to another independent HPPP Φ_S with density λ_S. Sensors are classified into two categories: regular sensors, with density λ_{SR} and cache-enabled D2D sensors, with density $\lambda_{SD} = \lambda_S - \lambda_{SR}$. The former do not have the caching ability, whereas the latter are cache-enabled. The transmit power of BSs of tier j, users and sensors, are assumed to be P_j^{T}, P_U^{T} and P_S^{T}, respectively. We also consider an open cellular network where a user can connect to any tier's BSs without any limitations [77]. Furthermore, it is assumed that the cellular user connects to the strongest BS, i.e., the one that provides the maximum average received power. Finally, a Rayleigh fading channel is considered for all the communications links in both the HetNet and WMSN. In the WMSN, a sensor can be connected to a CA via a designated communication channel or connected to another sensor via a D2D link (on a pre-defined nonexclusive D2D channel). A two-*tier macrocell*/femtocell *overlaid HetNet* with underlying cognitive D2D-enabled WMSN *is depicted in* Figure 4.2.

4.4.2 Spectrum allocation

One fundamental question is how to share the spectrum resources between cellular and D2D communications. D2D can be divided into two types, depending on the form of spectrum sharing: in-band and out-of-band [30,31]. The in-band refers to D2D using the cellular spectrum and the out-of-band corresponds to D2D using bands other than the cellular band (e.g., 2.4 GHz ISM band). The equivalent interpretation of out-of-band D2D is that a proportion η of the OFDMA resource blocks is assigned to D2D while the remaining resource blocks are utilized by cellular. In-band D2D can be divided into two more categories: overlay and underlay. Overlay means the usage of orthogonal time/frequency resources by cellular and D2D transmitters, while underlay means that D2D transmitters have opportunistic access to the time/frequency resources of cellular users. With this approach, our analysis is performed for two scenarios of D2D spectrum sharing: overlay and underlay in-band D2D. In addition, in underlay in-band D2D, we consider cognitive D2D communication. The term "cognitive D2D communication" is used in the sense that spectrum sensing is performed before transmission at each cache-enabled D2D sensor to ensure that the interference received from any adjacent cellular BS on a pre-defined non-exclusive D2D channel is lower than a pre-defined sensing threshold γ; otherwise, the channel is not available to be used by the D2D sensor. Notice that we do not disregard the imperfections associated with spectrum sensing at the D2D sides, which vary from other related studies. Errors in the form of false-alarms and miss-detections occur in spectrum sensing, and these errors can result in performance degradation. The total available bandwidth for each macrocell is divided into a set of orthogonal channels $C = \{c_1, c_2, \ldots, c_{|C|}\}$ where $|.|$ denotes the set of cardinality. Note that a macrocell is composed of one single macro BS (MBS) and a collection of SC BSs. Whereas a cellular user can be served over any channel $c_i \in C$ based on the channel availability

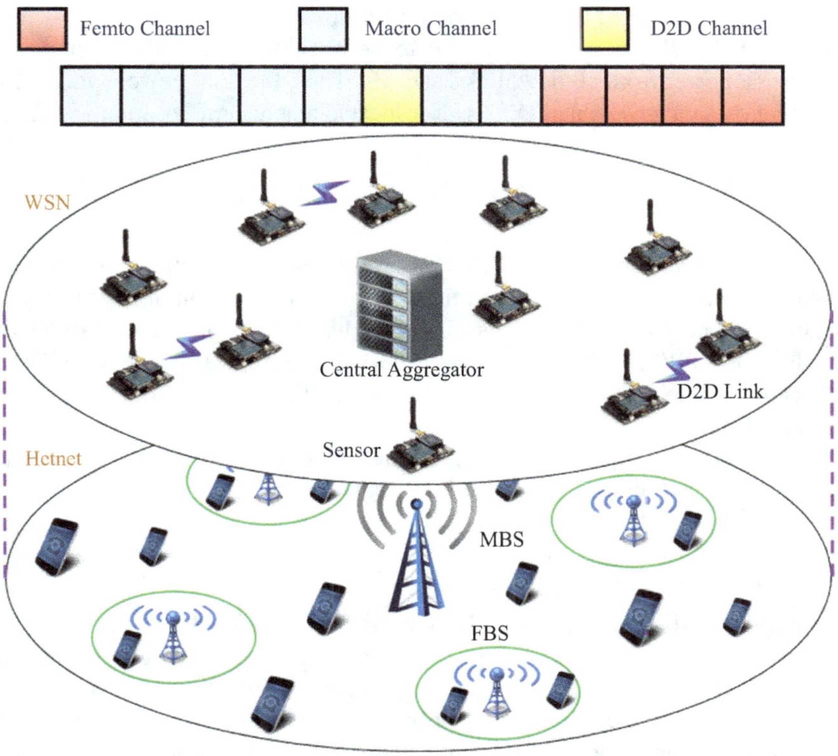

Figure 4.2 A two-tier macrocell/femtocell HetNet with underlying cognitive D2D-enabled WMSN

at the serving BS, all D2D communications between sensors take place on the same channel $c_d \in C$. As it will be discussed later, the channel c_d is not exclusive for sensors' D2D communications and can be used for cellular communication depending on the adopted spectrum access policy. It is worth mentioning that at any BS, each associated cellular user is served by only one channel at most. Also, in each tier, there is no intra-cell interference assuming that each BS serves no more than one user in each channel. In each macrocell, the set of orthogonal channels $C = \{c_1, c_2, \ldots, c_{|C|}\}$ is randomly allocated to BSs of different tiers in such a way that $\sum_{k=1}^{K} |C_k| = |C|$. Note that, in each macrocell, the number of assigned channels to each tier is considered to be fixed. Moreover, the tier to which the channel c_d belongs to (inside that macrocell) will be responsible for performing the adopted spectrum access policy discussed in Section 4.4.5.

4.4.3 EH model

All D2D transmitters (i.e., the cache-enabled D2D sensors) are powered by energy harvested from the ambient interference induced by the simultaneous cellular transmissions in the HetNet (i.e., downlink transmissions in this work) [78]. Each cache-enabled D2D sensor is presumed to be equipped with an EH circuit that harvests RF power from all downlink channels. Thus, the total power available for harvesting by a cache-enabled D2D sensor located at a generic location $y \in \mathbb{R}^2$ can be measured as

$$P_{EH}(y) = a \sum_{c \in C} \sum_{j \in \mathcal{K}} \sum_{x_i \in \Phi_j(c)} P_j^{\mathrm{T}} g_{x_i} |x_i - y|^{-\alpha} \quad (4.1)$$

where this term reflects the amount of RF power harvested from the simultaneous cellular downlink transmissions of all tiers. Note that the amount of power received from other concurrent D2D transmissions at the harvesting unit is not considered in P_{EH}. $\Phi_j(c)$ is a PPP of j-th tier with intensity $q_{c,j}\lambda_j$ that represents the set of BSs of tier j using channel $c \in C$, where $q_{c,j}$ is the probability that a BS in tier j uses this channel. In addition, $0 < a \leq 1$ is the efficiency of the conversion from RF to DC power [78,79]. α is the path-loss exponent, and g_{x_i} is the channel between a cellular BS of tier j located at $x \in \mathbb{R}^2$ and the D2D transmitter.

4.4.4 Spectrum sensing model of D2D sensors

As mentioned, all cache-enabled D2D sensors are considered to be cognitive (in underlay in-band D2D scenario) where each cache-enabled D2D sensor senses the state of the channel c_d before determining whether to use the channel for transmission or not. The key aim of spectrum sensing in this work is to avoid the interference resulting from the adjacent cellular transmissions on channel c_d. That is, the channel c_d is not used by a typical cache-enabled D2D sensor if the interference received on this channel from any neighboring cellular BS is higher than a predefined sensing threshold γ; otherwise, the channel is available to be used by that D2D sensor. Note that increasing the sensing threshold increases the likelihood of accessing channel c_d while at the same time increasing the aggregate interference. On the other hand, reducing the threshold of sensing gives more protection to the D2D transmission by reducing the aggregate interference; however, it decreases the possibility of accessing the c_d channel. In other words, cognition offers a protection region around each D2D transmitter in which that the D2D transmitter cannot use the c_d channel if this channel is used by at least one cellular BS within this region. In general, the protection region around a generic cache-enabled D2D sensor (located at location $y \in \mathbb{R}^2$) follows a random shape, i.e., to prevent interference from the neighboring cellular transmissions on channel c_d, we have [78]

$$P_j^{\mathrm{T}} g_{x_j} |x_j - y|^{-\alpha} > \gamma$$

$$|x_j - y| < r_{p_j}, \; r_{p_j} = \left(\frac{P_j^{\mathrm{T}} g_{x_j}}{\gamma} \right)^{1/\alpha}, j = 1, \ldots, K \quad (4.2)$$

where r_{p_j} is the random radius of the protection region related to tier j, g_{x_j} is the channel (with mean μ_j, $j \in \mathcal{K}$) between a neighboring cellular BS of tier j located at $x \in \mathbb{R}^2$ and the D2D sensors. The protection region centered around the cache-enabled D2D sensor has an average radius as follows

$$\overline{r_{p_j}} = \left(\frac{P_j^{\mathrm{T}}}{\mu_j \gamma}\right)^{1/\alpha} \Gamma\left(1 + \frac{1}{\alpha}\right), j = 1, \ldots, K \tag{4.3}$$

where the term above is obtained, taking into consideration that $\mathrm{E}\left[g^{1/\alpha}\right] = \left(\mu_j\right)^{-1/\alpha} \Gamma(1 + 1/\alpha)$. Therefore, to prevent the interference resulting from the neighboring cellular transmissions on channel c_d, we select the protection region as follows

$$\overline{r_p} = \max\left(\overline{r_{p_j}}\right), j = 1, \ldots, K. \tag{4.4}$$

The cache-enabled D2D sensors need to periodically detect the availability of channel c_d in the underlay in-band D2D. Each cache-enabled D2D sensor receives the interferer's radio signals independently during spectrum sensing. The basic energy detection method is used for spectrum sensing, and the energy statistics $\varnothing(y)$ of each D2D sensor's radio signal can be expressed as [80]

$$\varnothing(y) = \frac{1}{M} \sum_{i=1}^{M} |y(t)|^2, \tag{4.5}$$

where $y(t)$ is the signal received on channel c_d from the nearby cellular transmissions. Note that the signals received from other potential D2D transmissions (on c_d) are not considered because of their weak transmitting powers. $M = \tau f_s$ is defined as the sampling quality where f_s is the sampling frequency and τ denotes the sensing time. As stated earlier, the imperfections associated with the spectrum sensing at the D2D side must be considered, because errors in the form of false-alarms and miss-detections occur in spectrum sensing, and such errors may cause performance degradation. The probability of detection p_{DE} is the probability that a typical cache-enabled D2D sensor will correctly determine the existence of at least one nearby active cellular BS on c_d when there is at least one cellular BS. The probability of false alarm p_{FA} is the probability that a typical cache-enabled D2D sensor judges the existence of a nearby cellular BS on c_d while there is no cellular BS. The p_{FA} and p_{DE} can therefore be represented as follows [80]

$$p_{FA} = Q\left(\sqrt{2\eta_i + 1} Q^{-1}\left(\overline{P_{DE}}\right) + \sqrt{\tau f_s} \eta_i\right) \tag{4.6}$$

$$p_{DE} = Q\left(\frac{1}{\sqrt{2\eta_i + 1}}\left(Q^{-1}\left(\overline{P_{FA}}\right) - \sqrt{\tau f_s} \eta_i\right)\right), \tag{4.7}$$

where η_i denotes the sensing SNR of the i-th D2D transmitter sensor on the channel c_d. In the above expressions, $\overline{p_{DE}}$ and $\overline{p_{FA}}$, respectively, are the target p_{DE} and the target p_{FA}. $Q(.)$ represents the complementary distribution function of the standard

Gaussian, i.e.,

$$Q = \frac{1}{\sqrt{2\pi}} \int_{x}^{\infty} \exp\left(\frac{-t^2}{2}\right) \mathrm{d}t. \tag{4.8}$$

Thus, there are four possible cases in the underlay in-band D2D, depending on the absence and existence of active transmissions on channel c_d (within the D2D protection region):

- The D2D channel c_d is used by the neighboring cellular BSs, and a typical cache-enabled D2D sensor correctly detects the existence of cellular transmissions. The probability of such a case is $\left(1 - p_f^{\mathrm{D2D}}\right) p_{DE}$, where p_f^{D2D} is the probability that the D2D channel (i.e., c_d) is free, and therefore $\left(1 - p_f^{\mathrm{D2D}}\right)$ specifies the probability of existence of nearby cellular transmissions on channel c_d. In this situation, the typical cache-enabled D2D sensor cannot use the channel to transmit information.
- Nearby cellular BSs use the channel c_d, but the typical cache-enabled D2D sensor miss-detects the existence of cellular transmissions. The probability of this situation is $\left(1 - p_f^{\mathrm{D2D}}\right)(1 - p_{DE})$. The typical cache-enabled D2D sensor, in this situation, tries to use the occupied channel c_d for data transmission. However, the typical cache-enabled D2D sensor should not be permitted to perform data transmission to prevent any severe interference.
- Nearby cellular BSs do not use the channel c_d and the typical cache-enabled D2D sensor falsely detects the existence of cellular transmissions. The probability of this situation is $p_f^{\mathrm{D2D}} p_{FA}$. The typical cache-enabled D2D sensor also does not use the channel c_d to transmit data in this case.
- Nearby cellular BSs do not use channel c_d, and the typical cache-enabled D2D sensor correctly detects the absence of cellular transmissions. The probability of this situation is $p_f^{\mathrm{D2D}}(1 - p_{FA})$. The typical cache-enabled D2D sensor in this case uses c_d to transmit information.

It should be noted that we concentrate specifically on the fourth case for the underlay in-band D2D scenario, i.e., when the cellular network inside the D2D protection region is absent on c_d and the typical cache-enabled D2D sensor correctly senses the absence of cellular and therefore uses c_d for D2D transmission.

4.4.5 Spectrum access model for cellular transmissions

We consider two spectrum access policies as described out in [78], namely, random spectrum access (RSA) and prioritized spectrum access (PSA). These policies state how the spectrum is assigned to serve the cellular users for downlink transmissions. More explicitly, in RSA, in each macrocell, any channel $c_i \in C$ (including channel c_d which is used for sensors' D2D transmissions) can be assigned independently and randomly with the same probability to serve one of the cellular users (users connecting to either an MBS or SC BSs). On the other side, in PSA, any channel $c_i \in \mathrm{C} \backslash \{\mathrm{c_d}\}$ can be assigned independently and randomly to a cellular user as long as the number of cellular users is fewer than the number of accessible channels $|C|$. If the number of cellular users in each macrocell is greater than $|C| - 1$, only then, c_d will be assigned to a cellular user. As a result, for the RSA policy, the probability that a BS of tier k utilizes a generic channel $c_i \in C_k$ to serve one of its associated cellular users can be given by

$$q_{c,k}^{\mathrm{RSA}} = \left[1 - \sum_{n=0}^{|C_k|-1} \frac{|C_k| - n}{|C_k|} \mathbb{P}\{N_k = n\} \quad\right] \mathbb{P}\{c_d \in C_k\} \tag{4.9}$$

in which, $\sum_{k=1}^{K} |C_k| = |C|$, $\mathbb{P}\{c_d \in C_k\} = |C_k|/|C|$ is the probability that c_d belongs to the channels allocated to the k-th tier, and $\mathbb{P}\{N_k = n\}$ is the probability mass function (pmf) of the number of users served by a generic BS in tier k, which is as [81]

$$\mathbb{P}\{N_k = n\} = \frac{b^b}{\Gamma(b)} \frac{\Gamma(n+b)}{\Gamma(n+1)} \frac{(\mathcal{N}_k)^n}{(b + \mathcal{N}_k)^{n+b}}. \tag{4.10}$$

The above term is derived from an approximation of a Voronoi cell area by a Gamma-distributed random variable with a shape parameter $b = 3.575$ and a scale parameter $1/b\lambda_k$ [81]. Also, $\mathcal{N}_k$ denotes the average number of users associated with a BS in tier k. Notice that $q_d^{\mathrm{RSA}} = q_c^{\mathrm{RSA}}$ because, in each macrocell, any BS of tier k randomly and independently assigns any channel $c_i \in C_k$ with the same probability in the RSA policy. Let q_d denote the probability that a generic BS assigns the D2D channel c_d to serve one of its associated users. On the other hand, for the PSA policy, the following expression gives the probability that a BS of tier k uses a generic channel $c_i \in C_k \backslash \{c_d\}$ to serve one of its associated cellular users.

$$q_{c,k}^{\mathrm{PSA}} = \left[1 - \sum_{n=0}^{|C_k|-1} \frac{|C_k| - n - 1}{|C_k| - 1} \mathbb{P}\{N_k = n\} \quad\right] \mathbb{P}\{c_d \in C_k\}. \tag{4.11}$$

while the probability that a BS of tier k has to use c_d to serve one of its associated cellular users is denoted by

$$q_{d,k}^{\mathrm{PSA}} = \left[1 - \sum_{n=0}^{|C_k|-1} \mathbb{P}\{N_k = n\}\right] \mathbb{P}\{c_d \in C_k\}. \tag{4.12}$$

4.4.6 *Problem formulation*

4.4.6.1 Overlay mode: D2D coverage probability

We now provide analytical expressions of the D2D coverage probability for any typical sensor in the WMSN in the presence of the HetNet. As mentioned above, our evaluation is undertaken for two D2D spectrum sharing scenarios: overlay and underlay in-band D2D. We begin with the overlay in-band D2D, where cellular and D2D sensors utilize orthogonal time/frequency resources and therefore there is no interference between cellular and D2D transmissions. In other words, D2D links are not affected by cellular linksand vice versa. We now express the D2D coverage probability formula, which is defined as the probability that a typical sensor in the WMSN is in coverage, i.e., signal-to-interference-plus-noise ratio (SINR) is greater than a predetermined threshold τ_{D2D}. It should be noted that, in this chapter, the focus will be on the analysis of the WMSN in the presence of the HetNet, and the per-tier downlink cellular coverage probability in both the overlay and underlay in-band D2D is not studied. Please refer to [72–76] for the detailed analysis of cellular coverage probability.

Theorem 4.1: *The D2D coverage probability in overlay mode is expressed as*

$$\overline{p_{\mathrm{D2D}}^{cov}} = 2\pi\lambda_{UD}\int_0^d r_d \exp\left(-\pi\lambda_{SD}r_d^2\right)p_{\mathrm{D2D}}^{cov}(r_d)\mathrm{d}r_d \tag{4.13}$$

where for $p_{\mathrm{D2D}}^{cov}(r_d) = \mathbb{P}[\mathrm{SINR}_{\mathrm{D2D}} > \tau_{\mathrm{D2D}}] = \mathbb{P}\left[\frac{P_S^{\mathrm{T}} L r_d^{-\alpha}}{\sigma^2 + I_{\mathrm{D2D}}} > \tau_{\mathrm{D2D}}\right]$ *we have*

$$p_{\mathrm{D2D}}^{cov}(r_d) = e^{-\mu_{\mathrm{D2D}}\frac{\tau_{\mathrm{D2D}} r_d^{\alpha}}{P_S^{\mathrm{T}}}\sigma^2}\left[e^{-\pi\rho\lambda_{SD} r_d^2 (\tau_{\mathrm{D2D}})^{\frac{2}{\alpha}}\left(\frac{\frac{2\pi}{\alpha}}{\sin\left(\frac{2\pi}{\alpha}\right)}\right)}\right] \tag{4.14}$$

in which P_S^{T} denotes the D2D sensors' transmit power, r_d is a random variable indicating the D2D distance and d denotes the maximum allowable D2D distance. $\rho\lambda_{SD}$ reflects the portion of the cache-enabled D2D sensors that are engaged in active transmission, causing interference to the tagged D2D link, i.e., the one under consideration. Note that $\rho = (N_{\mathrm{D2D}} - 1)/\lambda_{SD}|\mathrm{S}|$,where N_{D2D}, which will be obtained later in Section 4.4.7, indicates the expected number of active D2D links in the network, and $|\mathrm{S}|$ represents the area of the cellular system. We also denote the predetermined D2D SINR threshold by τ_{D2D}, and L is the exponentially distributed channel (with mean μ_{D2D}) between the tagged D2D receiver sensor and its serving D2D transmitter sensor.

Remark 4.1: *In overlay mode, the aggregate interference (i.e., I_{D2D}) results only from other cache-enabled D2D sensors that have enough energy to communicate with their corresponding sensor receivers. The interfering D2D sensors therefore no longer constitute a homogeneous point process and analytical characterization of interference is not possible. For analytical tractability, therefore, we ignore the correlation between the locations of the interfering D2D sensors and approximate the point process by an HPPP with the same intensity $\rho\lambda_{SD}$. Note that the interfering D2D sensors can be arbitrarily close to the tagged D2D sensor. Indeed, in the derivation of I_{D2D}, the closest D2D interferer can be at a distance 0 from the tagged D2D connection, and thus, when calculating the aggregate interference, the integration limit will be from 0 to ∞.*

Corollary 4.1: *Consider $\sigma^2 \to 0$, then we will have*

$$\overline{p_{\mathrm{D2D}}^{cov}} = \frac{\lambda_{SD}\left(1 - e^{-\pi d^2\lambda_{SD} - \pi d^2\rho\lambda_{SD}(\tau_{\mathrm{D2D}})^{\frac{2}{\alpha}}\left(\frac{\frac{2\pi}{\alpha}}{\sin\left(\frac{2\pi}{\alpha}\right)}\right)}\right)}{\lambda_{SD} + \rho\lambda_{SD}(\tau_{\mathrm{D2D}})^{\frac{2}{\alpha}}\left(\frac{\frac{2\pi}{\alpha}}{\sin\left(\frac{2\pi}{\alpha}\right)}\right)}. \tag{4.15}$$

4.4.6.2 Underlay mode: D2D coverage probability

We now consider the underlay in-band D2D scenario, where the cache-enabled D2D sensors with cognition capability opportunistically access the time/frequency resource used by cellular users, i.e., the pre-defined nonexclusive D2D channel c_d, and thus cellular connections are influenced by D2D connections, and vice versa. As a result, interference will occur not only among cellular connections and within D2D connections but also between cellular and D2D connections. Note that at the cache-enabled D2D sensors we take into consideration the imperfections associated with the spectrum sensing.

Theorem 4.2: *The D2D coverage probability in underlay mode is expressed as*

$$\overline{p_{\mathrm{D2D}}^{cov}} = 2\pi\lambda_{SD}\int_0^d r_d \exp\left(-\pi\lambda_{SD} r_d^2\right) p_{\mathrm{D2D}}^{cov}(r_d)\mathrm{d}r_d \tag{4.16}$$

where $p_{\mathrm{D2D}}^{cov}(r_d) = \mathbb{P}[\mathrm{SINR}_{\mathrm{D2D}} > \tau_{\mathrm{D2D}}] = \mathbb{P}\left[\frac{P_S^{\mathrm{T}} f r_d^{-\alpha}}{\sigma^2 + I_{\mathrm{D2D}} + I_{\mathrm{C}}} > \tau_{\mathrm{D2D}}\right]$ is derived as follows

$$p_{\mathrm{D2D}}^{cov}(r_d) = e^{-\mu_{\mathrm{D2D}}\frac{\tau_{\mathrm{D2D}} r_d^{\alpha}}{P_S^{\mathrm{T}}}\sigma^2} e^{-\pi\rho\lambda_{SD} \quad r_d^2 (\tau_{\mathrm{D2D}})^{\frac{2}{\alpha}}\left(\frac{\frac{2\pi}{\alpha}}{\sin\left(\frac{2\pi}{\alpha}\right)}\right)}$$

$$\times \prod_{j\in\mathscr{K}}\left[e^{\pi B^2 r_d^2 q_{d,j}\lambda_{I_j} + \frac{2\pi B^2}{\alpha} q_{d,j}\lambda_{I_j} r_d^2 \Gamma\left(-\frac{2}{\alpha}\right)\mu_j \Omega\left(\frac{P_j^{\mathrm{T}}}{P_S^{\mathrm{T}} B^{\alpha}}\tau_{\mathrm{D2D}}, \mu_{\mathrm{D2D}}, \mu_j\right)}\right] \tag{4.17}$$

and

$$\Omega\left(\tau_{\mathrm{D2D}}, \mu_{\mathrm{D2D}}, \mu_j\right) = \left[\sum_{i=0}^{\infty} \frac{\Gamma(i+1)(\mu_{\mathrm{D2D}}\tau_{\mathrm{D2D}})^i}{\Gamma\left(i+1-\frac{2}{\alpha}\right)\left(\mu_j + \mu_{\mathrm{D2D}}\tau_{\mathrm{D2D}}\right)^{i+1}}\right]. \tag{4.18}$$

Remark 4.2: In underlay mode, each D2D sensor receiver suffers from two sources of interference, i.e., the cellular connections in the HetNet and D2D connections in the WMSN. For the cellular connections in the HetNet, the aggregate interference comes from all BSs of different tiers using the c_d channel. Thus, $q_{d,j}\lambda_{I_j}$ indicates the set of BSs in tier j that use channel c_d, where $q_{d,j}$ is given for both the RSA and PSA policies in Equations (4.9) and (4.12). For the D2D connections in the WMSN, the aggregate interference (i.e., I_{D2D}) comes only from other D2D transmitter sensors that have enough energy to communicate with their corresponding receiver sensors and can transmit on c_d. Thus, we assume the thinned process to be HPPP for tractable analysis with the same intensity $\rho\lambda_{SD}$ where $\rho = (N_{\mathrm{D2D}} - 1)/\lambda_{SD}|\mathrm{S}|$. For the cellular interference, i.e., I_{C}, since the cache-enabled D2D sensors perform spectrum sensing on $c_d$prior to transmission, the nearest interferer is then at least at a distance of $\overline{r_p}$ [given in (4.4)] from the tagged D2D receiver sensor. We assume, for better analytical tractability, that the nearest interferer is at least a distance of Br_d. r_d is the random variable representing the D2D distance, and $B = \overline{r_p}$. The integration limit when calculating the aggregate interference I_C will be therefore from Br_d to ∞. Note that to protect the D2D transmissions, the protection region is basically centered around the D2D receiver sensor rather than the D2D transmitter sensor. For simplicity, however, we assume the protection region centers around the D2D sensor. For I_{D2D}, the interfering D2D sensors can be arbitrarily close to the tagged D2D receiver sensor under consideration and there is no protection region, and therefore the integration limit for the calculation of the aggregate interference I_{D2D} will be from 0 to ∞.

Corollary 4.2: Consider $\sigma^2 \to 0$, then we will have

$$\overline{p_{\text{D2D}}^{cov}} = \frac{\pi\lambda_{SD}}{(Q)}\left(e^{Qd^2} - 1\right) \tag{4.19}$$

where

$$Q = -\pi\lambda_{SD} - \pi\rho\lambda_{SD}(\tau_{\text{D2D}})^{\frac{2}{\alpha}}\left(\frac{\frac{2\pi}{\alpha}}{\sin\left(\frac{2\pi}{\alpha}\right)}\right) + \sum_{j\in\mathcal{K}}\left(\pi B^2 q_{d,j}\lambda_{I_j} + \frac{2\pi B^2}{\alpha} q_{d,j}\lambda_{I_j}\Gamma\left(-\frac{2}{\alpha}\right)\mu_j\Omega\left(\frac{P_j^{\text{T}}}{P_S^{\text{T}}B^{\alpha}}\tau_{\text{D2D}}, \mu_{\text{D2D}}, \mu_j\right)\right) \tag{4.20}$$

in which, $B = \overline{r_p}$ and $\overline{r_p} = \max\left(\overline{r_{p_j}}\right)$ for $j = 1, \ldots, K$, as previously mentioned.

4.4.7 Cellular and D2D links coexistence

Sensor nodes nowadays are mainly equipped with memory modules of large storage capacity. This allows sensors to store popular multimedia files in their devices and exchange those files in a WMSN with other sensors upon request, resulting in the traffic offloading from the central system. In our cognitive D2D-enabled WMSN, sensors can request popular contents (files) from neighboring sensors via D2D communication, whereas they can also request contents from the CA through a designated wireless communication channel. It is assumed that sensors can request files from a set of m files, named "library," and unless stated otherwise, each cache-enabled D2D sensor has a cache memory of size $z = 1$, and m is set to 3. For instance, when a sensor node x requests a file from the "library" of size m, the sensor node x establishes a link with the CA. Then, the CA looks for the requested file in the certain area where the sensor is located at the center point. If the requested file is found, the channel c_d (if available) will be used for the D2D link between the sensor x and the sensor storing the file. Otherwise, sensor x receives the file from the CA. Note that a D2D link can be successfully established between a sensor requesting file i (i.e., D2D receiver sensor) and the sensor storing file i (i.e., D2D transmitter sensor) only when (1) there exists one cache-enabled D2D transmitter sensor storing file i in the area S ($|S| = \pi d^2$) centered around the D2D receiver sensor. As mentioned, d is the maximum permissible D2D distance.

(2) The D2D transmitter sensor harvests adequate energy to form a communication link with the receiver sensor, and (3) the D2D channel c_d is idle, i.e., the received interference from any adjacent cellular BS on the c_d is less than the pre-defined sensing threshold γ; otherwise, the sensor receives the file from the CA. As mentioned earlier, the sensors follow PPP with average density λ_S. Hence, the probability that l sensors exist in the area S is [82]

$$f(l; S) = \frac{(|S|\lambda_S)^l}{l!} e^{-|S|\lambda_S}, \tag{4.21}$$

where $|S|$ denotes the area of district S. We denote the probability of the file i being stored by a cache-enabled D2D sensor by p_i. Accordingly, the file i is distributed with the PPP model with average intensity $\lambda_{SD} p_i$. Furthermore, we denote by r_d the random distance between the reference sensor requesting file i and the nearest sensor that stores the file i. Obviously, there is no sensor storing file i in region S ($|S| = \pi d^2$) when $r_d > d$. Therefore, the probability that the reference sensor successfully find the requested file within the maximum permissible D2D distance to form a D2D link is as follows:

$$\Pr(r_d \leq d) = 1 - \Pr(r_d > d) = 1 - f(0; S) = 1 - e^{-\pi d^2 \lambda_{SD} p_i}. \tag{4.22}$$

Obviously, only sensors without file i will make a request to get file i. There are, therefore, on average, $\lambda_S |\mathrm{S}|(1 - p_i)$ sensors who may request file i in the network, where $|\mathrm{S}|$ indicates the area of the cellular system. According to [82], the probability of the sensors caching the ith ranked file can be expressed as (Zipf distributions)

$$p_i = \frac{(1/i)^{r_c}}{\sum_{j=1}^{m} (1/j)^{r_c}}, \quad i = 1, 2, \ldots, m, \tag{4.23}$$

where $r_c \geq 0$ is a parameter called skew coefficient and characterizes the distribution by controlling the relative popularity. In addition, let Q_i denote the probability that file i is requested by a sensor. Note that the sensor request probability Q_i also follows the Zipf distribution [82]. Furthermore, by p_t^{D2D} we denote the D2D transmission probability, i.e., the probability of c_d being available and the amount of harvested energy is adequate for D2D transmission.

$$p_t^{\mathrm{D2D}} = (1 - p_{FA}) p_f^{\mathrm{D2D}} p_s^{\mathrm{D2D}} \tag{4.24}$$

where, from the viewpoint of the cognitive D2D transmitter sensors, p_f^{D2D} is the probability that there is a channel available for D2D transmission, i.e., the probability that the D2D channel (i.e., c_d) is free inside the D2D protection region. In addition, p_s^{D2D} is the probability that a D2D transmitter sensor harvests enough energy to form a communication link with the receiver sensor, i.e., $p_s^{\mathrm{D2D}} = \mathbb{P}\left[P_{EH} > P_{EH}^{th}\right]$, where P_{EH}^{th} is a predetermined EH threshold. As explained earlier in this chapter, in the absence of the cellular network on channel c_d (within the D2D protection region), if the typical D2D transmitter sensor correctly detects the absence of cellular signaling, the probability of this situation is indicated by $(1-p_{FA})p_f^{\mathrm{D2D}}$. Clearly, $p_{FA}=0$ when the overlay mode is considered as the D2D spectrum sharing scenario. In summary, we can now obtain, based on the above explanations, the number of successfully activated D2D links for delivering file i as

$$n_i = \lambda_S |\mathrm{S}|(1-p_i)Q_i\left(1-e^{-\pi d^2 \lambda_{SD} p_i}\right)p_t^{\mathrm{D2D}}. \tag{4.25}$$

The expected number of active D2D links for all files within the WMSN can therefore be defined by

$$N_{\mathrm{D2D}} = \sum\nolimits_{i=1}^{m} \lambda_S |\mathrm{S}|(1-p_i)Q_i\left(1-e^{-\pi d^2 \lambda_{SD} p_i}\right)p_t^{\mathrm{D2D}}. \tag{4.26}$$

4.4.8 Simulation results

For simpler exposition and without loss of generality, a two-tier (femto-macro) HetNet (i.e., $K=2$) is henceforth considered. In the presented findings, unless otherwise specified, we consider the underlay in-band D2D mode where the D2D transmitter sensors with cognition capability have opportunistic access to the D2D channel c_d used by cellular users, and thus cellular connections are influenced by D2D connections and vice versa. Although the overlay in-band D2D can usually deliver better system performance without co-channel interference with dedicated resources, it is not as efficient as the underlay in-band D2D in terms of SE. Furthermore, we consider the PSA policy in what follows, unless otherwise explicitly specified. The performance metrics/policies, as well as the major simulation parameters, are summarized in Tables 4.1 and 4.2, respectively.

In Figure 4.3, we can see the effect of changing the total number of channels $|\mathrm{C}|$ accessible in each macrocell on the $q_{c,k}^{\mathrm{PSA}}$ and $q_{d,k}^{\mathrm{PSA}}$ in the underlay in-band D2D mode. It is apparent that increasing $|\mathrm{C}|$ would lead to a decrease of $q_{c,k}^{\mathrm{PSA}}$ and $q_{d,k}^{\mathrm{PSA}}$ in both the macro and femto tiers. This is because as the overall number of channels grows, the likelihood that a BS uses a generic channel $c_i \in C_k$ to serve one of its

Table 4.1 Performance metrics and policies

Metric/policy	Description
QoS of D2D links	Percentage of D2D receiver sensors with a throughput greater than a target threshold of 100 Mbps
D2D transmission probability	Probability of D2D channel c_d being available and the amount of harvested energy is adequate for D2D transmission
Prioritized spectrum access (PSA)	Any channel $c_i \in C\backslash\{c_d\}$ can be assigned independently and randomly to a cellular user as long as the number of cellular users is fewer than the number of accessible channels $\lvert C\rvert$
Random spectrum access (RSA)	Any channel $c_i \in C$ (including channel c_d) can be assigned independently and randomly with the same probability to serve one of the cellular users

Table 4.2 Major simulation parameters

Symbol	Value	Description
P^T_{MBS}	37 dBm	MBS transmit power
P^T_{FBS}	23 dBm	FBS transmit power
$P^T_U = P^T_S$	20 dBm	User transmit power
λ_{MBS}	$1 \times 10^{-7}\text{m}^{-2}$	MBS initial density
λ_{FBS}	$3\lambda_M$	FBS initial density
λ_U	$4 \times 10^{-5}\text{m}^{-2}$	User initial density
λ_S	λ_U	Sensor initial density
λ_{SD}	50% λ_S	Cache-enabled D2D sensors density
$\mu_{MBS} = \mu_{FBS} = \mu_{\text{D2D}}$	1	Rayleigh fading Parameter
d	200 m	Max. allowable D2D distance
$\lvert C \rvert$	200	Macrocell bandwidth
γ	−80 dBm	D2D sensing threshold
P^{th}_{EH}	−90 dBm	EH threshold
$\tau_{FBS} = \tau_{MBS} = \tau_{\text{D2D}}$	0 dB	Cell. and D2D SINR threshold

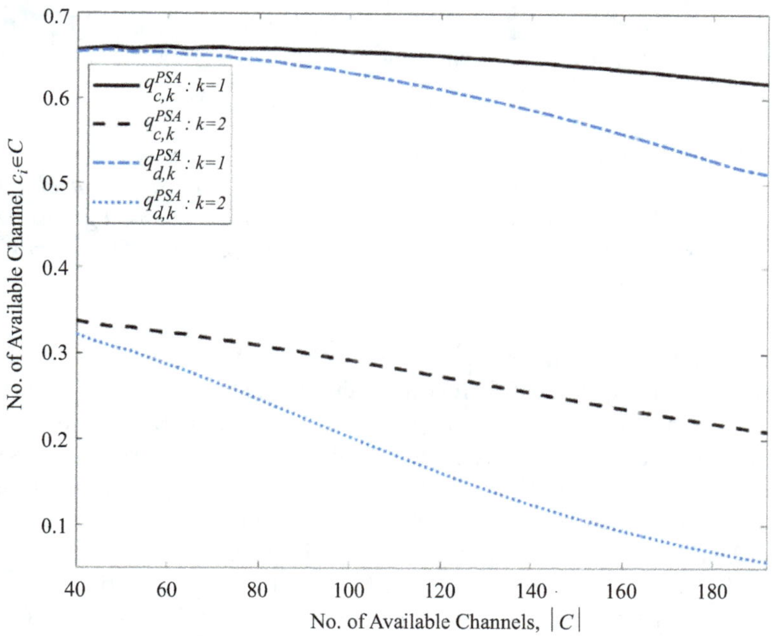

Figure 4.3 The $\mathbf{q}^{\mathbf{PSA}}_{c,k}$ and $\mathbf{q}^{\mathbf{PSA}}_{d,k}$ versus the total number of available channels $|\text{C}|$ *in each macrocell in the underlay in-band D2D mode*

associated cellular users will decrease. As stated, due to their high transmission power, more users are served by MBSs (as opposed to femto BSs (FBSs)). Thus, $q^{\text{PSA}}_{c,1}$ and $q^{\text{PSA}}_{d,1}$ are greater than $q^{\text{PSA}}_{c,2}$ and $q^{\text{PSA}}_{d,2}$, as seen in the figure.

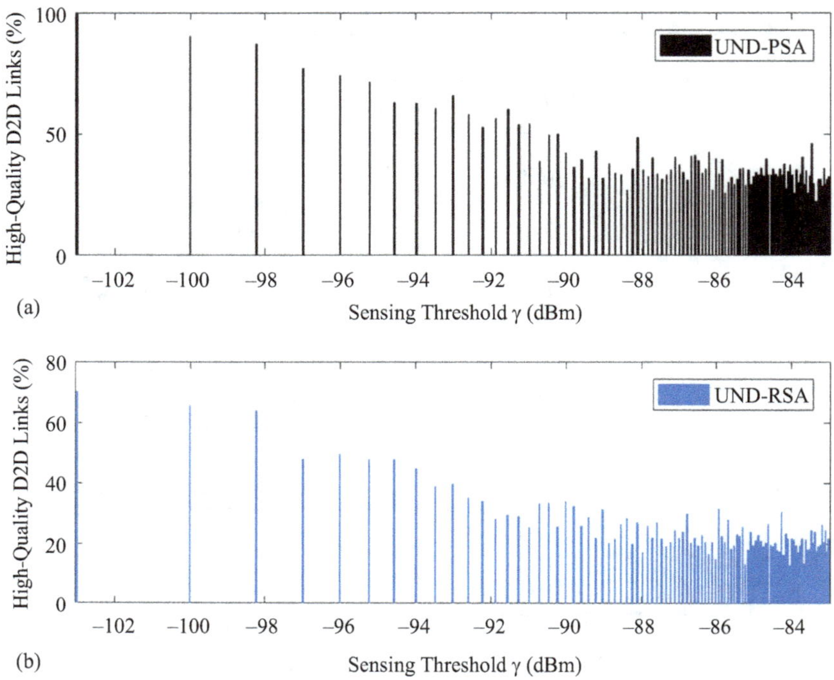

Figure 4.4 The proportion of D2D links with desired quality of service (QoS) versus the CR sensing threshold γ (in dBm) of sensors for both RSA and PSA policies. Higher γ: moving toward the non-cognitive D2D scenario. Lower γ: moving toward the cognitive D2D scenario

Figure 4.4 indicates the percentage of D2D links in the network with desired QoS. In this chapter, the QoS is related to the throughput a D2D receiver sensor achieves. In specific, the QoS is the percentage of D2D receiver sensors with a throughput greater than a target threshold of 100 Mbps. Increasing the sensing threshold γ raises the probability of accessing the channel c_d while at the same time rising the aggregate interference for both the PSA and RSA policies in the underlay in-band D2D. This would result in a reduction in the proportion of D2D connections in the network with desirable QoS. On the other side, lowering the sensing threshold provides greater protection for D2D transmissions by reducing aggregate interference on channel c_d. This will obviously increase the fraction of D2D connections with desirable QoS. In comparison, the percentage of D2D connections with desirable QoS is always higher when using the PSA policy (for all values of γ). It is worth noting that increasing the spectrum sensing threshold excessively very high can be viewed as a scenario where the D2D transmitter sensors of the network are all non-cognitive, such that they can access the c_d channel without taking account of the cellular network presence. This implies a strong probability of

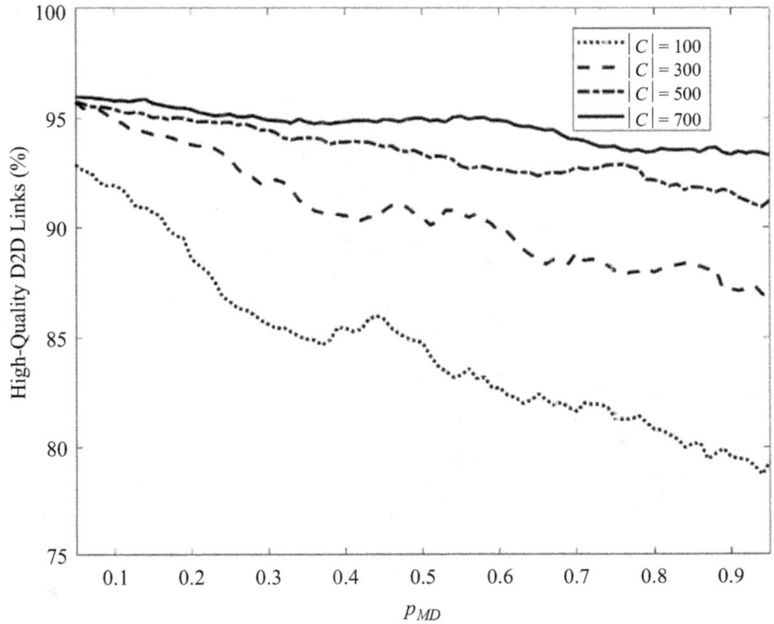

Figure 4.5 The percentage of D2D links with desired quality of service (QoS) versus the miss-detection probability (i.e., $\mathbf{p_{MD}}$*) for different values of* |C|

interference between the cellular network and D2D links of the WMSN. Therefore, cognition plays a major role in assessing all measures relevant to QoS.

Simulations are used in Figure 4.5 to display the fraction of D2D connections in the WMSN with the desired QoS as a function of the probability of miss-detection (i.e., $p_{MD} = 1 - p_{DE}$). As previously explained, miss-detection is known to be the situation when the cognitive D2D transmitter sensors wrongly identify the channel c_d to be idle for the transmission. In this scenario, with a miss-detection probability $(1 - p_{DE})$, the D2D and cellular transmissions may interfere severely with each other (on channel c_d). As a consequence, the greater the probability of miss-detection, the smaller the percentage of D2D connections with desired QoS will become. Also, increasing the number of channels accessible in each macrocell (i.e., |C|) reduces the likelihood that c_d may be used by cellular communications, which in effect decreases the aggregate interference on this channel. Thus, a greater proportion of the WMSN's D2D connections will obtain a higher throughput than the target throughput.

As can be noticed in (4.24) and seen in Figure 4.6, p_t^{D2D}, i.e., the D2D transmission probability, is a function of both p_{FA} and the predefined EH threshold P_{EH}^{th}. As seen in the simulation, when p_{FA} is decreased, p_t^{D2D} increases linearly. Moreover, the greater the value of P_{EH}^{th}, the lower the D2D transmission probability. This is because the probability of a D2D transmitter sensor harvesting

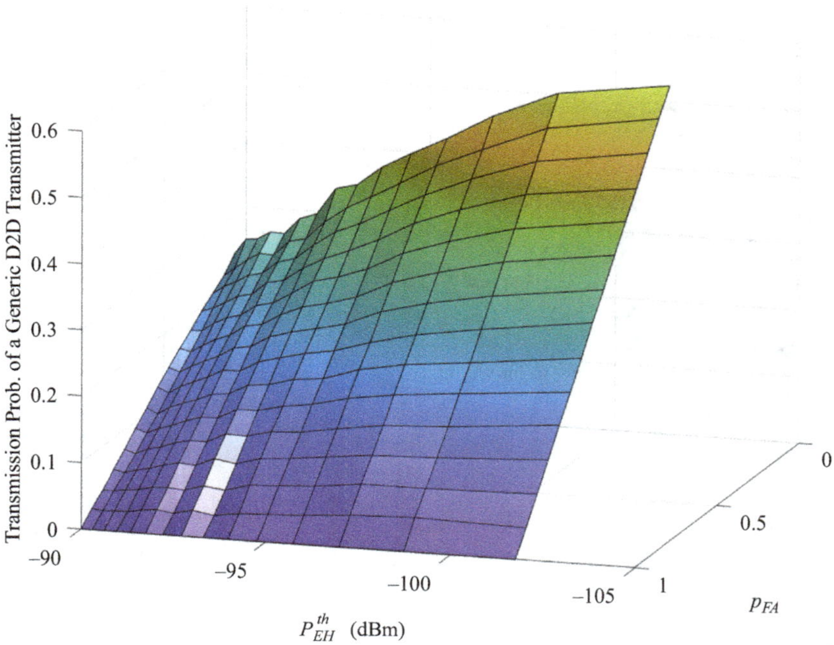

Figure 4.6 The D2D transmission probability ($\mathbf{p}_t^{\mathbf{D2D}}$) versus the probability of false alarm (i.e., $\mathbf{p}_{FA}$) and the predetermined EH threshold (i.e., $\mathbf{P}_{EH}^{th}$) of the cognitive D2D transmitter sensors

enough energy (i.e., p_s^{D2D}) decreases as P_{EH}^{th} increases. As a consequence, we will observe a lower D2D transmission probability.

4.5 Future challenges and research opportunities

As for the specific research directions in the future, besides what have been discussed and presented in this chapter, there are a number of other issues that need to be carefully investigated in this research study. One important point is the energy-and-spectral efficiency trade-off when implementing the proposed WMSN-based architecture. The SE and EE are among the most important venues for technological advances in current and emerging wireless networks. The past decade has witnessed tremendous efforts and advancement made by both the academic and industrial communities for improving SE. Due to the steadily rising energy cost and environmental concerns, energy and power efficiencies of wireless networks have become more crucial over the last few years. While there has been a paradigm shift from improving SE to reducing energy consumption, a dilemma also arises as some EE criteria are in conflict with the SE objectives [83–85]. Hence, there is an urgent need to address key challenges and state-of-the-art solutions for joint spectrum and energy-efficient design and optimization of wireless networks.

Existing works on the energy-and-spectral efficiency trade-off can be divided into two categories: the first one is to characterize the mathematical relationship between the EE and the SE as accurately as possible; the other one is to maximize the EE with a minimum SE requirement or to maximize the SE with a minimum EE requirement. In future, we will consider the spectrum-energy trade-off utility function design with different concerns, and use the decomposition methodology to analyze it. This is the important thing to investigate the trade-off relationship between the energy and spectral efficiency when implementing the proposed cognitive D2D-enabled WMSN over a multi-tier cellular network. Another issue to consider is that power control can be used for interference management of the concurrent D2D connections between sensors aiming at further EE improvements [86–88].

4.6 Conclusion

Multimedia data should be transmitted in real time in the WMSN. Supporting high throughput multimedia data from sensors is crucial. A centralization problem can arise when all the data is sent to a CA. D2D communication should therefore be deployed into the WMSN. In this chapter, the objective was to model and analyze a cognitive D2D-enabled WMSN based on a multi-tier and multi-channel cellular network using stochastic geometry. We considered a framework for wireless multimedia sharing, where some sensors are cache-enabled such that they can store popular multimedia files and share files via D2D communication. Furthermore, each cognitive cache-enabled D2D sensor can harvest energy from the concurrent downlink cellular transmissions and opportunistically use one of the channels allocated to cellular users to communicate with the corresponding sensor node. It was shown that EH can efficiently power D2D communications. We also observed that cognitive channel access can help improve the QoS of D2D links in the WMSN.

List of acronyms

Acronym	Explanation
BS	Base Station
CA	Central aggregator
CR	Cognitive radio
CRN	CR network
D2D	Device-to-device
E2E	End-to-end
EE	Energy efficiency
EH	Energy harvesting
FBS	Femto base station
HetNet	Heterogeneous network
HPPP	Homogeneous Poisson point process
IoT	Internet of Things

(Continues)

(*Continued*)

Acronym	Explanation
MBS	Macro base station
MIMO	Multiple-input and multiple-output
PSA	Prioritized spectrum access
QoS	Quality of service
RET	Renewable energy technology
RF	Radio frequency
RSA	Random spectrum access
SC	Small cell
SE	Spectral efficiency
SISO	Single-input single-output
SWIPT	Simultaneous wireless information and power transfer
WMSN	Wireless multimedia sensor network
WSN	Wireless sensor network

References

[1] Alsharif, M.H., Nordin, R., Abdullah, N.F., and Kelechi, A.H.: 'How to make key 5G wireless technologies environmental friendly: A review', *Transactions on Emerging Telecommunications Technologies*. 2018, 29, e3254.

[2] Elkhodr, M., Shahrestani, S., and Cheung, H.: 'Emerging wireless technologies in the Internet of Things: A comparative study', *International Journal of Wireless and Mobile Computing*, vol. 8, no. 5, pp. 67–82, Oct. 2016.

[3] Sinha, R.S., Wei, Y., and Hwang, S.-H.: 'A survey on LPWA technology: LoRa and NB-IoT,' *ICT Express* 2017, 3, 14–21.

[4] Panahi, F.H. and Falahati, A.: 'Spectral efficient impulse radio-ultra-wideband transmission model in presence of pulse attenuation and timing jitter', *IET Communications*, vol. 6, no. 11, pp. 1544–1554, 24 July 2012.

[5] Ahmed, N., Rahman, H., and Hussain, M.I.: 'A comparison of 802.11 ah and 802.15.4 for IoT', *ICT Express* 2016, 2, 100–102.

[6] Panahi, F.H.Farhadi, P., and Panahi, Z.H.: 'Spectral-Efficient Green Wireless Communications via Cognitive UWB Signal Mode', *Automatika*, vol. 57, br. 3, str. 793–809, 2016.

[7] Panahi, F.H. and Falahati, A.: 'Joint IR-UWB Power Spectral Lines and Interference Suppression Based on Coded Auxiliary Independent Signaling in Presence of Pulse Attenuation and Timing Jitter', *Wireless Personal Communications*, vol.69, no 4, pp. 1241–1260, 2013.

[8] Falahati, A. and Panahi, F.H.: 'Spectrally efficient ultra-wideband transmission over wireless sensor networks', IET Conference on Wireless Sensor Systems (WSS 2012), 2012.

[9] Chu, X., and Sethu, H.: 'Cooperative topology control with adaptation for improved lifetime in wireless sensor networks', *Ad Hoc Networks* 2015, 30, 99–114.

[10] Lin, S., Miao, F., Zhang, J., Zhou, G., Gu, L., and He, T.: 'ATPC: Adaptive transmission power control for wireless sensor networks', *ACM Transactions on Sensor Networks (TOSN)* 2016, 12, 6.

[11] Cui, S., Goldsmith, A.J., and Bahai, A.: 'Energy-efficiency of MIMO and cooperative MIMO techniques in sensor networks', *IEEE Journal on Selected Areas in Communications.* 2004, 22, 1089–1098.

[12] Jayaweera, S.K.: 'Virtual MIMO-based cooperative communication for energy-constrained wireless sensor networks', *IEEE Transactions on Wireless Communications* 2006, 5, 984–989.

[13] Cui, S., Goldsmith, A.J., and Bahai, A.: 'Energy-constrained modulation optimization', *IEEE Transactions on Wireless Communications* 2005, 4, 2349–2360.

[14] Costa, F.M., and Ochiai, H.: 'A Comparison of Modulations for Energy Optimization in Wireless Sensor Network Links', In Proceedings of the IEEE in Global Telecommunications Conference (GLOBECOM 2010), Maiami, FL, USA, 6–10 December 2010, pp. 1–5.

[15] Misra, S., Kumar, M.P., and Obaidat, M.S.: 'Connectivity preserving localized coverage algorithm for area monitoring using wireless sensor networks', *Computer and Communications* 2011, 34, 1484–1496.

[16] Karasabun, E., Korpeoglu, I., and Aykanat, C.: 'Active node determination for correlated data gathering in wireless sensor networks', *Computer Networks* 2013, 57, 1124–1138.

[17] Panahi, F.H., Panahi, F.H., Heshmati, S., and Ohtsuki, T.: 'Optimal Sleep & Wakeup Mechanism for Green Internet of Things,' 2019 27th Iranian Conference on Electrical Engineering (ICEE), Yazd, Iran, 2019, pp. 1659–1663.

[18] Carrano, R.C., Passos, D.G., Magalhães, L.C.S., and Vinicius, N.C.: 'Survey and taxonomy of duty cycling mechanisms in wireless sensor networks', *IEEE Communications Surveys and Tutorials* 2014, 16, 181–194.

[19] De Paz Alberola, R., and Pesch, D.: 'Duty cycle learning algorithm (DCLA) for IEEE 802.15.4 beacon-enabled wireless sensor networks', *Ad Hoc Networks* 2012, 10, 664–679.

[20] Wan, Z., Tan, Y., and Yuen, C.: 'Review on Energy Harvesting and Energy Management for Sustainable Wireless Sensor Networks', In Proceedings of the IEEE 13th International Conference on Communication Technology (ICCT), Jinan, China, 25–28 September 2011, pp. 362–367.

[21] Sudevalayam, S., and Kulkarni, P.: 'Energy harvesting sensor nodes: Survey and implications', *IEEE Communications Surveys and Tutorials* 2011, 13, 443–461.

[22] Panahi, F.H., Hajimirzaee, P., Erfanpoor, S., Panahi, F.H., and Ohtsuki, T.: 'Smart Image-Processing based Energy Harvesting for Green Internet of Things,' 2018 Smart Grid Conference (SGC), Sanandaj, Iran, 2018, pp. 1–5.

[23] Panahi F.H., and Panahi F.H.: 'Energy Harvesting Technologies and Market Opportunities', In: Nojavan S., Zare K. (eds) *Electricity Markets*. Springer, Cham. https://doi.org/10.1007/978-3-030-36979-8_1.

[24] Shi, Y., Xie, L., Hou, Y.T., and Sherali, H.D.: 'On Renewable Sensor Networks with Wireless Energy Transfer', In Proceedings of the IEEE in INFOCOM, Shanghai, China, 10–15 April 2011, pp. 1350–1358.

[25] Erol-Kantarci, M., and Mouftah, H.T.: 'Suresense: Sustainable wireless rechargeable sensor networks for the smart grid', *IEEE Transactions on Wireless Communications* 2012, 19.

[26] Pan, G., Lei, H., Yuan, Y., and Ding, Z.: 'Performance analysis and optimization for SWIPT wireless sensor networks,' *IEEE Transactions on Communications*, vol. 65, no. 5, pp. 2291–2302, May 2017, doi: 10.1109/TCOMM.2017.2676815.

[27] Guo, S., Wang, F., Yang, Y., and Xiao, B.: 'Energy-efficient cooperative for simultaneous wireless information and power transfer in clustered wireless sensor networks,' *IEEE Transactions on Communications*, vol. 63, no. 11, pp. 4405–4417, Nov. 2015.

[28] Zhang, R., Maunder, R.G., and Hanzo, L.: "Wireless information and power transfer: From scientific hypothesis to engineering practice,' *IEEE Communications Magazine*, vol. 53, no. 8, pp. 99–105, Aug. 2015.

[29] Kisseleff, S., Chen, X., Akyildiz, I., and Gerstacker, W.H.: 'Efficient charging of access limited wireless underground sensor networks,' *IEEE Transactions on Communications*, vol. 64, no. 5, pp. 2130–2142, May 2016.

[30] Lin, X., Andrews, J.G., and Ghosh, A.: 'Spectrum sharing for device-to-device communication in cellular networks,' *IEEE Transactions on Wireless Communications*, vol. 13, no. 12, pp. 6727–6740, Dec. 2014.

[31] Paramonov, A., Hussain, O., Samouylov, K., Koucheryavy, A., Kirichek, R., and Koucheryavy, Y.: 'Clustering optimization for out-of-band D2D communications', Wireless Communications and Mobile Computing, (2017).

[32] Panahi, F.H. and Ohtsaki, T.: 'Optimal channel-sensing policy based on Fuzzy Q-Learning process over cognitive radio systems,' 2013 IEEE International Conference on Communications (ICC), Budapest, 2013, pp. 2677–2682, doi: 10.1109/ICC.2013.6654941.

[33] Panahi, F.H. and Ohtsaki, T.: 'Optimal channel-sensing scheme for cognitive radio systems based on fuzzy Q-learning', *IEICE Transactions on Communications*, Vol. E97-B, No. 2, pp. 283–294, 2014.

[34] Hu, F., Chen, B., and Zhu, K.: 'Full spectrum sharing in cognitive radio networks toward 5G: a survey', *IEEE Access*, vol. 6, pp. 15754–15776, 2018, doi:10.1109/ACCESS.2018.2802450.

[35] Rault, T., Bouabdallah, A., and Challal, Y.: 'Energy efficiency in wireless sensor networks: A top-down survey', *Computing Networks.*, vol. 67, pp. 104–122, Jul. 2014.

[36] Alsharif, M.H., Sunghwan, K., and Nuri, K.: 'Energy harvesting techniques for wireless sensor networks/radio- frequency identification: A review', *Symmetry* 2019, 11, 865. https://doi.org/10.3390/sym11070865.

[37] Correia, L.H., Macedo, D.F., Santos, A.L.d., Loureiro, A.A., and Nogueira, J.M.S.: 'Transmission power control techniques for wireless sensor networks', *Computer Networks* 2007, 51, 4765–4779.

[38] Zhou, P., Liu, W., Yuan, W., and Cheng, W.: 'Energy-Efficient Joint Power and Rate Control Via Pricing in Wireless Data Networks', In Proceedings of the IEEE Wireless Communications and Networking Conference, Las Vegas, NV, USA, 31 March–3 April 2008, pp. 1091–1096.

[39] Tsiropoulou, E.E., Vamvakas, P., and Papavassiliou, S.: 'Joint utility-based uplink power and rate allocation in wireless networks: A non-cooperative game theoretic framework', *Physics Communications* 2013, 9, 299–307.

[40] Musku, M.R., Chronopoulos, A.T., Popescu, D.C., and Stefanescu, A.: 'A game-theoretic approach to joint rate and power control for uplink CDMA communications', *IEEE Transactions on Communications* 2010, 58, 923–932.

[41] Anastasi, G., Conti, M., and Di Francesco, M.: 'An Adaptive Sleep Strategy for Energy Conservation in Wireless Sensor Networks', (2009).

[42] Anastasi, G., Conti, M., Di Francesco, M., and Passarella, A.: 'Energy conservation in wireless sensor networks: A survey', *Ad Hoc Networks* 2009, 7, 537–568.

[43] Chang, H., and Rubin, I.: 'Optimal downlink and uplink fractional frequency reuse in cellular wireless networks', *IEEE Transactions on Vehicular Technology* 2016, 65, 2295–2308.

[44] Alsharif, M.H., and Kim, J.: 'Optimal solar power system for remote telecommunication base stations: a case study based on the characteristics of South Korea's solar radiation exposure', *Sustainability* 2016, 8, 942.

[45] Zhang, F., Hackworth, S.A., Liu, X., Chen, H., Sclabassi, R.J., and Sun, M.: 'Wireless Energy Transfer Platform for Medical Sensors and Implantable Devices', In Proceedings of the Annual International Conference of the IEEE in Engineering in Medicine and Biology Society (EMBC 2009), Minneapolis, MN, USA, 3–6 September 2009, pp. 1045–1048.

[46] Jonah, O., and Georgakopoulos, S.V.: 'Efficient Wireless Powering of Sensors Embedded in Concrete Via Magnetic Resonance', In Proceedings of the IEEE International Symposium on Antennas and Propagation (APSURSI), Boston, MA, USA, 18–13 July 2011, pp. 1425–1428.

[47] Griffin, B., and Detweiler, C.: 'Resonant: Wireless Power Transfer to Ground Sensors from a UAV', In Proceedings of the IEEE International Conference on Robotics and Automation (ICRA), St. Paul, MN, USA, 14–18 May 2012, pp. 2660–2665.

[48] Ho, S., Wang, J., Fu, W., and Sun, M.: 'A comparative study between novel witricity and traditional inductive magnetic coupling in wireless charging', *IEEE Transactions on Magnetics.* 2011, 47, 1522–1525.

[49] Kurs, A., Karalis, A., Moffatt, R., Joannopoulos, J.D., Fisher, P., Soljačić, M.: 'Wireless power transfer via strongly coupled magnetic resonances', *Science* 2007, 317, 83–86.

[50] Kline, M., Izyumin, I., Boser, B., and Sanders, S.: 'Capacitive Power Transfer for Contactless Charging', in Proceedings of the Twenty-Sixth Annual IEEE in Applied Power Electronics Conference and Exposition (APEC), FortWorth, TX, USA, 6–11 March 2011, pp. 1398–1404.

[51] Hui, S.: 'Planar: wireless charging technology for portable electronic products and Qi', *Proceedings of the IEEE* 2013, 101, 1290–1301.

[52] Lu, X.,Wang, P., Niyato, D., Kim, D.I., and Han, Z.: 'Wireless charging technologies: Fundamentals, standards, and network applications', *IEEE Communications Surveys and Tutorials* 2016, 18, 1413–1452.

[53] Lu, X., Wang, P., Niyato, D., Kim, D.I., and Han, Z.: 'Wireless networks with RF energy harvesting: A contemporary survey', *IEEE Communications Surveys and Tutorials* 2015, 17, 757–789.

[54] Xie, L., Shi, Y., Hou, Y.T., and Lou, A.: 'Wireless power transfer and applications to sensor networks', *IEEE Transactions on Wireless Communications* 2013, 20, 140–145.

[55] Lu, X., Wang, P., Niyato, D., and Han, Z.: 'Resource Allocation in Wireless Network with RF Energy Harvesting and Transfer,' IEEE Network, to appear.

[56] Lu, X., Wang, P., Niyato, D., and Hossain, E.: 'Dynamic spectrum access in cognitive radio networks with RF energy harvesting,' *IEEE Wireless Communications*, vol. 21, no. 3, pp. 102–110, June 2014.

[57] Nishimoto, H., Kawahara, Y., and Asami, T.: 'Prototype implementation of ambient RF energy harvesting wireless sensor networks', in Proceedings of IEEE Sensors, Kona, Hawaii, USA, November 2010.

[58] Nishimoto, H., Kawahara, Y., and Asami, T.: 'Prototype implementation of ambient RF energy harvesting wireless sensor networks', *IEEE Sensors Journal*, pp. 1282–1287, Nov. 2010.

[59] Huang, K., and Zhou, X.: 'Cutting the last wires for mobile communications by microwave power transfer', *IEEE Communications Magazine* 2015, 53, 86–93.

[60] Akyildiz, I.F., Melodia, T., and Chowdhury, K.R.: 'A survey on wireless multimedia sensor networks', *Computer Networks.* 2007, 51, 921–960.

[61] Liu, W., Lu, K., Wang, J., Xing, G., and Huang, L.: 'Performance analysis of wireless sensor networks with mobile sinks', *IEEE Transactions on Vehicular Technology*, vol. 61, no. 6, pp. 2777–2788, July 2012.

[62] Xie, L., Shi, Y., Hou, Y.T., *et al.*: 'A mobile platform for wireless charging and data collection in sensor networks', *IEEE Journal on Selected Areas in Communications*, vol. 33, no. 8, pp. 1521–1533, Aug. 2015.

[63] Wang, C., Guo, S., and Yang, Y.: 'An optimization framework for mobile data collection in energy-harvesting wireless sensor networks', in IEEE Transactions on Mobile Computing, doi: 10.1109/TMC.2016.2533390.

[64] Park, J., Seo, H., and Choi, C.: 'Simple contending-type MAC scheme for wireless passive sensor networks', *IEEE Trans. Smart Process. Comput.* 2017, 6, 299–304.

[65] Lee, S., Koh, J., and Jung, C.: 'An energy-efficient QoS-aware routing algorithm for wireless multimedia sensor networks', *International Journal of Multimedia and Ubiquitous Engineering* 2014, 9, 245–252.

[66] Usman, M., Yang, N., Jan, M., He, X., Xu, M., and Lab, K.: 'A joint framework for QoS and QoE for Video transmission over wireless multimedia sensor networks', *IEEE Transactions on Mobile Computing* 2018, 17, 746–759.

[67] Panahi, F.H., Moshirvaziri, S., Momene, A., Panahi, F.H., and Ohtsuki, T.: 'D2D-aided Optimal Utilization of Renewable Energies for Green Powered Base Stations', 2018 Smart Grid Conference (SGC), Sanandaj, Iran, 2018, pp. 1–6.

[68] Panahi, F.H., Panahi, F.H., Ghaderzadeh, M., and Mohammadisarab, A.: 'M2M Communications as a Promising Technique to Support Green Powered Base Stations', 27th Iranian Conference on Electrical Engineering (ICEE), Yazd, Iran, 2019, pp. 1654–1658.

[69] Saurabh, S., Pais, A., and Chatterjee, S.: 'Efficient key management in IoT using mobile aggregator', *Security in Computing and Communications.* 2016, 625, 161–172.

[70] Vijay, G., Ben Ali Bdira, E., and Ibnkahla, M.: 'Cognition in wireless sensor networks: a perspective', *IEEE Sensors Journal* 2011, 11, 582–592.

[71] Akan, O.B., Karli, O.B., Ergul, O.: 'Cognitive radio sensor networks', *IEEE Networks* 2009, 23, 34–40.

[72] Panahi, F.H., Panahi, F.H., Hattab, G., Ohtsuki, T., and Cabric, D.: 'Green heterogeneous networks via an intelligent sleep/wake-up mechanism and D2D communications', *IEEE Trans. on Green Communications and Networking*, vol. 2, no. 4, pp. 915–931, Dec. 2018.

[73] Panahi, F.H., and Ohtsuki, T.: 'Stochastic geometry modeling and analysis of cognitive heterogeneous cellular networks', *EURASIP Journal on Wireless Communications and Networking*, vol. 141, pp. 1–19, Dec. 2015.

[74] Panahi, F.H., and Ohtsuki, T.: 'Stochastic geometry based analytical modeling of cognitive heterogeneous cellular networks', in Proc. IEEE Int. Conf. on Commun. (ICC), pp. 5281–5286, June 2014.

[75] Panahi, F.H., Panahi, F.H., and Ohtsuki, T.: 'Energy efficiency analysis in cache-enabled D2D-aided heterogeneous cellular networks', *IEEE Access*, vol. 8, pp. 19540–19554, 2020, doi:10.1109/ACCESS.2020.2968455.

[76] Panahi, F.H., Panahi, F.H., Hattab, G., Ohtsuki, T., and Cabric, D.: 'Green heterogeneous networks via an intelligent power control strategy and D2D communications', IEEE 28th Annual International Symposium on Personal, Indoor, and Mobile Radio Communications (PIMRC), Montreal, QC, 2017, pp. 1–8.

[77] Jo, H., Sang, Y., Xia, P., and Andrews, J.: 'Heterogeneous cellular networks with flexible cell association: A comprehensive downlink SINR analysis', *IEEE Transactions on Wireless Communications*, vol. 11, pp. 3484–3495, Oct. 2012.

[78] Sakr, A.H., and Hossain, E.: 'Cognitive and EH-based D2D communication in cellular networks: Stochastic geometry modeling and analysis', *IEEE Transactions on Communications*, vol. 63, no. 5, pp. 1867–1880, May 2015.

[79] Sakr, A.H., and Hossain, E.: 'Analysis of multi-tier uplink cellular networks with EH and flexible cell association', in Proc. of 2014 IEEE Global Telecommunications Conference (GLOBECOM), 2014, pp. 4712–4717.

[80] Liang, Y., Zeng, Y., Peh, E.C.Y., and Hoang, A.T.: 'Sensing-throughput tradeoff for cognitive radio networks', *IEEE Transactions on Wireless Communications*, vol. 7, no. 4, pp. 1326–1337, April 2008, doi: 10.1109/TWC.2008.060869.

[81] ElSawy, H., Sultan-Salem, A., Alouini, M., and Win, M.Z.: 'Modeling and analysis of cellular networks using stochastic geometry: a tutorial', *IEEE Communications Surveys and Tutorials*, vol. 19, no. 1, pp. 167–203, First quarter 2017, doi: 10.1109/COMST.2016.2624939.

[82] Malak, D., Al-Shalash, M., and Andrews, J.G.: 'Spatially correlated content caching for device-to-device communications', arXiv preprint arXiv:1609.00419, 2016.

[83] Panahi, F.H., Ohtsuki, T., Jiang, W., Takatori, Y., and Nakagawa, T.: 'Joint interference alignment and power allocation for multi-user MIMO interference channels under perfect and imperfect CSI', *IEEE Transactions on Green Communications and Networking*, vol. 1, no. 2, pp. 131–144, June 2017, doi: 10.1109/TGCN.2017.2694061.

[84] Zhang, J., Zhang, J., Zhou, Y., Ji, H., Sun, J. and Al-Dhahir, N.: 'Energy and spectral efficiency tradeoff via rate splitting and common beamforming coordination in multicell networks', *IEEE Transactions on Communications*, vol. 68, no. 12, pp. 7719–7731, Dec. 2020, doi: 10.1109/TCOMM.2020.3023736.

[85] Al-Obiedollah, H.M., Cumanan, K., Thiyagalingam, J., *et al.*: 'Spectral-energy efficiency trade-off-based beamforming design for MISO non-orthogonal multiple access systems', *IEEE Transactions on Wireless Communications*, vol. 19, no. 10, pp. 6593–6606, Oct. 2020, doi:10.1109/TWC.2020.3004292.

[86] Panahi, F.H., Ohtsuki, T., Jiang, W., Takatori, Y., and Uehara, K.: 'Interference alignment and power allocation for multi-user MIMO interference channels', 2016 IEEE International Conference on Communications (ICC), Kuala Lumpur, 2016, pp. 1–7, doi: 10.1109/ICC.2016.7510851.

[87] Zeb, J., Hassan, A., and Nisar, M.D.: 'Joint power and spectrum allocation for D2D communication overlaying cellular networks', *Computer Networks*, vol. 184, 2021, 107683, https://doi.org/10.1016/j.comnet.2020.107683.

[88] Zabetian, N., Mohammadi, A., and Masoudi, M.: 'Energy-efficient power allocation for device-to-device communications underlaid cellular networks using stochastic geometry', *Transactions on Emerging Telecommunications Technologies* 30.12 (2019): e3768.

Chapter 5

A systematic study on the metamaterial microstrip antenna design for self-powered wireless systems

Taha A. Elwi[1]

Abstract

This chapter discusses comprehensively the possibility of desigining a low-profile metamaterial (MTM) and electromagnetic bandgap structures for self-powered wireless systems. The designed antenna would be mounted on a flexible solar panel substrate for wearable applications. For this, an antenna based on a Hilbert-shaped MTM is exemplified to performance over the frequency ranging from a few MHz up to 6 GHz to satisfy the frequency bands of modern 5G networks and wireless sensor networks/Internet of Things applications. In this study, the antenna is mounted on a flexible solar panel as a substrate to ensure solar energy harvesting as well as harvesting the RF energy. For this, the key design points for such work are eliminating the shadowing effects of the antenna on the solar panel to maintain efficiency and increasing the antenna bandwidth to harvest the RF energy over the entire frequency band of interest. Nevertheless, the negative effects of solar cell metallization on the antenna performance must be reduced significantly. The antenna must be circularly polarized to ensure RF energy reception based on any polarization. Therefore, the harvested RF energy for the proposed design is found to be about 83% at 3.5, 4.2, 5.4, and 6 GHz. Next, the proposed study is extended to measure the output DC voltages from the solar energy harvesting to be about the same output DC voltage before introducing the antenna structure. The antenna polarization is found to be circular where the excel ratio (AR) is less than 3 dB over the frequencies of interest. Such a study is conducted to realize the most effective reason for maximizing the harvesting outputs. The key challenge of such research is found to be how to maximize such antenna efficiency with omni-directional radiation patterns to support the RF energy harvesting with minimum negative effects on the solar energy harvesting efficiency. On top of that, developing a low-cost, low-profile, lightweight, and ultra-wideband antenna that can be integrated into self-powered wireless systems could be another challenge for such research goals. Finally, this chapter will also identify the major advances for each design

[1]Communication Engineering Department, Al-Ma'moon University College, Baghdad, Iraq

step to develop cost-effective, efficient, and reliable energy harvesting systems to maximize energy management.

5.1 Introduction

Recently, our environment has been filled with electromagnetic waves due to the wide expansion in modern wireless communication services [1]. A great portion of the radiated energy could be wasted in the air while a small amount of it would be received [2]. If such energy, radio frequency (RF), has been used wisely, a reusable benefit could be gained with the natural harvested energy including solar resources [2]. In many low-power assumption devices, including sensor networks [3], radio-frequency identification [4], Internet-of-Things (IoT) [5], and electric clock, the harvested RF energy is considered the main power resources. With the aid of a super capacitor, the harvested RF energy can provide sufficient power for the entire life span [5]. Several research studies conducted their research on developing single-band rectennas [6], multiband rectennas [7], and arrays [8]. Because, the harvested energy usually comes from different directions and sources, an omni-directional antenna with multibands is preferable receptors [5]. Therefore, a rich multi-path signal environment combined with an uncertainty of the source position makes an omnidirectional antenna a preferred choice for ambient RF energy harvesting [6]. For this, the harvested energy efficiency could be increased widely when antenna-based fractal geometries are realized [7].

Since their introduction in the 1950s, microstrip antennas have been extensively investigated due to their low profile, lightweight, low cost, conformability to mounting hosts, and ease of manufacturing [8]. It is well known that the conventional microstrip antenna exhibits a relatively small bandwidth and moderate gain which limits its use in modern ultra-wideband (UWB) communication systems [9]. Recently, the Federal Communications Commission (FCC) allocated the frequency spectrum from 3.1 to 10.6 GHz for UWB applications [10]. Numerous challenges arise to overcome the inherent performance limitations of the traditional microstrip antenna to allow a 10-dB return loss bandwidth over the entire band allocated for UWB applications. A myriad of design methods has been proposed to increase the bandwidth such as aperture coupling, parasitic elements, multilayered patches, and modifications of the feed [11].

Recently, the performance of microstrip antennas has been enhanced with the introduction of metamaterials (MTM) and electromagnetic bandgap (EBG) structures [12,13]. For example, split-ring resonators (SRRs) were applied to miniaturize the size, increase the bandwidth, and enhance the radiation patterns of printed circuit antennas [12,14]. EBG structures patterned either on the metal patches or etched on the ground plane were proposed [15]. MTMs possess several attractive features such as negative effective constitutive parameters, surface wave reduction, and distinctive stopbands [16]. However, resonant MTM structures suffer from two fundamental limitations: narrow bandwidth over which the desired properties occur and losses due to the conducting inclusions [17]. Miniaturized antennas made of fractal-shaped space-filling curves were proposed in [18] with enhanced bandwidth. A microstrip antenna composed of six spiral unit cells of MTMs and a printed dipole element backed with a truncated ground plane of a rectangular slot was proposed in [19]. In [20], a printed meander RFID tag antenna on open complementary SRRs was introduced. A spiral-shaped antenna printed on a dielectric substrate containing SRR inclusions to radiate over the frequency ranging from 8 to 18 GHz was explored in [21]. A multi-resonant microstrip antenna based on the Hilbert curve fractal patch excited with a co-planar waveguide was reported in [22]. A rectangular patch antenna consisting of a periodic array of MTM structures fed through strip-line and backed by a defected ground plane was reported in [13] to achieve broad bandwidth and high gain.

5.2 Motivation and objectives

5.2.1 Motivation

Due to the rapid need for green energy harvesting as new resources for self-powered wireless systems, several research groups around the world are attracted to harvest RF energy [7]. For this, the proposed study is conducted to summarize the main effective antenna parameters including efficiency, gain, bandwidth, matching impedance, and material losses on the harvested voltage. This study prompts such researchers to develop novel antenna designs for RF energy harvesting, which could increase power wireless devices capability for both 5G and IoT systems [18].

In the last few decades, attention is increased toward wireless power transfer due to the ability of seamless usage as well as the ability of wireless charging for different applications including modern cars based electrical DC motors [12]. This need is recently shifted toward self-powered wireless devices to avoid daylight and night limitations as in photovoltaic panels [3]. This interest has been coupled with the importance of controlling the data collection and positioning navigation surfaces [5]. Nevertheless, the most important need is the national borders security against smugglers on both land and sea [20]. Therefore, monitoring such surfaces in real time is not an easy matter with all these input parameters that complicate the parameters compromisation by itself [2]. All these requirements motivated researchers to investigate and develop low-cost, low-profile, and lightweight antenna structures that can be integrated into different wireless systems.

5.2.2 Objectives

In light of the problem statement challenge that is presented in the abstract, the main research objective is to investigate and develop a low-cost, low-profile, and lightweight antenna structures that can be integrated into self-powered wireless devices. Many systems are available that use energy from external natural/ non-natural sources which can be captured and stored. There is a need for such systems because: they can provide sufficient power as and when it is needed, they offer long lifetime with little or no maintenance needs, and they are easy and cheap to install with enough flexibility [9]. RF and microwave energy are among the external non-natural sources that can be captured and used to power the electronic systems of wireless sensor nodes. Therefore, it is the subject of this research to develop microstrip antennas based on a printed circuitry to receive and transmit the collected data by the sensor node and at the same time to be used for harvesting the RF and microwave energy.

To overcome the antenna limitations, several printed circuit antennas were discussed in this chapter to show the following advantages:

- The use of fractal geometry can provide an excellent size reduction in comparison with other techniques by increasing the current path within a limited area. Moreover, fractal antennas provide multi-bands at different frequency resonances.
- Partial ground plane antennas may be a subject for providing a wide frequency band with relatively small size.
- The introduction of the MTM could enhance the antenna gain and polarization stability.
- The EBG structure can be introduced to enhance the antenna radiation efficiency and radiation patterns.

Finally, these designs will be studied extensively using two different commercial software packages including CST, MWS, and HFSS. The fabricate process of the proposed antenna and the effects on the harvesting efficiency will be summarized in this chapter based on experimental measurements.

5.3 Recent related previous approaches

Most integrated wireless systems rely on batteries that suffer from different drawbacks such as charging losses, low charge and discharge cycle life, overheat failure, substantial maintenance costs, and intelligent power management systems, an urgent need is increased to invoke the green energy techniques as alternatives [1]. Therefore, many scientists applied their research studies to couple antennas with solar panels for self-powered devices at low energy [2]. Mainly, the proposed approaches can be classified into five main approaches:

1. Placing the antenna structure under the solar panel directly or on the ground plane with sufficient space around the antenna to allow the antenna radiation leaving the unutilized surface space. Such technique is the simplest type; however, the antenna metallization affects negatively on the solar panel efficiency. Moreover, the solar panel bus bar metallic has negative effects on the antenna performance. Therefore, to minimize these effects, the ground plane of antennas is mounted on the back panel side of the solar panel. For example, a 920-MHz coplanar waveguide inverted-F antenna (IFA) integrated on a-Si sun-oriented cells [22].

2. Integrating solar panels to the metallic parts of antennas as a reflector, for example. In such a configuration, the antenna would operate normally without the solar cells. For example, a dipole antenna was proposed in [23] and backed with a solar panel reflector.
3. Using the solar cell as a radiating element patch of the microstrip antenna. Such a method may reduce the overall system weight and size; however, a low front-to-back ratio could take place. In [26], a GPS antenna vehicular system was constructed based on four solar cells as array patch antennas. This approach overcomes the efficiency reduction due to shadowing effects; however, a significant enlargement in the antenna ground plane has occurred.
4. Using the solar cell as a ground plane for microstrip antennas was proposed as in [27]. However, the effects of shadowing on the solar cell impact negatively on the output efficiency and the antenna impedance matching.
5. Integrating the solar panel fully with the antenna structure to form the radiating element with no need for additional metallization. In [10], a solar panel was concentrated as a parabolic reflector to enhance the dipole gain at 1.5 GHz and focus the solar light to a certain point.

5.4 Antenna geometry

The configuration of the proposed antenna is shown in Figure 5.1. The patch consists of an array of 3 $\times$ 5 planar Hilbert MTM elements printed on a 0.394-mm-thick FR4 substrate with a relative permittivity of $\varepsilon_r = 4.4$ and a loss tangent of $\tan\delta = 0.02$. The dimensions of the fractal geometry are chosen in order to increase the effective electrical length of the unit cell within a 4 $\times$ 4-mm footprint and to control the input impedance matching at the desired resonant frequencies through capacitive coupling using the T-shaped stub, as shown in Figure 5.1(b). Moreover, the unit cell provides a small shadowing area that is essential for self-powered wireless devices [22]. To ensure that the driving current excites each unit cell, the transmission-like network shown in Figure 5.1(b) couples energy to the unit cells, as shown in Figure 5.1(b). The feed network is excited with a 0.4-mm wide, 50-Ω off-centered transmission line with a length of 8 mm, as depicted in Figure 5.1(a) and (b). The defected ground plane and unit cells on the patch direct the radiation pattern in the direction parallel to the patch plane. In Figure 5.1(c), the width of the partial ground plane, the width of the feed line, and the position of the feed line have been chosen to optimize the bandwidth of the equivalent solid patch [13]. The rest of the ground plane has been loaded with a 7 $\times$ 6 periodically etched EBG cells, as shown in Figure 5.1(c). The details of the geometry of the unit cell structure are given in Figure 5.1(d) to (f).

The elementary radiator is inspired by the approach outlined in [14] and different Hilbert orders are presented to achieve an EBG structure with a direction-dependent bandgap. In this study, a patch structure based on the Hilbert MTM unit cell of the third order is proposed. The width of the Hilbert line is selected so as not to cause unwanted cross-lines between the inner and outer traces because of the limited resolution of the printing and etching processes.

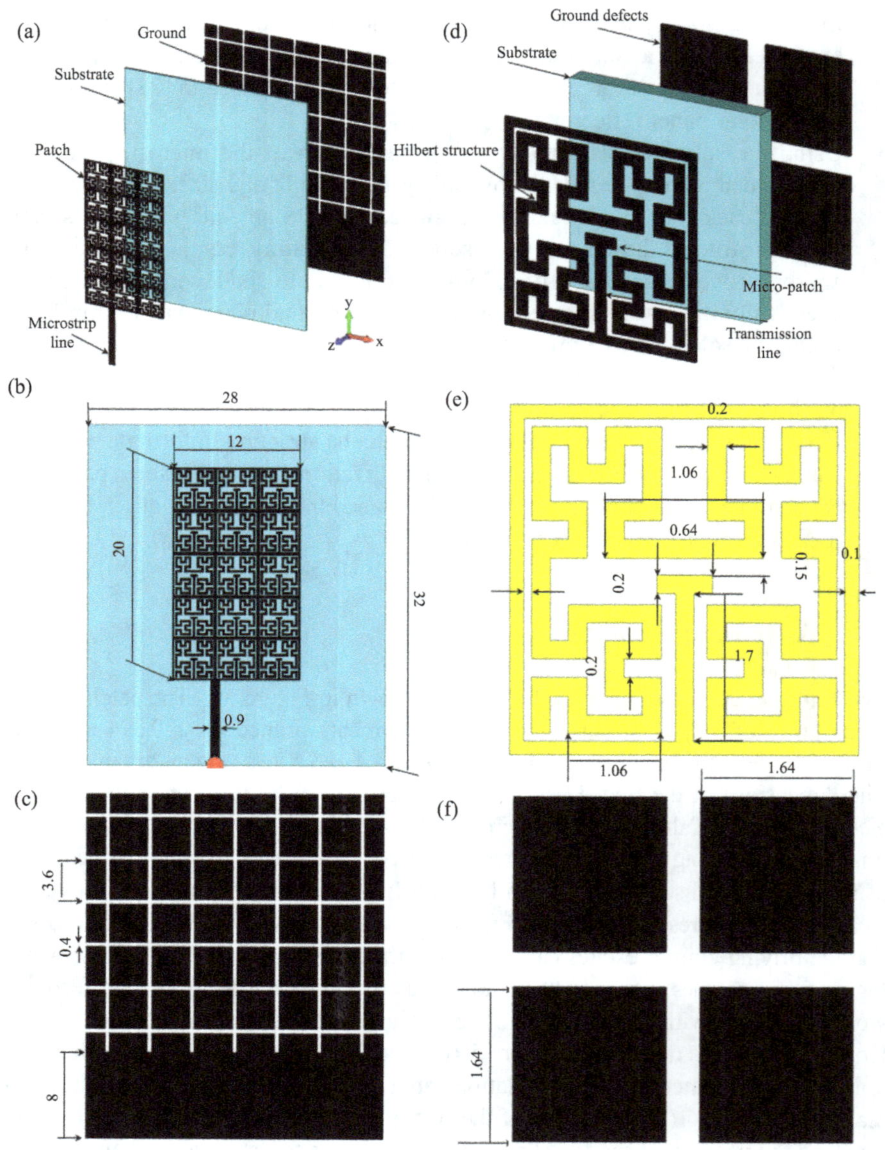

Figure 5.1 Geometry of the proposed antenna: (a) 3-D view, (b) 3-D decomposed view, (c) front view, (d) front view without the square EBG, (e) front view with square EBG, and (f) 3-D view with square EBG. Note: All numbers are in mm

5.5 Performance characterization

The performance characterizations are discussed in this section. The given study is performed to achieve the following performance matrix as follows:

1. Maximize the antenna bandwidth to cover UWB matching impedance below −10 dB,
2. Increase the antenna gain ultimately over the entire frequency band of interest,
3. Realize omni-directional patterns at different frequencies,
4. Reduce the antenna size ultimately without affecting the performance of the antenna,
5. Minimize the shadowing effects to maintain the solar panel efficiency ultimately, and
6. Maintain the antenna performance against bending effects and mechanical forces.

Therefore, the goal of this study is to achieve the design specifications by conducting printing technology based on conductive nanomaterial ink and flexible solar panels.

5.6 Design methodology

In this section, the author offers the procedure followed to arrive at the final design given in Figure 5.1. This section is also intended to explain the principle of operation of the antenna based on simulations conducted using CST MWS [24]. The design starts with a solid rectangular patch backed by a full ground plane, denoted as case 1, as shown in Figure 5.2(a). Next, the solid patch in case 1 is changed to a fishnet geometry, denoted as case 2, see Figure 5.2(c). Further, a Hilbert conductor pattern, which makes the current antenna design different from the approach reported in [21], is introduced inside the fishnet geometry as seen in Figure 5.2(e) and is referred to as case 3. Further modification considered in the design includes the introduction of a partial ground plane common to cases 1, 2, and 3 and denoted as cases 4, 5, and 6, respectively. The geometrical details are summarized in Table 5.1. In all cases, the patch is fed by a 50-Ω microstrip transmission line, 0.4 mm in width, positioned on top of the substrate and backed with the partial ground plane, as shown in Figure 5.1(c).

5.6.1 Design development

The S_{11} spectra evaluated from CST MWS are presented in Figure 5.3. It is found that the full ground plane, cases 1 to 3, provides narrow bandwidths around multiple frequency bands as can be seen in Figure 5.3(a). However, when the partial ground plane is introduced in cases 4 to 6, the antenna shows wider bandwidths as seen in Figure 5.3(b). However, in the cases of the full ground plane, 1 to 3, the number of frequency resonances is increased as the solid patch is changed to the proposed fractal geometry. It should be noted that the parameters gp_l, TL_x, and TL_y

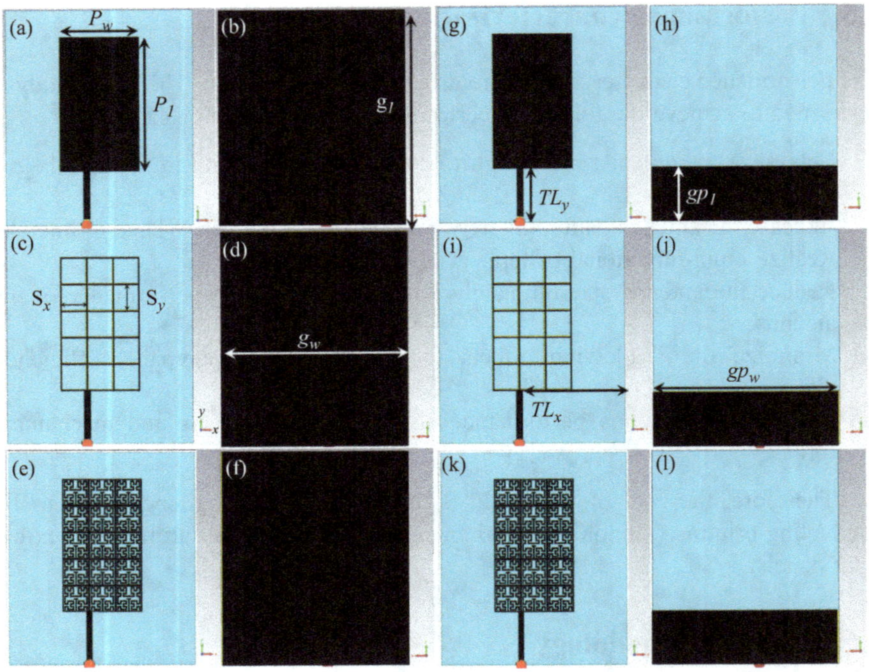

Figure 5.2 Numerical model of the proposed cases: (a), (c), (e), (g), (i), and (k) are the front views of cases 1–6, respectively. (b), (d), (f), (h), (j), and (l) are the back views of cases 1–6, respectively

Table 5.1 Geometrical details of the proposed cases

Part name	Symbol	Dimension (mm)
Patch width	p_w	12
Patch length	p_l	20
Full ground plane width	g_w	28
Full ground plane length	g_l	32
Partial ground plane width	gp_w	28
Partial ground plane length	gp_l	8
TL length	TL_y	8
TL position	TL_x	16
Slot width	S_x	3.68
Slot length	S_y	3.68

have been fixed in all cases and their values have been determined in [13] are determined to optimize the bandwidth resulting in the S_{11} spectrum of case 4 shown in Figure 5.3(b). It should be mentioned that although the same geometries and dimensions of the substrate, partial ground plane, and patch reported in [13] were used in [21]; however, they were not referenced in [21].

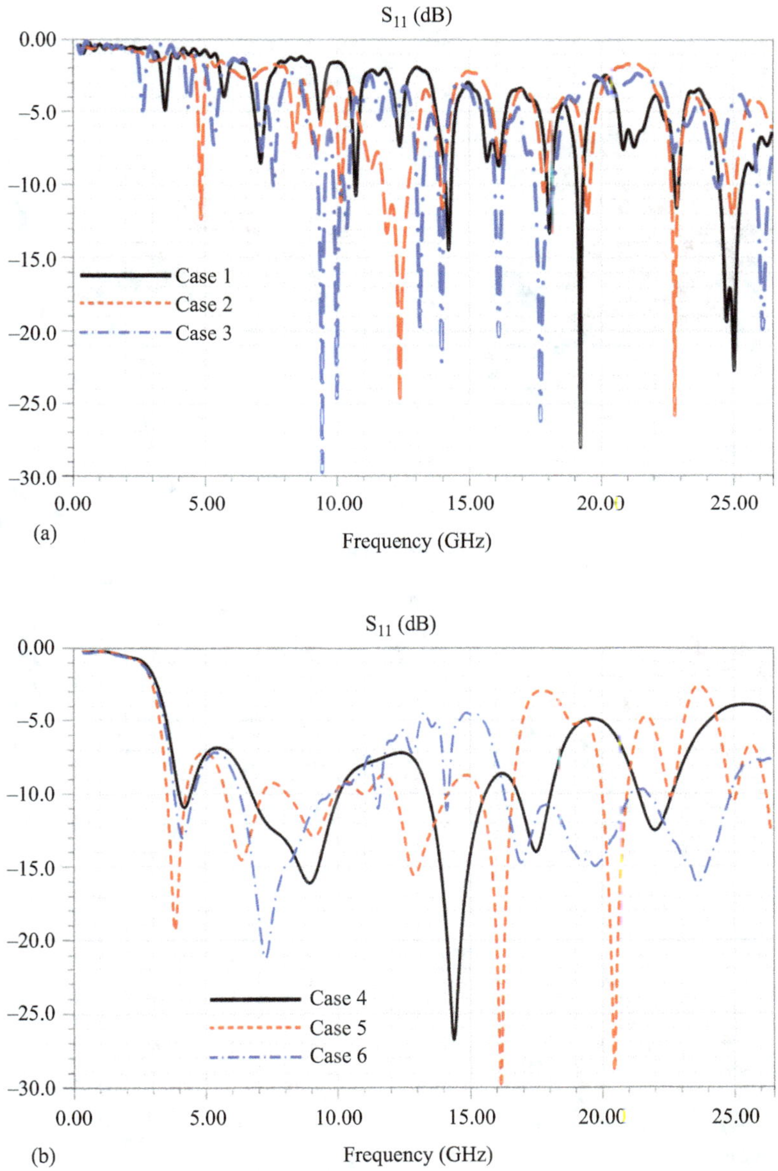

Figure 5.3 S_{11} spectra: (a) 1 to 3 and (b) 4 to 6

The radiation patterns are also affected by the change in the geometries of the ground plane and the patch. For example, the 3-D radiation patterns are evaluated at 5.8 GHz, where the antenna possesses excellent matching. Case 1 shows broadside radiation along the *z*-axis, as depicted in Figure 5.4(a). However, changing the patch to case 2, the boresight direction deviates slightly, while in case 3 the

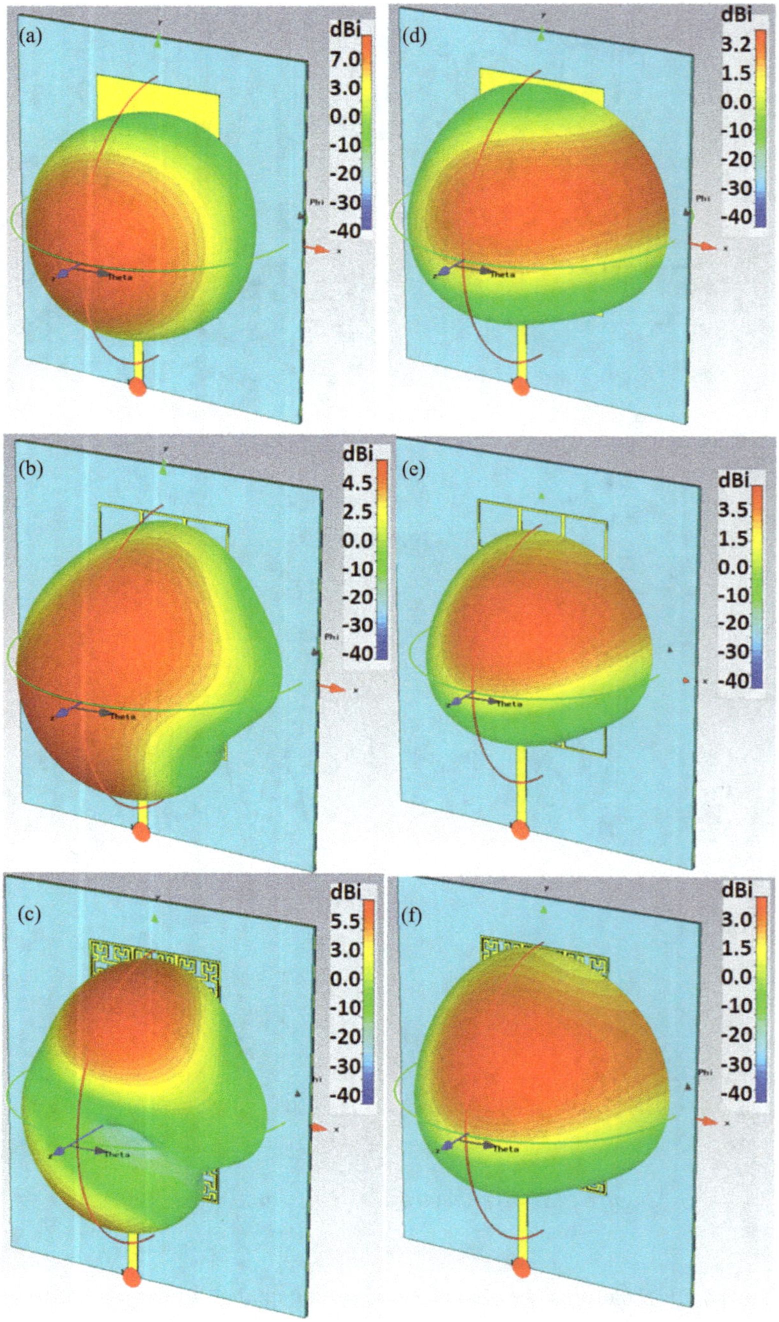

Figure 5.4 3-D radiation patterns at 5.8 GHz: (a) to (f) for cases 1 to 6, respectively

deviation increases further, as shown in Figure 5.4(b) and (c), respectively. In general, the boresight directions for cases 4 to 6 are along the patch side, as seen in Figure 5.4(d) to (f).

5.6.2 Principles of operation

In this section, the principle of operation of the proposed antenna shown in Figure 5.5 and the method used to enhance the bandwidth are provided. The boresight direction at 5.8 GHz has been changed from the x-axis to the y-axis. This is accomplished by changing the resonant mode at 5.8 GHz from TM_{020} to TM_{001}, by choosing a patch length (p_l), which is less than the width (p_w). A microstrip antenna similar to case 4 but with a rectangular patch of length, $p_l = 4$ mm, as shown in Figure 5.6(a) and (b), and width, $p_w = 12$ mm, is considered. The S_{11} spectrum and radiation pattern are presented in Figure 5.6(c) to (e). As seen in Figure 5.6(d), the radiation pattern of the proposed antenna is along the y-axis, in which the antenna is an end-fire radiator. Surface current propagating along the y-axis is subjected to high attenuation due to the stopband encountered at 5.8 GHz as shown in Figure 5.5. However, the proposed unit cell in Figure 5.1(d) allows propagation with minimum attenuation along the x-axis. Hence, the proposed unit cell is an anisotropic EBG structure, which allows controlling the p_l/p_w ratio without changing their physical lengths.

The EBG structure, see Figure 5.1(d), consists of defects on the ground plane with an array of a periodic square array to achieve a bandgap from 5 to 8.5 GHz when it is placed below the FR4 substrate with Hilbert MTM patch as seen in Figure 5.7.

5.6.3 Bandwidth

To specify the effects of the coupling gap (G), see Figure 5.1(d), S_{11} spectra of the proposed antenna are calculated by changing G to 0.05 mm, 0.1 mm, and 0.15 mm as seen in Figure 5.8. It is found that the coupling gap exhibits a significant effect on the magnitude of the return loss, bandwidth, and resonant frequency. As the gap distance increases, the bandwidth increases, while the return loss starts to degrade. Hence, the coupling gap is fixed at 0.15 mm. An observable effect is found in the antenna matching by changing the length insets of T-shaped stubs inside the Hilbert-shaped MTM unit cells.

5.6.4 Radiation patterns

The far-field radiation patterns are evaluated numerically around the WLAN and WiMax bands. The antenna shows a steerable end-fire radiation pattern from 3.3 to 3.9 GHz, which is suitable for WiMax applications, as seen in Figure 5.9. As shown in Figure 5.9(a), the realized gain varies from 0 to 2.5 dB over the frequency band from 3.3 to 3.9 GHz. In addition, the antenna provides a steering rate of 0.12°C/MHz, as depicted in Figure 5.9(b) and (c). The 3-dB beamwidth varies from 110° and 140° as presented in Figure 5.9(c). The maximum realized gain is 4.75 dBi at 5.5 GHz with an insignificant change in the gain and the main lobe direction over the frequency band from 4.4 to 6 GHz as shown in Figure 5.9(d) to (f).

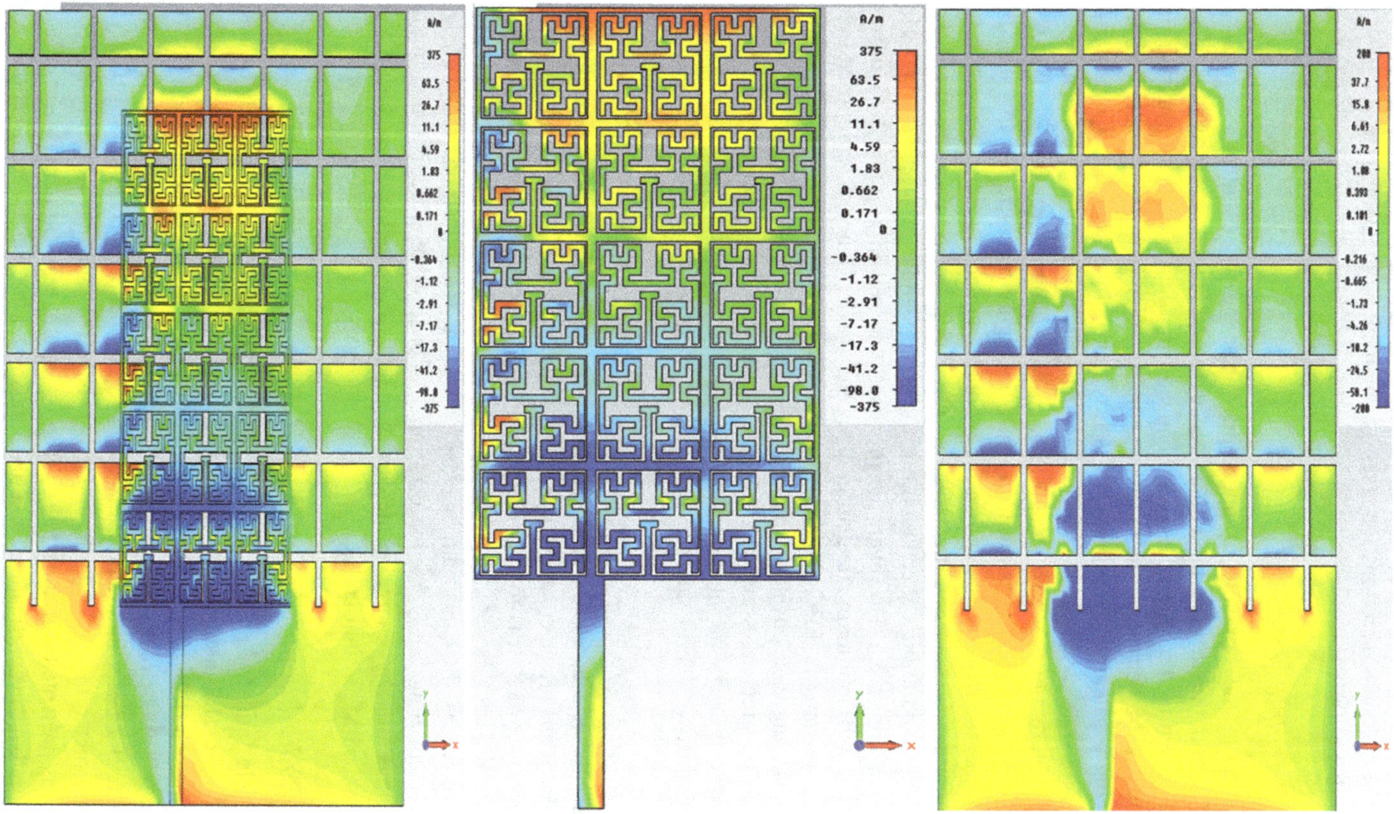

(a) (b) (c)

Figure 5.5 Surface current distribution at 5.8 GHz: (a) on both patch and EBG structure, (b) on the patch, and (c) on the EBG structure

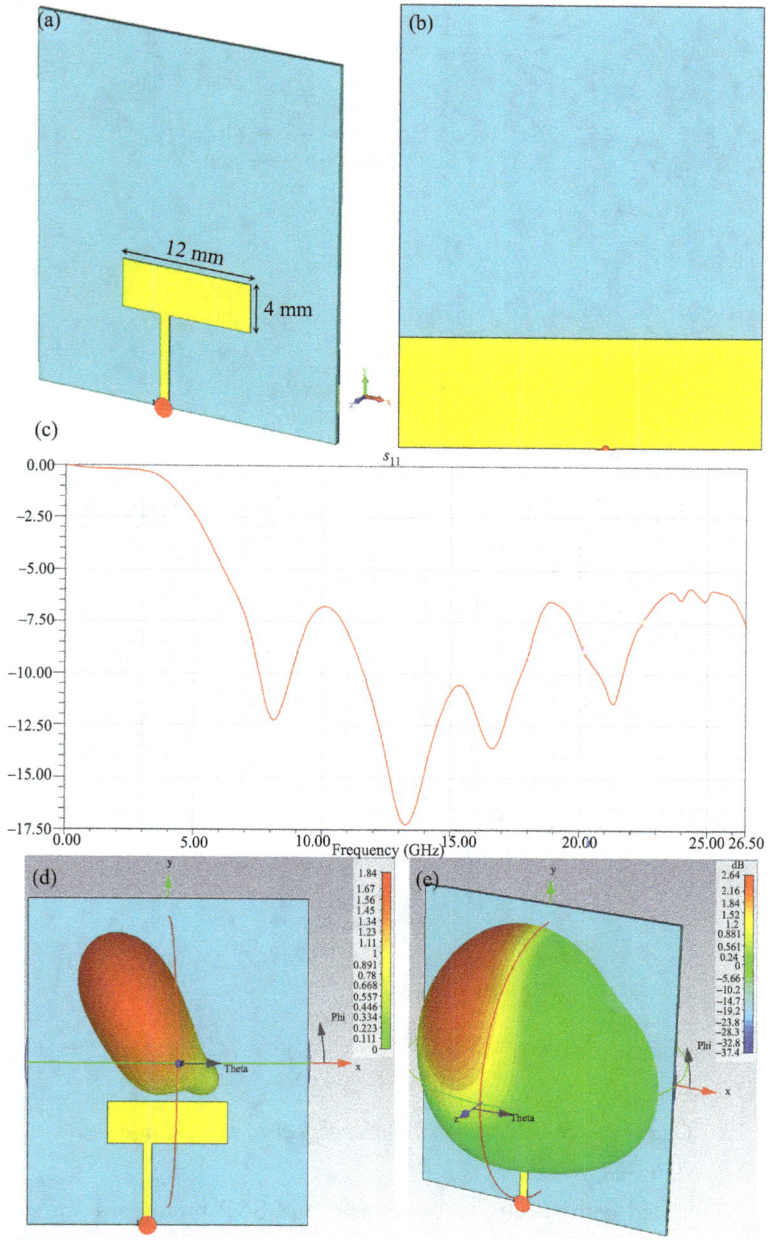

Figure 5.6 The microstrip antenna geometry and performance: (a) and (b) geometry, (c) S_{11}, (d), and (e) 3-D radiation pattern

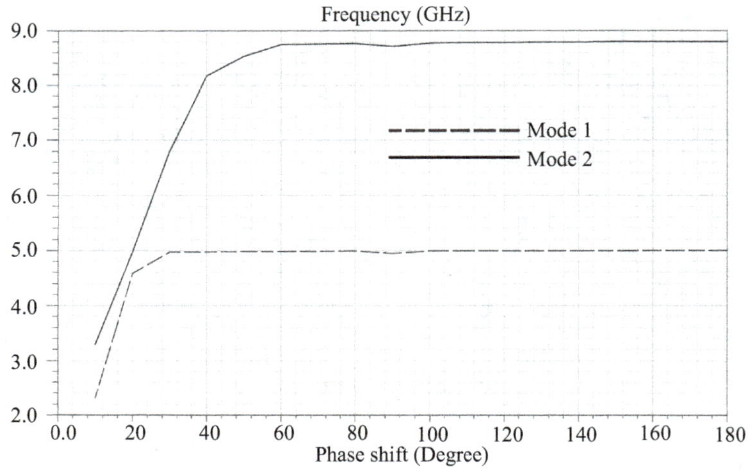

Figure 5.7 Dispersion diagram of the proposed unit cell

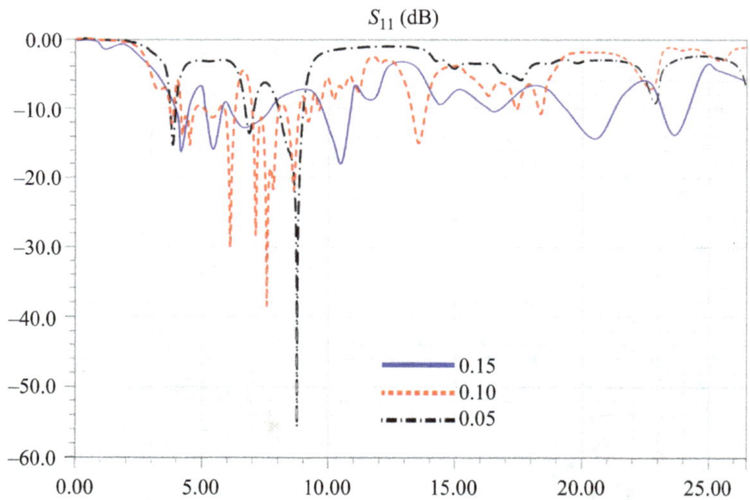

Figure 5.8 Change in the S_{11} spectra with respect to the change in the coupling gap

The simulated gain spectra of the copper and SNP prototypes are shown in Figure 5.10. The copper prototype shows a variation in the realized gain from −5 to 5 dBi as frequency changes from 3 to 18 GHz, as shown in Figure 5.10(a). This is very desirable for wireless applications in the medium and long ranges. The SNP prototype shows a fluctuation in the realized gain from −10 to 1 dBi within the same frequency band, as depicted in Figure 5.10(b). This can serve wireless applications from short to medium range. The results presented in Figure 4.3.10 are validated against those obtained from HFSS [25].

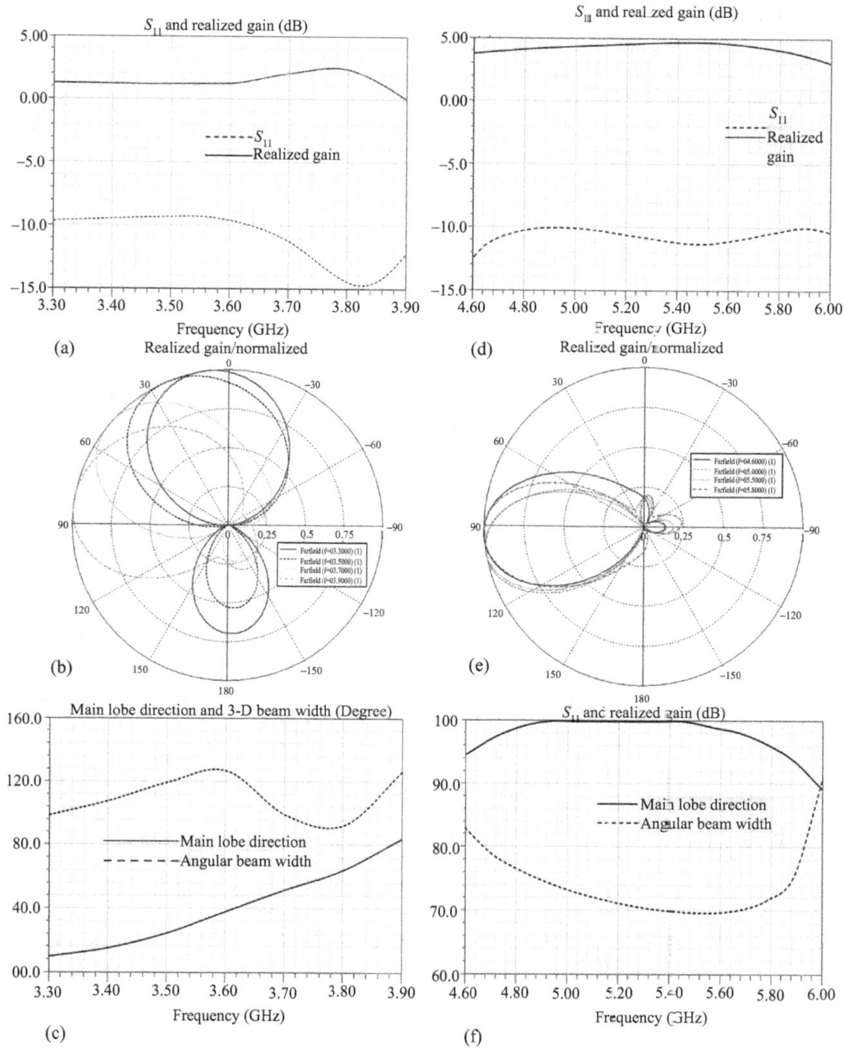

Figure 5.9 Antenna performance from 3.3 to 3.9 GHz: (a) S_{11} and realized gain versus frequency, (b) radiation patterns in a linear scale, and (c) main lobe direction and beamwidth. Antenna performance from 4.4 to 6 GHz, (d) S_{11} and realized gain versus frequency, (e) radiation patterns in a linear scale, and (f) main lobe direction and beamwidth

5.7 Printing process

5.7.1 Fabrication of the antenna on FR4 substrate

The DMP 2800 printer is used to print the antenna shown in Figure 5.1 on an FR4 substrate with a 10 pl nominal drop volume cartridge. The substrate is set on a

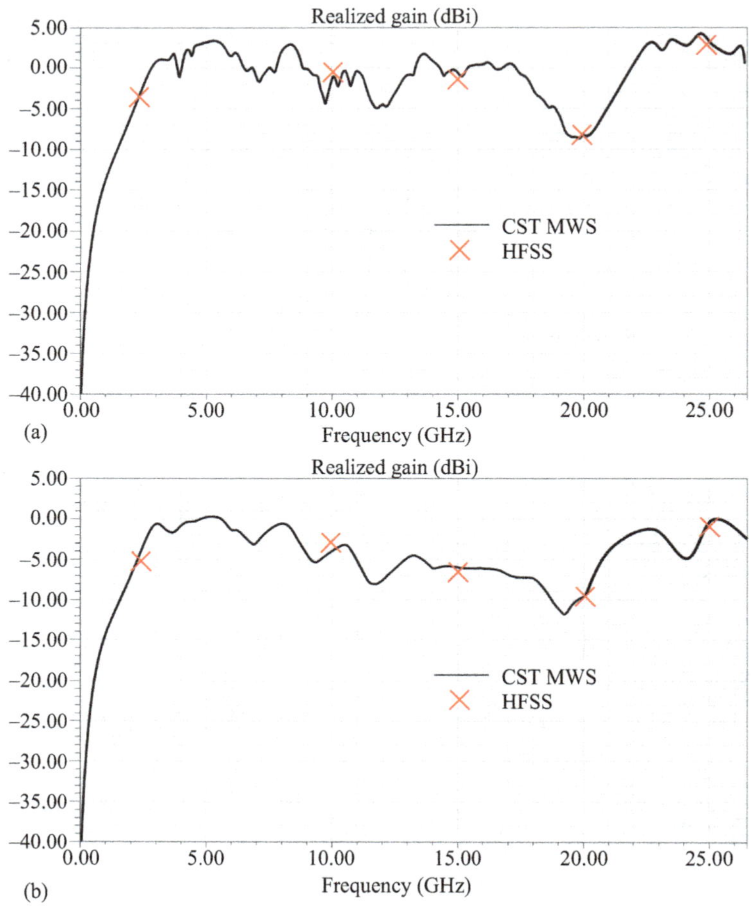

Figure 5.10 Comparison of the simulated realized gain of the proposed antenna: (a) copper electrical conductivity, $\sigma = 5.7 \times 10^7$ S/m, and (b) SNP electrical conductivity, $\sigma = 8.8 \times 10^4$ S/m

vacuumed plate, with a printing area of 200 × 300 mm. The printer is integrated with fiducial and drop watcher cameras. The vacuumed plate is heated up to 60 °C. A piezo-electric controller is connected to the head of the printer through 16 jets.

A single inkjet with 20 mV for the piezo-driven jetting device is used to print the proposed antenna. Five layers are printed on the FR4 substrate after heating the vacuumed platen to 60 °C. The head printer is set to 0.5 mm in height. The separation time between two consecutive printing passes is fixed at 20 s. The SNP ink is heated to 45 °C inside the cartridge. The printed antenna is annealed inside the ProtoFlow LPKF's convection oven for 10 h at 120 °C. The 50-Ω SMA port is attached to the ground plane and the patch using silver paste. The antenna is annealed for 15 min at 120 °C after attaching the SMA port. The printed SNP

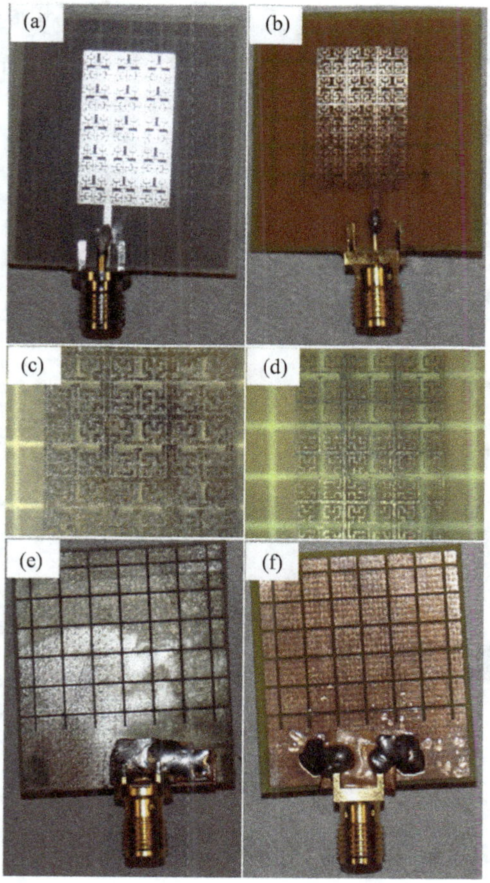

Figure 5.11 *The fabricated prototypes: (a) and (b) are the front and (c) the back views based on SNP, respectively. (d) and (e) are the front, and (f) the back views based on copper, respectively*

antenna is displayed in Figure 5.11(a) to (c). The four-point technique is used to measure the DC electrical conductivity of the printed antenna using a Keithley 224 current source and a Keithley 617 electrometer and is found to be $8.8 \pm 0.2 \times 10^4$ S/m. The measured conductivity is used in CST MWS and HFSS to evaluate the performance of the SNP prototype. An identical prototype is fabricated from copper using a chemical etching process. The 50-Ω SMA port is soldered to the patch and the ground plane. The prototype is shown in Figure 5.11(d) to (f).

5.7.2 Fabrication of the antenna on a solar panel

A flexible polycrystalline silicon solar panel with an active area of 8.4 × 6.0 cm, 2.5 mm thickness, is used as the substrate. The solar panel is made of hydrophobic materials, the surface of which is not adhesive; hence, inkjet printing without surface

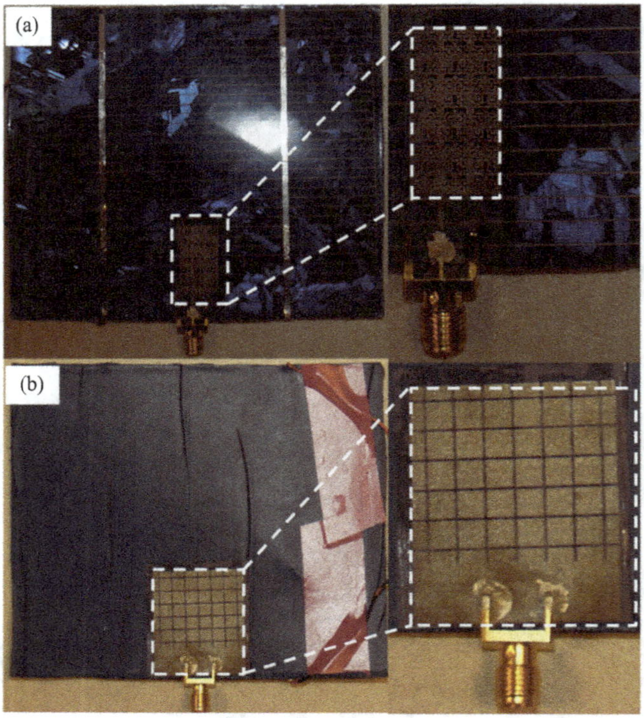

Figure 5.12 Fabricated antenna prototypes based on the EBG structure on the solar panel: (a) front view and (b) back view

treatment is very difficult. Nevertheless, the attempts of treating the surface using plasma or chemical surface treatment may lead to efficiency reduction with time in outdoor usage. Furthermore, the high temperature of the curing process may damage the material of the solar cell. Due to these reasons, the patch and ground plane are printed on a clear sticker paper, 50 μm in thickness, the surface of which is transparent and highly adhesive. The patch layer is stacked on the upper surface of the solar cell and the ground plane layer on the lower surface as can be seen in Figure 5.12(a) and (b), respectively. Silver paste is used to connect the SMA 50 Ω port to the ground plane and the patch through the transmission line.

5.8 Antenna performance characterization

The SNP and copper prototypes are tested and the measured results are compared against numerical simulations as seen in Figure 5.13. The S_{11} spectrum for the copper prototype shows reasonable agreement with the measured results as depicted in Figure 5.13(a). The S_{11} spectrum measurements between 300 kHz and 7.5 GHz reveal excellent agreement with results obtained from CST, MWS, and HFSS. The measured and simulated S_{11} for the SNP prototype are presented in Figure 5.13(d). An

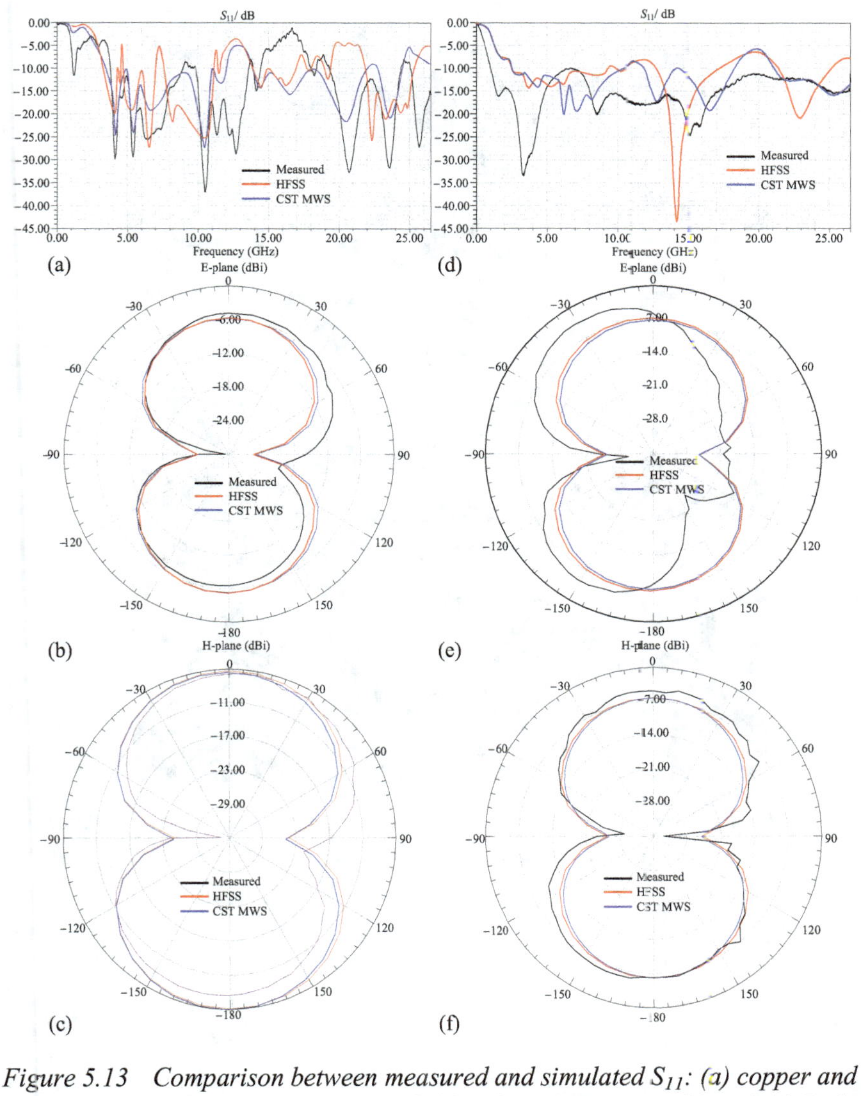

Figure 5.13 Comparison between measured and simulated S_{11}: (a) copper and (d) SNP, respectively. Far-field radiation patterns: (b) and (e) E-plane (y–z) and H-plane (x–z), respectively, for copper patch and (c) and (f) E-plane (y–z) and H-plane (x–z), respectively, for SNP

acceptable agreement has been achieved between measured and simulated S_{11}. The SNP prototype shows a 10-dB return loss bandwidth from 1 to 26.5 GHz as seen in Figure 5.13(d). In Figure 5.13, the far-field radiation patterns of the copper and SNP prototypes in the *E*- and *H*-planes are provided at 2.45 GHz. The copper prototype shows a realized gain of −5 dBi which is in excellent agreement with the simulated results, as can be seen in Figure 5.13(b) and (c). The simulated realized gain of the

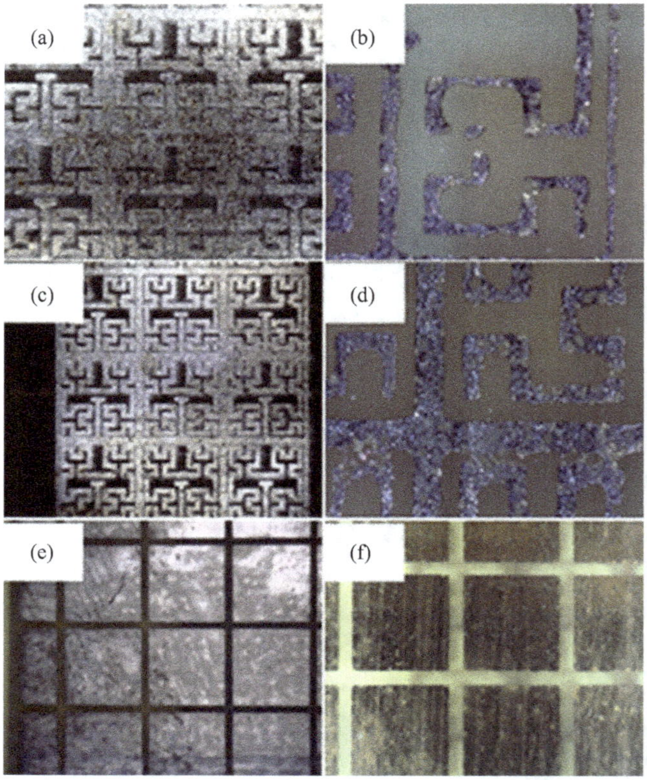

Figure 5.14 Antenna fabrication defects: for SNP antenna (a), (b) patch and (c) ground, for copper antenna (d), (e) patch and (f) ground

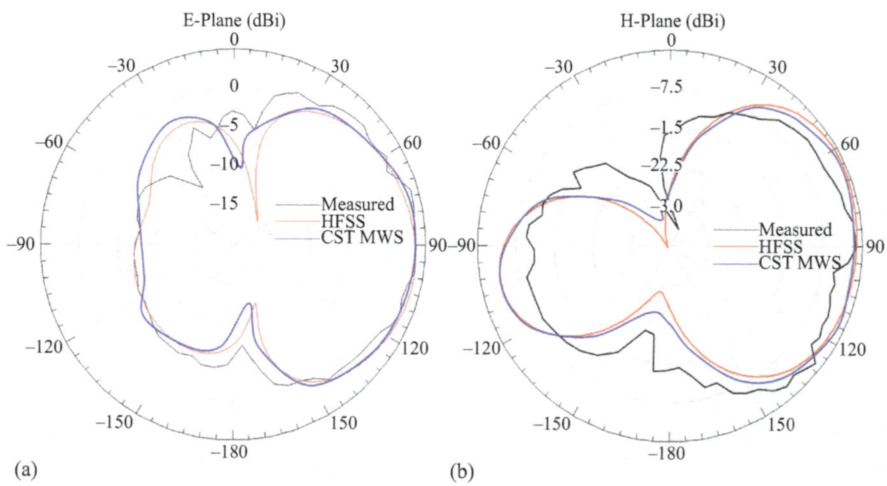

Figure 5.15 Comparison between measured and simulated far-field radiation patterns of the copper prototype at 5.8 GHz: (a) E-plane (y–z) and (b) H-plane (x–z)

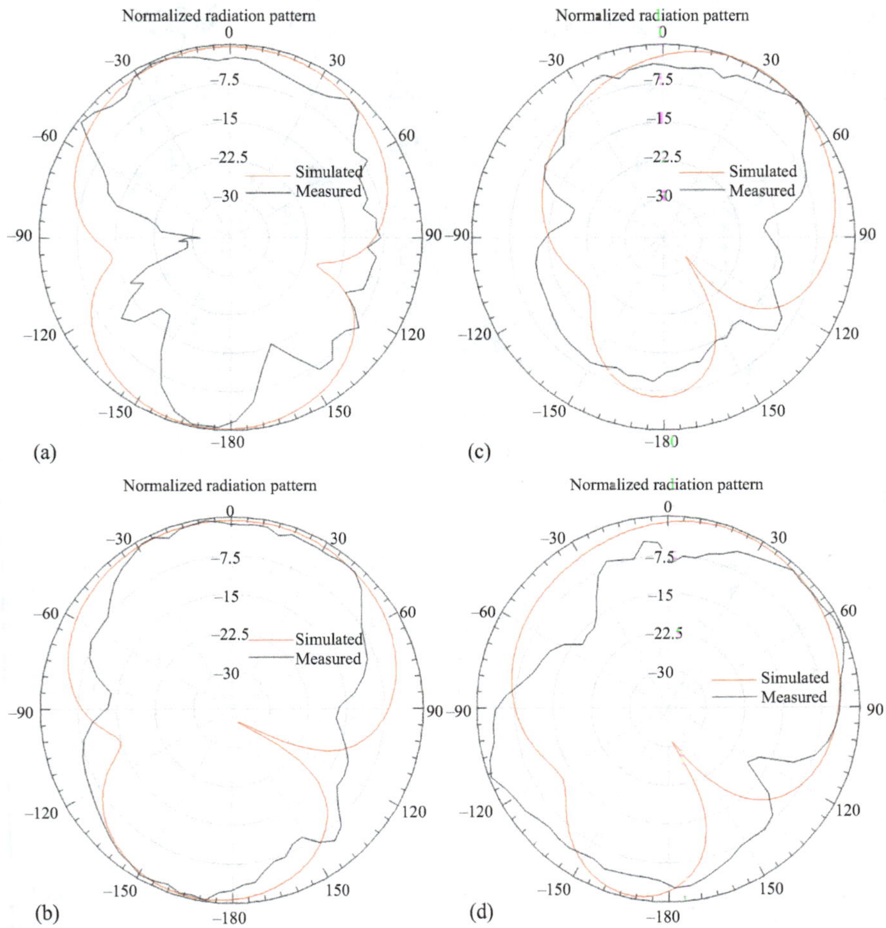

Figure 5.16 Comparison between measured and simulated far-field radiation patterns of the copper prototype E-plane (y–z); (a) 3.3 GHz, (b) 3.5 GHz, (c) 3.7 GHz, and (d) 3.9 GHz

SNP prototype is −7 dBi which is in agreement with the measured value (−7.5 dBi). The *E*-plane pattern shows a good agreement against the simulated results, as shown in Figure 5.13(e). Also, the simulated *H*-plane pattern shows excellent agreement with the measured results, as can be seen in Figure 5.13(f).

The discrepancies between the measured and simulated S_{11} are attributed to manufacturing defects, in particular, the patch structure as seen in Figure 5.14 which is magnified by a factor of 50 using a Dino-Lite Digital Microscope. The SNP patch, shown in Figure 5.14(a)–(c), has few defects as seen in Figure 5.14(a) and (b). These defects are due to substrate morphology and adhesion. The copper patch shown in Figure 5.14(d) and (e) also suffers from etching defects that are due to the tolerance in the thickness of the copper layer and photo-resistor layers in addition to the substrate

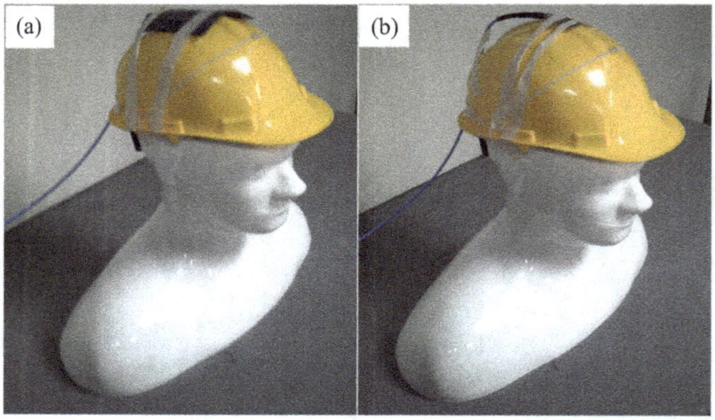

Figure 5.17 Fabricated prototypes based SNP on the safety helmet on the phantom structure: (a) on the solar cell substrate and (b) on the FR4 substrate

morphology. Furthermore, the observed defects in the SNP ground plane were caused by the micro-crakes as can be seen in Figure 5.14(c). The copper ground structure has no observable defects as depicted in Figure 5.14(f).

In Figure 5.15, the far-field radiation patterns in the *E*- and *H*-planes of the copper prototype are displayed at 5.8 GHz and compared to the simulated results.

In Figure 5.16, the radiation patterns, at 3.3, 3.5, 3.7, and 3.9 GHz of the copper prototype and displayed only in the *E*-plane.

5.9 Antenna integration on a flexible solar panel

The integration of printed circuit antennas into solar panels has been investigated in [23]. In [22], a microstrip antenna based on a solant design was presented, while a square patch antenna was reported for satellite communication in [23]. In [13], a dipole antenna mounted on a solar panel was reported.

5.9.1 Antenna performance

The performance of the SNP antenna has been tested when printed on a flexible solar cell and compared against an identical prototype printed on an FR4 substrate. Experimental measurements of S_{11} for both prototypes have been conducted in air and on a phantom model as can be seen in Figure 5.17. S_{11} is measured over the frequency range from 300 kHz to 26.5 GHz, as shown in Figure 5.18(a) and (d). The radiation patterns are measured on the phantom model in the *E*- and *H*-planes at 2.45 GHz as presented in Figure 5.18(b), (c), (e), and (f).

5.9.2 Fabrication of the antenna on a solar panel

The measured *I–V* characteristics of the solar cell with and without the proposed antenna are shown in Figure 5.19. These measurements have been performed in an

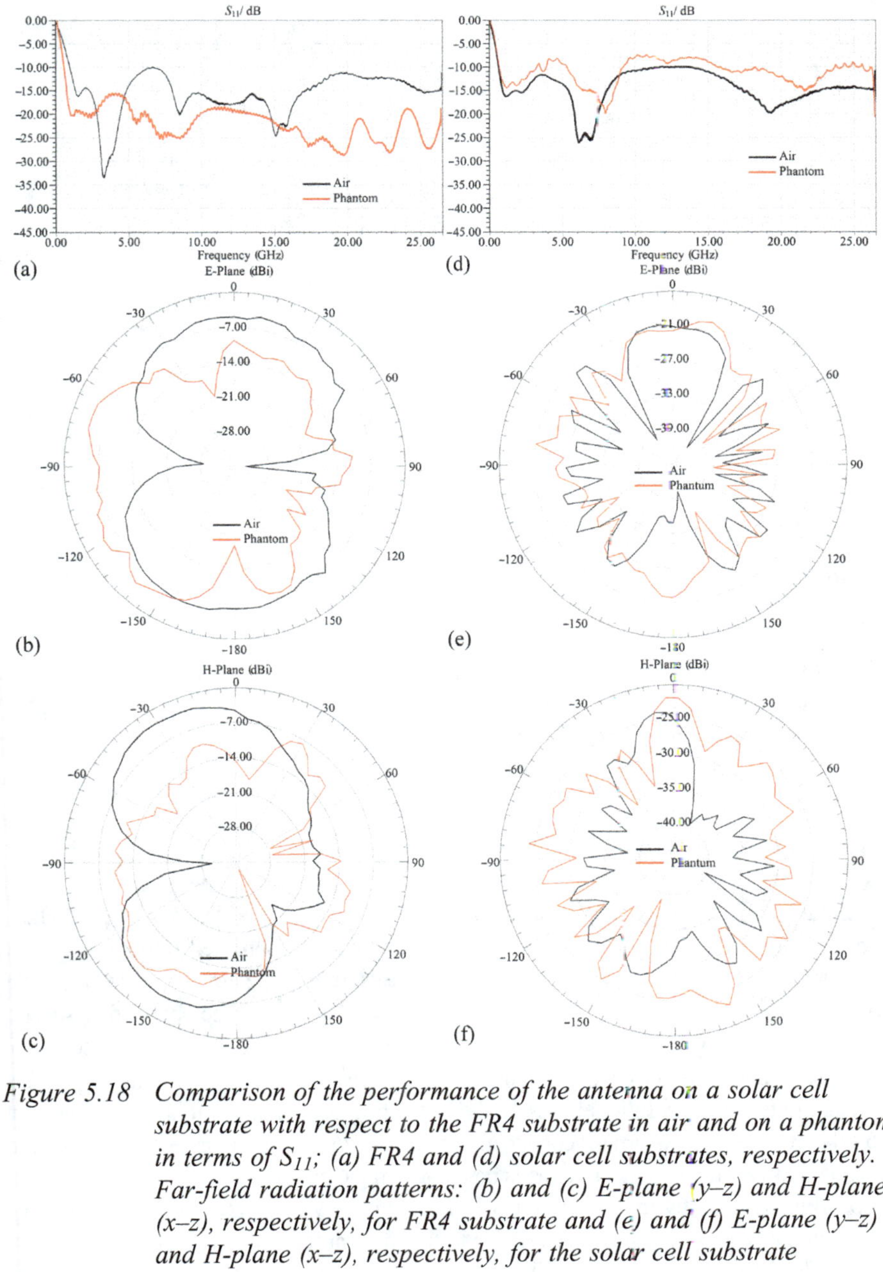

Figure 5.18 Comparison of the performance of the antenna on a solar cell substrate with respect to the FR4 substrate in air and on a phantom in terms of S_{11}; (a) FR4 and (d) solar cell substrates, respectively. Far-field radiation patterns: (b) and (c) E-plane (y–z) and H-plane (x–z), respectively, for FR4 substrate and (e) and (f) E-plane (y–z) and H-plane (x–z), respectively, for the solar cell substrate

outdoor environment during a sunny day. The measured short circuit current when the input voltage is equal to zero is 0.175 A in both cases. The measured open-circuit current, where the voltage is maximum, is 1.1 mA. Between the maximum and minimum values of the input voltage, the current produced from the solar cell

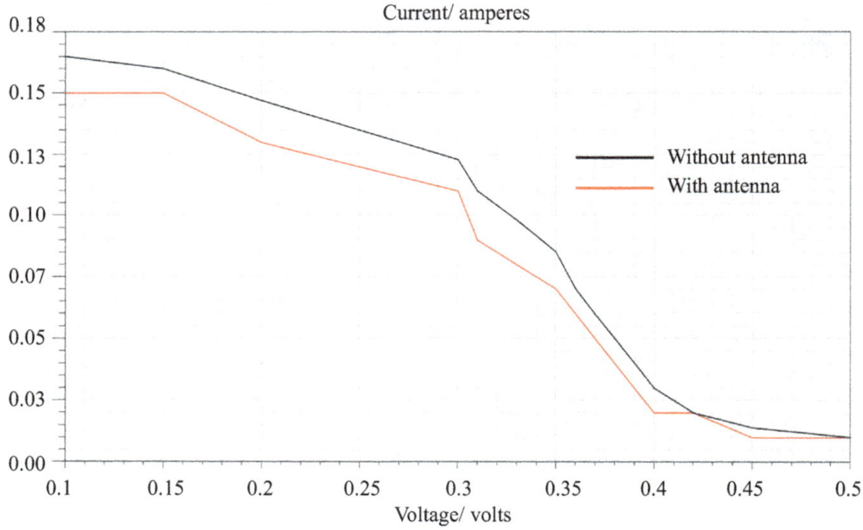

Figure 5.19 I–V characteristics measurements

with and without the proposed antenna is measured at a voltage step of 0.05 V. It is found that the proposed antenna does not show a significant reduction in the produced current as can be seen in Figure 5.19. Therefore, the aim of minimizing the shadowing effects due to the patch structure has been demonstrated.

5.10 RF energy harvesting measurements

Now, the harvested RF energy performance in terms of η_c at 3.5, 4.2, 5.4, and 6 GHz is tested for the proposed antenna. The proposed antenna is connected through an SMA port with 50Ω to an RFD102A test board. The measured η_c spectra are realized using a UWB standard horn antenna that covers the entire band of inters. The obtained η_c spectra are presented in Figure 5.20 for the proposed antenna. Therefore, the measurements are conducted after considering the antenna in two scenarios as a flat structure and bent structure to test the flexibility effects on the harvesting performance. It is found that η_c at 3.5 GHz reaches the maximum enhancement up to 83% for the flat case of the proposed antenna. However, when the antenna is bent over on the patch length, the enhancement is dropped insignificantly to be 81% for 5.4 GHz. It is very important to mention that the conversion efficiency is calculated relative to the transmitter source power variation from −20 to 30 dBm. Generally, the enhancement in η_c is attributed to μ_r component of the INP substrate that reduces the surface wave diffractions from the substrate edges. It is good to mention that η_c at all frequencies of interest are obtained from measurements in open-field environments according to the recommendations in [20].

Finally, the authors summarized their achievements from the proposed antenna-based INP substrate with respect to other published work to provide the

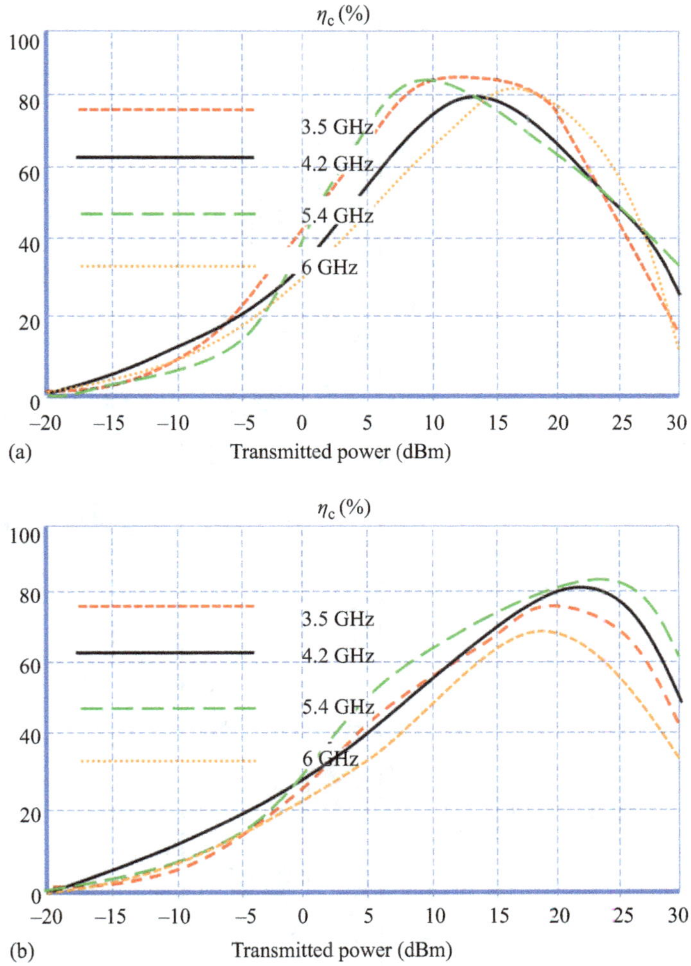

Figure 5.20 η_c measurements at 3.5, 4.2, 5.4, and 6 GHz. (a) flat and (b) bent

achieved advancements. It is found that only the proposed work provided a full study on the effects of bending on harvesting efficiency at different frequencies. Moreover, the used INP mixture supported the proposed design to achieve a circularly polarized antenna with a pure resistive part of 50Ω from 3 up to 10 GHz for the matching impedance. The proposed antenna in this work conducted the use of a microstrip line to provide a circular polarization, while the proposed design in [21] used a coplanar waveguide to achieve a linear polarization. For this, the proposed work performance in terms of harvesting efficiency, listed in Table 5.2, is compared to their similar works that were published in the literature [21].

Table 5.2 Comparison with other published results

Ref.	BW (GHz)	Max. gain (dBi)	Size (mm^2)	Frequencies (GHz)	η_c (%)	η_c after bending (%)	Polarization	AR (dB)
14	1	4	10 × 28	18.8	42	Not applied	Linear	–
15	0.67	3	10 × 25	12	41	Not applied	Linear	–
20	2.4–10	3.4	32 × 28	2.45, 5.8	78, 88	Not applied	Linear	–
The proposed work	3–10	4.6	36 × 22	3.5, 4.2, 5.4, 6	84, 80, 83, 82	78, 80, 81, 74	Circular	<3

5.11 Conclusion

In this section, a wearable antenna consisting of a 3 × 5 array of Hilbert-shaped MTM unit cells on the FR4 substrate backed by a partial ground plane and square EBG defects has been presented. The antenna exhibits a steerable radiation pattern from 3.3 to 3.9 GHz with a steering rate of 0.12°/MHz. The antenna shows an end-fire radiation pattern at 5.8 GHz. The realized gain varies from 0 to 2.5 dBi over the frequency band from 3.3 to 3.9 GHz and from 0 to 4.5 dBi over the frequency band from 4.4 to 6 GHz. A good agreement is found between the measured and simulated S_{11} within the frequency range from 0.1 to 26.5 GHz. The SNP prototype exhibits a 10 dB return loss bandwidth from 1 to 26.5 GHz where the realized gain varies from −10 to 1 dBi. Experimental results demonstrated that the current generated from the solar panel has not been significantly affected by the antenna since the shadowing area of the patch is very small.

List of acronyms

3D	Three dimensional
CST MWS	Computer simulation technology microwave studio
FDTD	Finite-difference time domain
FEM	Finite-element method
FIT	Finite integration technique
f_o	Resonant frequency
GHz	Gigahertz
HFSS	High-frequency structure simulator
MICs	Microwave integrated circuits
mm	millimeter
MMIC	Microwave and millimeter-wave integrated circuits
PCB	Printed circuit board
SEM	Scanning electron microscopy
SMA	Sub-miniature version A connector
WLAN	Wireless local area network
WiMax	Worldwide interoperability for microwave access
ε_r	Dielectric constant of the substrate
λ_o	Free space wavelength

References

[1] C. A. Balanis, *Antenna Theory: Analysis and Design*. John Wiley & Sons, 2006.

[2] M. T. P.-López and M. G. Forner, *Powering Autonomous Sensors: An Integral Approach with Focus on Solar and RF Energy Harvesting*. Springer, London, 2011.

[3] S. Priya and D. J. Inman, *Energy Harvesting Technologies*. Springer, Boston, MA, 2009.

[4] D. Nalin, K. Jayakody, S. K. Sharma, and S. Chatzinotas, *Wireless Information and Power Transfer: A New Paradigm for Green Communications*. Springer, 2018.

[5] T. Soyata, L. Copeland, and W. Heinzelman, "RF Energy Harvesting for Embedded Systems: A Survey of Tradeoffs and Methodology," *IEEE Circuits and Systems Magazine*, volume 16, no. 1, pp. 22–57, 2016.

[6] T. A. Elwi, "A Miniaturized Folded Antenna Array for MIMO Applications," *Wireless Personal Communications*, volume 98, no. 2, 2018.

[7] T. A. Elwi, "Electromagnetic Band Gap Structures Based an Ultra-Wideband Microstrip Antenna," *Microwave and Optical Letters*, volume 59, no. 4, pp. 827–834, 2017.

[8] T. A. Elwi, M. A. Rasheed, L. W., Anber, and M. Q. Fahad, "Gain enhancement of a miniaturized inverted λ-dipole antenna," *IEEE ICICT19 International Conference for Information and Communication Technology*, April 2019.

[9] Z. Al-Dulaimi, T. A. Elwi, and D. C. Atilla, "Design of a Meander Line Monopole Antenna Array Based Hilbert-Shaped Reject Band Structure for MIMO Applications," *IETE Journal of Research*, volume 66, no.1, pp. 1–10, 2020.

[10] H. M. Al-Sabbagh, T. A. Elwi, Y. Al-Naiemy, and H. M. Al-Rizzo, "A Compact Triple-Band Metamaterial-Inspired Antenna for Wearable Applications," *Microwave and Optical Technology Letters*, volume 62, no. 2, pp. 763–777, 2020.

[11] T. A. Elwi, "Printed Microwave Metamaterial-Antenna Circuitries on Nickel Oxide Polymerized Palm Fiber Substrates," *Nature Scientific Reports*, volume 9, no. 2174, pp. 1–14, 2019.

[12] T. A. Elwi and B. A. Ahmed, "A Fractal Metamaterial Based Printed Dipoles on a Nickel Oxide Polymer Palm Fiber Substrate for Wi-Fi Applications," *AEU-International Journal of Electronics and Communications*, volume 96, pp. 122–129, 2018.

[13] T. A. Elwi, O.-A. Tawfeeq, Y. Alnaiemy, H. S. Ahmed, and N. Lajos, "An UWB Monopole Antenna Design Based RF Energy Harvesting Technology," *IEEE 3rd Scientific Conference of Electrical Engineering SCEE 2018*, December 2018.

[14] T. A. Elwi, "Metamaterial Based a Printed Monopole Antenna for Sensing Applications," *International Journal of RF and Microwave Computer-Aided Engineering*, volume 28, no. 7, 2018.

[15] A. I. Imran and T. A. Elwi, "A Cylindrical Wideband Slotted Patch Antenna Loaded with Frequency Selective Surface for MRI Applications," *Engineering Science and Technology, An International Journal*, volume 20, no. 3, pp. 990–996, 2017.

[16] A. Takacs, H. Aubert, S. Fredon, L. Despoisse, and H. Blondeaux, "Microwave Power Harvesting for Satellite Health Monitoring," *IEEE Transactions on Microwave Theory and Techniques,* volume 62, no. 4, pp. 1090–1098, 2014.

[17] A. Okba, S. Charlot, P-F. Calmon, A. Takacs, and H. Aubert, "Multiband rectenna for microwave applications," *IEEE Wireless Power Transfer Conference*, Aveiro, Portugal, 5–6 May 2016.

[18] T. A. Elwi, O.-A. Tawfeeq, Y. Alnaiemy, H. S. Ahmed, and N. Lajos, "An UWB monopole antenna design based RF energy harvesting technology," *IEEE 3rd Scientific Conference of Electrical Engineering SCEE 2018*, December 2018.

[19] T. A. Elwi and A. M. Al-Saegh, "Further Realization of a Flexible Metamaterial Based Antenna on Indium Nickel Oxide Polymerized Palm Fiber Substrates for RF Energy Harvesting," *International Journal of Microwave and Wireless Technologies, Cambridge*, volume 5, no. 4, pp. 1–9, 2020.

[20] T. A. Elwi, D. A. Jassim, and H. H. Mohammed, "Novel Miniaturized Folded UWB Microstrip Antenna-Based Metamaterial for RF Energy Harvesting," *International Journal of Communication Systems*, volume 1, no. 2, 2020.

[21] T. A. Elwi, Z. A. AL-Hussain, and O. Tawfeeq, "Hilbert Metamaterial Printed Antenna Based on Organic Substrates for Energy Harvesting", *IET Microwaves, Antennas and Propagation*, volume 12, no. 4, pp. 1–8, 2019.

[22] T. A. Elwi, "Novel UWB Printed Metamaterial Microstrip Antenna Based Organic Substrates for RF-Energy Harvesting Applications," *AEU – International Journal of Electronics and Communications*, volume 101, pp. 1–10, 2019.

[23] T. A. Elwi, "A Further Realization of a Flexible Metamaterial-Based Antenna on Nickel Oxide Polymerized Palm Fiber Substrates for RF Energy Harvesting," *Wireless Personal Communications*, volume 10, no. 12, pp. 1–15, 2020.

[24] Computer Simulation Technology/Microwave Studio CST MWS. 2010. http://www.cst.com.

[25] Ansoft's High Frequency Structure Simulator HFSS. 2010. 13th version, http://www.ansoft.com.

[26] D. Ali, T. Elwi, and S. Özbay, "Metamaterial-based Printed Circuit Antenna for Biomedical Applications", *Avrupa Bilim ve Teknoloji Dergisi*, no. 26, pp. 12–15, 2021, doi:10.31590/ejosat.950318.

[27] A. I. Imran, T. A. Elwi, and A. J. Salim, "On the Distortionless of UWB Wearable Hilbert-Shaped Metamaterial Antenna for Low Energy Applications", *Progress In Electromagnetics Research M*, volume 101, pp. 219–239, 2021.

Chapter 6

On the trade-off of RF energy harvesting and transmission intervals in cognitive IoT network using fuzzy logic

Najam Ul Hasan[1], Prajoona Valsalan[1], and Waleed Ejaz[2]

Abstract

This study explores an optimal mode selection strategy for radio frequency (RF)-powered Internet of Things (IoT) devices based on cognitive radios (CRs). The RF energy harvesting allows the IoT devices to operate for a possibly much longer lifespan. It is assumed that the IoT devices harvest RF energy obtained from the primary user. However, IoT devices cannot carry out RF energy harvesting and opportunistic spectrum access at the same time. The IoT system can then determine whether to access the spectrum or harvest RF energy at each time slot to optimize the expected total throughput. This chapter canvasses over methods for selecting the suitable mode for IoT devices using a fuzzy logic-based decision-making process. Numerical results have also been discussed, showing that the developed policy balances obtaining the immediate throughput and harvesting the RF energy for future use. The simulation results show the improved performance of cognitive IoT networks using an optimal model selection strategy.

6.1 Introduction

The Internet of Things (IoT) paradigm connects objects to share information over the Internet from one location to another. These objects perceive and collect information from the real world and transmit the collected data to remote users. It allows the information to be available from everywhere at any time. There are a range of IoT applications including eHealth, surveillance, transportation, and environment monitoring and logistics management [1,2]. IoT devices are generally energy constrained and one of the main objectives is its long-term functionality. Although substantial

[1]Department of Electrical and Computer Engineering, Dhofar University, Salalah, Oman
[2]Department of Electrical Engineering, Lakehead University, Barrie, ON, Canada

work has been done to optimize the life of IoT devices, the energy issue remains a significant one. Energy harvesting (EH) technology is seen as a viable solution for providing adequate energy for energy-intensive IoT applications [3,4].

For harvesting, different energy sources, such as illumination, vibration, or heat, are available. However, with the unprecedented growth in wireless technologies in recent years electromagnetic (EM) waves have become more attractive choice for EH applications in particular [5]. EM signals have the qualities of being conveniently attainable and efficiently utilizable in a wireless communication. This opens the door to develop battery-free IoT devices. In this study, we consider a cognitive radio (CR)-based IoT network that consists of IoT devices that can harvest RF energy from the primary users [6–8]. RF energy storage allows the node to run for a potentially longer life without periodic battery replacements or an external power source. Besides, opportunistic spectrum access is a way to resolve interference issues.

This chapter presents the mode selection strategy by considering the trade-off between the harvested energy and throughput the CR-based IoT devices. The goal is to maximize the residual energy of the IoT devices while satisfying throughput requirements. A fixed-time slot network architecture is considered. Each IoT device can be in one out of the following four modes: (1) full harvesting mode, (2) full transmission mode, (3) semi-harvesting mode, and (4) sleep mode. Selection among these modes is obvious and trivial. However, the point of concern is when the IoT device is in the semi-harvesting or semi-transmission mode for how long it should harvest. To solve this problem a fuzzy logic-based algorithm is developed for IoT devices' residual energy and the amount of data it has in its buffer that needs to be transmitted.

The remaining chapter is organized as follows: Section 6.2 highlights the related work and how it differs from our work. Section 6.3 presents the system model. Section 6.4 presents the fuzzy logic technique for choosing the mode and determining the IoT device harvesting time. The simulation results and discussion are presented in Section 6.5, and the chapter is eventually concluded in Section 6.6.

6.2 Related work

Increasing the life of the network is one of the key design goals in battery-operated IoT devices, primarily by minimizing energy usage while trying to maintain the network's end-to-end performance. The energy harvesting from the ambient environment can be used, and battery power as a complement prolongs the IoT nodes' lives. Several alternatives can be used for energy harvesting, such as wind, solar, or radio signals. [9,10]. Energy harvesting based on ambient radio frequency is a secure and convenient energy source because it carries energy and information simultaneously. These days, simultaneous information reception and energy harvesting can be conveniently accomplished by using power splitting devices, and the radio signals are converted into usable DC power. That is why energy harvesting based on RF in wireless networks is becoming more common. Another factor that has promoted it to the next level is the wide range of radio sources, such as Wi-Fi networks, radio towers, base stations, and cell phones. Recent developments have also allowed radio frequency-based energy harvesting on CR networks (CRNs) as a greening solution.

In CR-based IoT networks, the nodes must coexist with other primary user networks. These nodes can share the spectrum with the primary user in a vertical manner in which primary user is the privileged user, whereas the CR-based devices are the secondary user. The CR-based devices can opportunistically access the spectrum while considering that it does not harm the primary user transmission. The CR-based devices can share the spectrum either using an overlay or underlay approach. In the overlay approach, the CR devices can use the spectrum if and only if the primary users use the spectrum. Moreover, in the underlay approach, the CR can use the same spectrum used by primary user by appropriate power adjustment. However, it is compulsory to use one of these approaches; the CR devices can also follow a hybrid approach. In the overlay approach, the CR devices must sense the spectrum before using it. While sensing, these nodes can also harvest energy. If the primary user's signal is detected, they can utilize that. If not, then they can harvest using the noise signal present in the sense channel. In the underlay approach, as the CR-based devices transmit in the same spectrum where the primary user is operating so they can harvest and transmit without sensing.

Keeping in view the advantage offered by energy harvesting for various wireless network devices, many researchers considered CRNs with energy harvesting as a new means to recharge the battery of CR nodes [11–13]. In [14], energy harvesting for CRN and non-CRN has been considered. The analysis shows that energy harvesting allows increasing CRN lifetime significantly. In [15], the authors suggested that energy harvesting can be more beneficial if we perform joint optimization of the energy harvesting and sensing process. Park *et al.* [16] investigated the trade-off between harvesting and sensing throughput and increased the estimated manageable secondary throughput while protecting primary users. Lee *et al.* [17] represented mobile devices in the secondary network opportunistically either by harvesting radio energy from the transmission of nearby primary users or by transmitting data using an overlay approach. Besides, the proposed technique maximizes the throughput of the CRN by optimizing the transmitting power and the number of CR devices supported under an outage probability constraint. Arroca *et al.* [18] studied the cognitive wireless body area network with RF energy harvesting capabilities and presented physical, medium-to-medium-access, and network layer problems and discussed some potential solutions. They proposed some useful ideas for CR-based energy harvesting devices for combined data reception and RF energy harvesting. The authors in [19] investigate maximum harvesting energy and energy efficiency for cooperative multichannel CR where half the number of CRs senses the spectrum, and the other half harvest.

In [20], the CR divides the channel into two sub-channel sets: one for primary user sensing and the other for primary user energy harvesting. In the transmission slot, the energy harvested will be used to compensate for energy loss due to spectrum sensing to guarantee the throughput. The problem is formulated to evaluate the best sub-channel set for sensing and harvesting to optimize the throughput, harvested energy, and energy efficiency across all sub-channels. However, dynamic spectrum access for RF-driven CRNs has not been thoroughly explored in an underlay approach. The work discussed in this chapter is one of the first of its kind. In this work, the authors have made use of the trade-off between energy harvesting and transmission intervals. Endeavoring to find the best possible time interval for energy harvesting under the provided condition of the battery nodes and the data to be transmitted. The approach offers reduced algorithm complexity, design simplicity, and logical consistency.

6.3 System model and problem form Ulation

6.3.1 System model

We assume an energy-harvested CR-based IoT network, which consists of several IoT devices equipped with CR capabilities that have to coexist with the licensed users (also known as primary users) network, as shown in Figure 6.1. Both networks operate in a time-slotted structure and share a common structure. The primary users are privileged users, whereas CR-based IoT devices can unscrupulously access the spectrum using an underlay approach. It is assumed that the CR-based IoT devices are self-powered, whereas the CR-based IoT devices have no fixed power supplies (e.g., batteries) and get power from the primary user. Assuming that CR-based IoT devices cannot harvest and transmit together due to duplex-constrained hardware. In each time slot, the CR-based IoT devices use a fraction of the time slot to harvest energy and the rest portion of the slot for the transmission. The CR-based IoT devices operate in the following four different modes (also depicted in Figure 6.1).

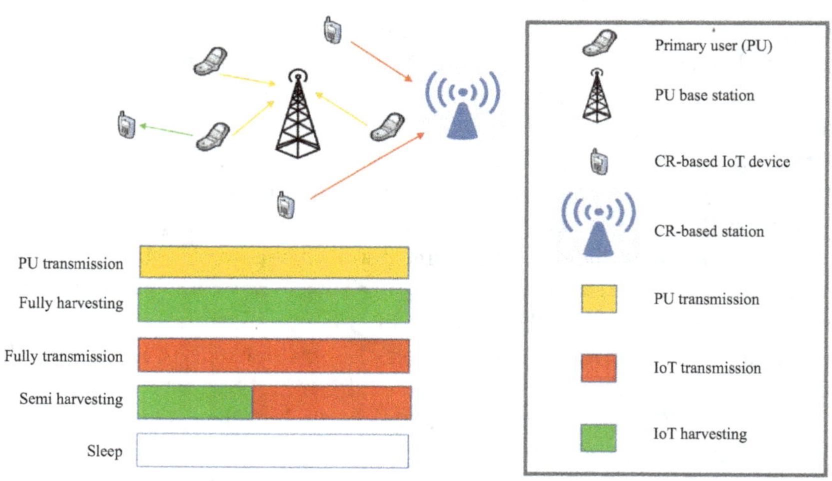

Figure 6.1 Coexistence of primary users and CRNs with slotted structure and different modes of operation for CR-based IoT devices

Fully energy harvesting mode: In this mode, the IoT devices harvest energy from primary users for the complete duration of a time slot and no transmission. This mode is selected when the battery of an IoT device is very low and has data to send. Also, its battery capacity is not full.

Fully transmission mode: In this mode, the IoT devices does not harvest energy, and the complete duration of a time slot is used for transmission. This mode is selected when the battery is full, and the IoT device has data to send.

Semi-harvesting mode: In this mode, the IoTs device will harvest for some portion of the time slot, and for the remaining portion, it will transmit; this mode occurs if battery is neither full nor empty and the device has some data to send.

Sleep mode: In this mode, the IoT device neither harvests nor it transmits any data. This mode is selected when the battery is full, and the IoT device has no data to be transmitted.

6.3.2 Problem formulation

We consider that the primary user and the CR-based IoT devices co-use the channel in an underlay fashion. However, CR-based IoT devices are dependent on energy harvesting to keep alive and perform transmission. Moreover, the transmission time of CR-based IoT device depends on the harvesting time. If harvesting is done for more time, it may result in lower spectrum utilization and may not satisfy the throughput demand of the IoT device. On the other hand, if it is done for a shorter time, then it may deplete the battery and not transmit its data. Therefore, a balance needs to be maintained between the energy harvesting time and the data transmission time based on the residual energy and the amount of data the device has to send. By doing so, the device can be functional for a much longer time and the overall spectrum utilization will also be improved.

6.4 Proposed solution

The residual energy and the amount of data a CR-based IoT device needs to transmit are fed to the fuzzification process to determine the duty cycle, as shown in Figure 6.2.

6.4.1 Input membership functions

For each input in the process of fuzzification, a membership function is developed. The membership function of residual energy is divided into three ranges: low, medium, and high. Similarly, the membership function of data rate is divided into low, medium, and high. The fuzzification process for both inputs is shown in Figures 6.3 and 6.4.

The values of the residual energy membership function are as follows:

$$\text{Low} = \begin{cases} 1, & x < 10. \\ \dfrac{15 - x}{15 - 10}, & 10 < x < 15 \\ 0, & x \geq 15 \end{cases} \tag{6.1}$$

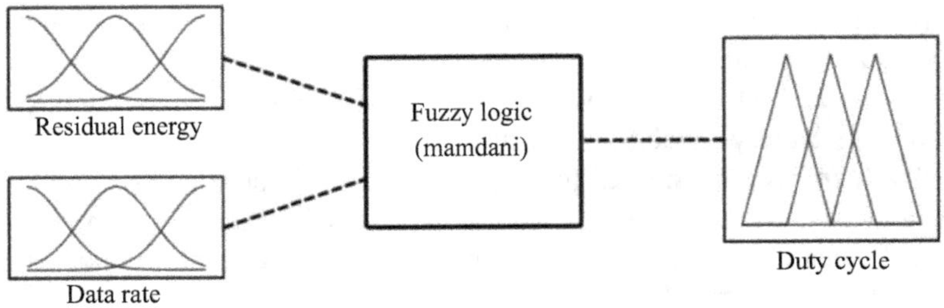

Figure 6.2 Overview of fuzzy logic system

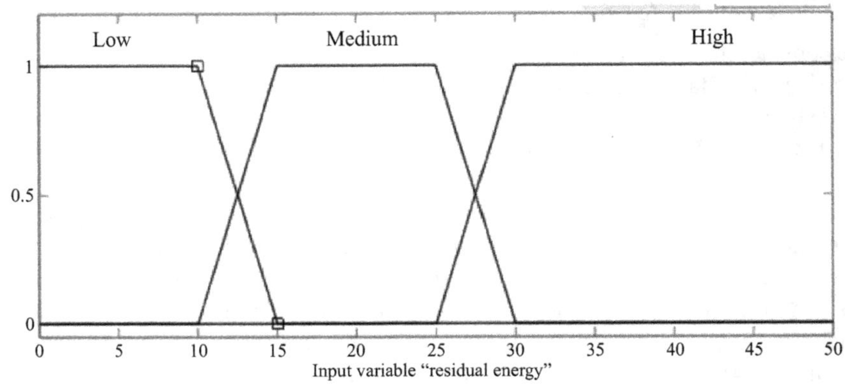

Figure 6.3 Fuzzy membership function for residual energy in Joule

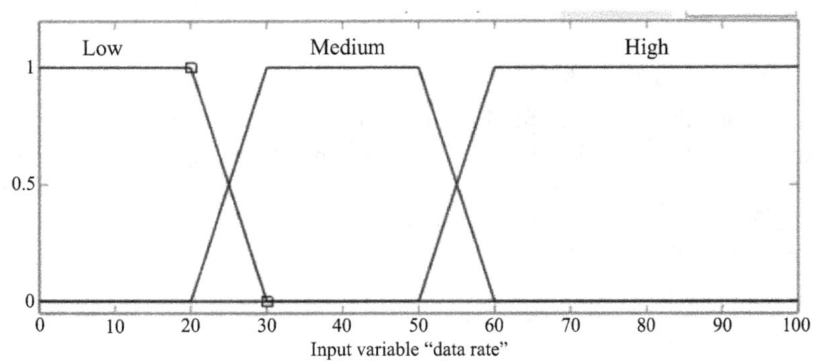

Figure 6.4 Fuzzy membership function for data

$$\text{Medium} = \begin{cases} 1, & 15 < x < 25. \\ \dfrac{15 - x}{15 - 10}, & 10 < x < 15 \\ \dfrac{x - 30}{30 - 25}, & 25 < x < 30 \\ 0, & 10; x > 30 \end{cases} \tag{6.2}$$

$$\text{High} = \begin{cases} 0, & x \leq 25. \\ \dfrac{30 - x}{30 - 25}, & 10 < x < 15 \\ 1, & x \geq 30 \end{cases} \tag{6.3}$$

The values of the data membership function are as follows:

$$\text{Low} = \begin{cases} 1, & x \leq 20. \\ \dfrac{30 - x}{30 - 20}, & 20 < x < 30 \\ 0, & x \geq 30 \end{cases} \tag{6.4}$$

$$\text{Medium} = \begin{cases} 1, & 30 < x < 50. \\ \dfrac{30 - x}{30 - 20}, & 20 < x < 30 \\ \dfrac{x - 60}{60 - 50}, & 50 < x < 60 \\ 0, & x < 20; x > 60 \end{cases} \tag{6.5}$$

$$\text{High} = \begin{cases} 0, & x \leq 50 \\ \dfrac{60 - x}{60 - 50}, & 50 < x < 60 \\ 1, & x \geq 60 \end{cases} \tag{6.6}$$

The overlapping ranges for both residual energy and the amount of data are $10 < x < 15$, $20 < x < 30$ and $25 < x < 30$, $50 < x < 60$, respectively.

6.4.2 *Fuzzy rules*

The fuzzy rules are used to infer the output of the fuzzified inputs. Based on the specified membership function of the two input-residual energy and data to be transmitted, a set of fuzzy rules for combining two inputs, the fuzzification rules are given, as follows in Table 6.1.

The combination for the given ranges of the two input results in nine rules in total. The output of these rules is the duty cycle or the duration for the slot. A CR-based IoT devices can harvest RF energy from the primary user. To have a better understanding, these rules are further expressed in Table 6.2.

Table 6.1 Input combinations for the two inputs

Residual energy	**Amount of data**
Low	Low
Low	Medium
Low	High
Medium	Low
Medium	Medium
Medium	High
High	Low
High	Medium
High	High

Table 6.2 Rules for Defuzzification

No.	**Rule**
1	(ResidualEnergy==LOW) & (DataRate==LOW) => (DutyCycle==MEDIUM) (1)
2	(ResidualEnergy==LOW) & (DataRate==MEDIUM) => (DutyCycle==MEDIUM) (1)
3	(ResidualEnergy==LOW) & (DataRate==HIGH) => (DutyCycle==HIGH) (1)
4	(ResidualEnergy== MEDIUM) & (DataRate==LOW) => (DutyCycle==MEDIUM) (1)
5	(ResidualEnergy== MEDIUM) & (DataRate==MEDIUM) => (DutyCycle==MEDIUM) (1)
6	(ResidualEnergy== MEDIUM) & (DataRate==HIGH) => (DutyCycle==HIGH) (1)
7	(ResidualEnergy== HIGH) & (DataRate==LOW) => (DutyCycle==LOW) (1)
8	(ResidualEnergy== HIGH) & (DataRate== MEDIUM) => (DutyCycle== MEDIUM) (1)
9	(ResidualEnergy== HIGH) & (DataRate==HIGH) => (DutyCycle== MEDIUM) (1)

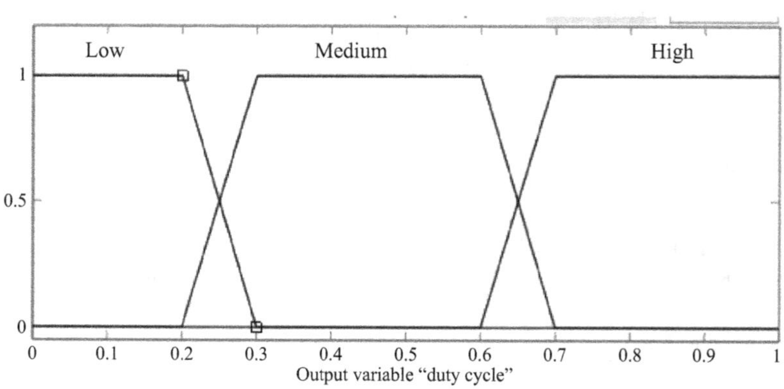

Figure 6.5 Membership function for duty cycle

6.4.3 Output membership functions

For the defuzzification process, the membership function for the output variable duty cycle is defined, as shown in Figure 6.5. The membership function ranges from 0 to 1, indicating the duration of the timeslot for which the energy is harvested.

The values of the duty cycle membership function are as follows:

$$\text{Low} = \begin{cases} 1, & x \leq 0.2. \\ \dfrac{0.3 - x}{0.3 - 0.2}, & 0.2 < x < 0.3 \\ 0, & x \geq 0.3 \end{cases} \tag{6.7}$$

$$\text{Medium} = \begin{cases} 1, & 0.3 < x < 0.5. \\ \dfrac{0.3 - x}{0.3 - 0.2}, & 0.2 < x < 0.3 \\ \dfrac{x - 0.6}{0.6 - 0.5}, & 0.5 < x < 0.6 \\ 0, & x < 0.2; x > 0.6 \end{cases} \tag{6.8}$$

$$\text{High} = \begin{cases} 0, & x \leq 0.5 \\ \dfrac{0.6 - x}{0.6 - 0.5}, & 0.5 < x < 0.6 \\ 1, & x \geq 0.6 \end{cases} \tag{6.9}$$

6.5 Simulation results

The proposed fuzzy logic-based optimal mode selection strategy is simulated in MATLAB. The fuzzy logic toolbox in MATLAB contains Simulink blocks, apps, and functions that can be used to analyze and model complex fuzzy inference systems using simple logic rules.

Figure 6.6 shows the resultant duty cycle for the given inputs, including the residual energy and data rate. For illustration purposes, we selected the value of residual energy equal to 25 (i.e., 25% of battery left for the IoT device) and the data rate requirement of the IoT device is 50 Mbps. The duty cycle is calculated as 45 using the proposed fuzzy logic-based scheme. This means that the IoT device will harvest energy for 45% of the time slot duration. On the other hand, 55% of the time slot duration will be utilized for the transmission. In the given scenario, the IoT device is in semi-harvesting mode. Similarly, the mode of IoT devices can be full harvesting mode, full transmission mode, and sleep mode, depending on the residual energy and data rate requirement of the IoT device. The proposed fuzzy logic-based scheme enabled IoT devices to find optimal duty cycle in a given mode.

Figure 6.7 shows the duty cycle calculated using the proposed fuzzy logic-based scheme for residual energy and data rate. It is important to note that the duty cycle decreases when the residual energy increases (i.e., the IoT devices will spend more time in transmission and less time in harvesting). On the other hand, when the residual energy is less (e.g., 10% or 20%) and the required data rate increases, then the resultant duty cycle increases. This illustrates the importance of optimal duty cycle selection for the given residual energy and data rate. Similarly, more input parameters can be added to find the optimal duty cycle, such as latency, reliability, etc.

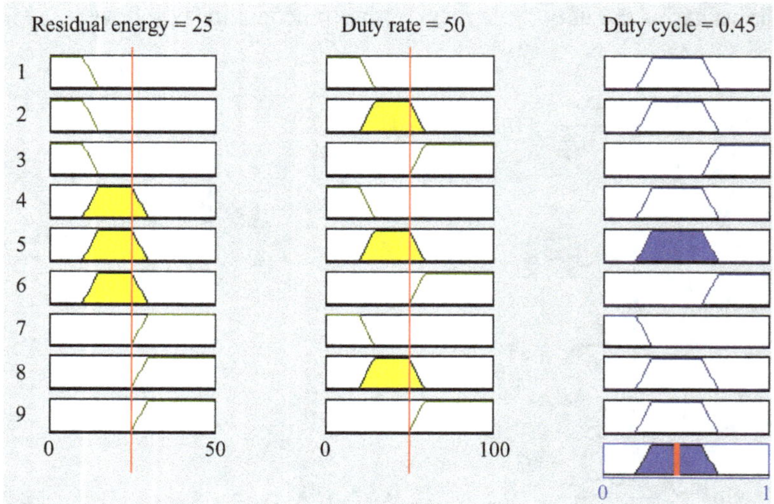

Figure 6.6 Resultant duty cycle for the given inputs including the residual energy and data rate

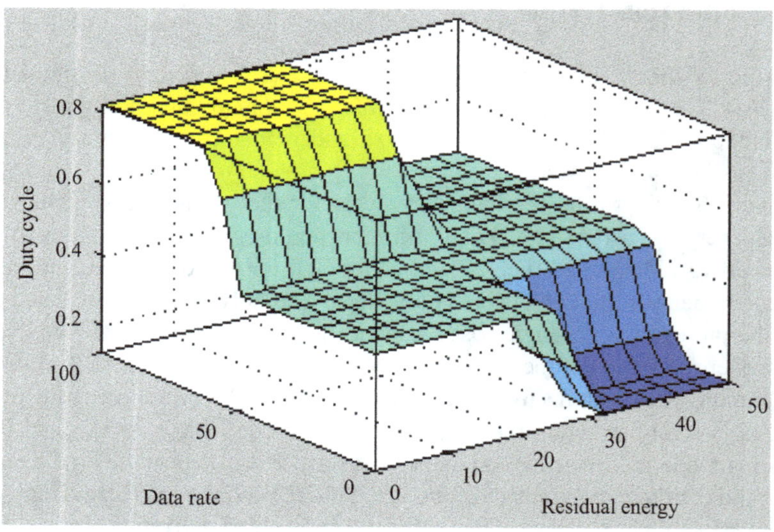

Figure 6.7 Surface view of output in reference to the inputs

6.6 Conclusion

This work presented a system to estimate the appreciated energy harvesting interval for CR-based IoT devices operating in an underlay approach. This system takes two inputs: the residual energy of the IoT devices and the required data rate of each IoT device. Based on these inputs, the output is calculated that is named as the duty cycle. This duty cycle will determine that for a given duration of a slot, for how long, the IoT device will sense/harvest energy or transmit. For the calculation of the duty cycle, we opted the fuzzy logic approach. This work can be further extended to the decision making of channel selection that can result in higher throughput and the amount of harvested energy. However, since the energy is harvested from the same primary user, the spectrum is shared. Thus, it gives rise to a new challenge that needs to be addressed, i.e., to select a channel in which the primary user is transmitting with high power may result in a higher amount of energy harvesting; on the other hand, the CR-based IoT must reduce its power, which results in lowering the throughput. Keeping this challenge in view, a smart solution can be developed to take care of primary user transmission power while selecting a channel for the CR-based IoT devices per their requirements. As a concluding remark, it can be said that further exploration is needed in this research direction.

References

[1] V. Nallarasan and K. Kottilingam, "Spectrum management analysis for cognitive radio IoT," In *2021 International Conference on Computer Communication and Informatics (ICCCI)*, pp. 1–5.

[2] P. V. Dudhe, N. V. Kadam, R. M. Hushangabade, and M. S. Deshmukh, "Internet of Things (IOT): an overview and its applications," In *2017 International Conference on Energy, Communication, Data Analytics and Soft Computing (ICECDS)*, Chennai, India, 2017, pp. 2650–2653.

[3] H. Elahi, K. Munir, M. Eugeni, S. Atek, and P. Gaudenzi, "Energy harvesting towards self-powered IoT devices," *Energies*, vol. 13, no. 21, pp. 5528, 2020.

[4] S. Zeadally, F. K. Shaikh, A. Talpur, and Q. Z. Sheng, "Design architectures for energy harvesting in the Internet of Things," *Renewable and Sustainable Energy Reviews*, vol. 128, pp. 109901, 2020.

[5] S. Sharma, V. Panwar, and S. C. Yadav, "Different sources of energy harvesting: a survey," In *2017 International Conference on Computing, Communication and Automation (ICCCA)*, Greater Noida, India, 2017, pp. 1370–1373, doi: 10.1109/CCAA.2017.8230039.

[6] J. Singla, R. Mahajan, and D. Bagai, "A survey on energy harvesting cognitive radio networks," In *2018 6th Edition of International Conference on Wireless Networks Embedded Systems (WECON)*, Rajpura (near Chandigarh), India, 2018, pp. 6–10, doi: 10.1109/WECON.2018.8782061.

[7] S. Aslam, W. Ejaz, and M. Ibnkahla, "Energy and spectral efficient cognitive radio sensor networks for Internet of things," *IEEE Internet of Things Journal*, vol. 5, no. 4, pp. 3220–3233, 2018.

[8] B. Alzahrani and W. Ejaz, "Resource management for cognitive IoT systems with RF energy harvesting in smart cities," *IEEE Access*, vol. 6, pp. 62717–62727, 2018.

[9] W. Khalid and H. Yu, "Residual energy analysis with physical-layer security for energy-constrained UAV cognitive Radio Systems," In *Proceedings of the 2020 International Conference on Electronics Information, and Communication (ICEIC)*, Barcelona, Spain, 19–22 January 2020; pp. 1–3.

[10] F. Zhang, T. Jing, Y. Huo, and K. Jiang, "Outage probability minimization for energy harvesting cognitive radio sensor networks," *Sensors*, vol. 17, no. 2, pp. 224, 2017.

[11] T. Kalluri, M. Peer, V. A. Bohara, D. B. da Costa, and S. U. Dias, "Cooperative spectrum sharing-based relaying protocols with wireless energy harvesting cognitive user," *IET Communications*, vol. 2, no. 7, pp. 838–847, 2018.

[12] D. Xie, X. Lai, X. Lei, and L. Fan, "Cognitive multiuser energy harvesting decode-and-forward relaying system with direct links," *IEEE Access*, vol. 6, pp. 5596–5606, 2017.

[13] Z. Yan, S. Chen, X. Zhang, and H.-L.Liu, "Outage performance analysis of wireless energy harvesting relay-assisted random underlay cognitive networks," *IEEE Internet of Things Journal*, vol. 5, no. 4, pp. 269102699, 2018.

[14] S. T. John, "Performance measure and energy harvesting in cognitive and non-cognitive radio networks," In *2015 International conference on innovations in information, embedded and communication systems (ICIIECS)*, 2015, pp. 1–5.

[15] S. Zhang, H. Zhao, AS Hafid, and S. Wang, "Joint optimization of energy harvesting and spectrum sensing for energy harvesting cognitive radio," In *2016 IEEE 84th Vehicular technology conference (VTC-Fall)*, 2016, pp. 1–5.

[16] S. Park, J. Heo, B. Kim, W. Chung, H. Wang, and D. Hong, "Optimal mode selection for cognitive radio sensor networks with RF energy harvesting," In *Proceedings of the IEEE 23rd International Symposium on Personal, Indoor and Mobile Radio Communications-(PIMRC)*, Sydney, NSW, Australia, September 2012; pp. 2155–2159.

[17] S. Lee, R. Zhang, and K. Huang, "Opportunistic wireless energy harvesting in cognitive radio networks," *IEEE Trans. Wireless Communications*, vol. 12, no. 9, pp. 4788–4799, 2013.

[18] N. Arroca, J. M. Ferro, L. M. Borges, J. Tavares, and F. J. Velez, "Electromagnetic energy harvesting for wireless body area networks with cognitive radio capabilities," In *Proceedings of the URSI Seminar Portuguese Committee*, Lisbon, Portugal, 1–4 Nov. 2012.

[19] A. Alsharoa, N. M. Neihart, S. W. Kim, and A. El Kamal, "Multi-band RF energy and spectrum harvesting in cognitive radio networks," In *Proceedings of the IEEE International Conference on Communication (ICC)*, May 2018, pp. 1–9.

[20] X. Liu, F. Li, and Z. Na, "Optimal resource allocation in simultaneous cooperative spectrum sensing and energy harvesting for multichannel cognitive radio," *IEEE Access*, vol. 5, pp. 3801–3812, 2017.

Part III

Emerging trends in energy harvesting

Chapter 7
UAV-assisted energy harvesting for WSNs/IoT networks

Madiha Amjad[1], Maham Waheed[2], Syed Ali Hassan[2], and Hassaan Khaliq Qureshi[2]

Abstract

The stringent energy constraint imposed by the small size of sensing nodes is a critical issue in realizing the perpetual operation of wireless sensor networks/ Internet of Things (WSNs/IoT) networks. The unmanned aerial vehicle (UAV) network is a promising technology that is already being used for several applications. They can be a substantial and convenient source of energy harvesting for WSNs/IoT applications, especially in far-flung areas. However, due to mobile nature and constraints such as budget and airspace congestion, and different capabilities of different UAVs, many factors need to be considered in designing UAV-assisted energy harvesting schemes. In this chapter, we present a comprehensive review of various schemes that are proposed to improve energy harvesting directly or indirectly by improving some dependent performance objectives in UAV-assisted WSNs/IoT networks.

7.1 Introduction

Unmanned aerial vehicles (UAVs) are gaining huge attention these days due to the vast range of applications they support. The UAV market is expected to show an annual increase of EUR 10 billion by 2035 and this will reach EUR 15 billion by the year 2035 [1]. As a UAV can get access to the places where structure-based Internet connectivity is impossible, it has been proposed for facilitating wireless sensor networks/Internet of Things (WSNs/IoT) networks in many ways [2–5]. As sensor nodes have limited energy, a cost-effective and energy-efficient way of

[1]Department of Information Technology, Khwaja Fareed University of Engineering and Information Technology (KFUEIT), Rahim Yar Khan, Pakistan
[2]School of Electrical Engineering and Computer Science (SEECS), National University of Sciences and Technology (NUST), Islamabad, Pakistan

collecting information is required. To achieve this goal, it is important to manage minimum path loss between sensing devices and the base station (BS). The mobility and dynamic positioning of UAVs, compared to a stationary terrestrial BS. Accomplishes this task by providing smaller path loss leading to better channel conditions and energy efficiency [2]. Furthermore, UAVs can provide resilience in WSNs/IoTs networks by dealing with critical node issues in a better way [3,6]. UAVs are also used as a platform for sensor nodes/Internet of Things devices (SNs/IoTDs), by hosting sensor nodes and flying over the areas that need to be monitored [4]. They can also assist in processing WSNs' information. Instead of putting heavy computational tasks on the sensor node or transferring the whole data to the Internet for analysis, UAVs perform necessary computation and reduce the amount of transferred information [5].

In addition to assisting WSNs/IoT networks in their operation, UAVs can power small sensor devices using wireless power transfer [7]. The magnetic resonant coupling technique for energy harvesting is preferred for this arrangement as this scheme allows energy harvesting over large distances [8]. However, UAVs have limited energy-carrying capabilities, it is important to design efficient schemes for power transfer and operation of UAVs. Furthermore, depending on the type of UAV, different functional capabilities of the UAVs need to be considered while designing energy harvesting schemes. Several schemes in the literature focus on improving energy harvesting by optimizing different factors such as spectrum allocation [9], power allocation [10], user's offloading time, and user's transmit power [11], time of hovering and time of wireless power transfer and workflow [12], and node association [13].

In this article, we provide a comprehensive review of the UAV-assisted energy harvesting schemes for WSNs/IoT networks. This thorough analysis will help to identify approaches that are efficient for particular use cases. An overview of various factors impacting energy harvesting in UAV-assisted WSNs/IoTs along with schemes used for optimization and involved parameters is depicted in Figure 7.1.

The rest of the chapter is arranged as follows. In Section 7.2, system model is discussed in detail. The impact of trajectory on energy harvesting is analyzed in Section 7.3 with a focus on trajectory optimization along with the methods of powering UAV to complete its assigned tasks efficiently.

In Section 7.4, the impact of the node association, its various optimization schemes, and its impact on EH are discussed. The influence of different resource allocation schemes on the performance of UAV-assisted energy harvesting is explored in Section 7.5. Furthermore, throughout the paper, the theory of related analytical and statistical approaches for optimizing resources are discussed for making discussed schemes comprehensible.

7.2 System model of UAV-assisted WSNs/IoT networks

UAV-assisted WSNs/IoTs consists of several devices and components including single/multiple UAVs, SNs/IoTDs, terrestrial/satellite BS involving multiple

Trajectory

• Objective:
Maximize Harvested Energy, Maximize Coverage, Maximize Energy Transfer, Maintain Minimum Data Requirement, Maintain Fairness.
• Parameters:
UAV's Path, Hovering Points, Hover Time, Flight Time, Power Transfer Time, Energy Harvesting Time, Distance between Hovering Points, Distance between UAV and Sns/IoTDs, Distance between Hovering Point and Sns/IoTDs, UAV's Velocity and Acceleration, Harvested Power, Uplink Transmit Power, Number of Transmission Slots, Subcarrier Allocation, Interference among UAVs, Load Balancing in Multiple UAVs.
• Optimization Alogrithms:
SCA, SCAIA, CVX, CCCP, Dual Method, Taylor Series, Taylor Expansion, Lagrange Multipliers, Geometry-based Alogrithms, MAVNS,MA, MTSP, TSP, VNS.

Association

• Objective:
Maximizing Minimum Data Rate, Minimizing Packet Loss, Minimizing Energy Consumption of UAV, Satisfaction Rate for Energy Harvesting.
• Parameters:
Association, Charging Time, Transmit Power, Trajectory, Hovering Time, Operation Schedule, Offloading Capacity, Beam Pattern.
• Optimization Alogrithms:
MDPs and K-means Clustering, Decomposition of Non-convex Optimization, Iterative Alogrithm, RL, DQN, DDPG.

Resource Allocation

• Objective:
Minimize Packet Loss, Maximize Minimum Data Rate, Maximize Minimum Achievable Rate for Information Transfer, Minimize Energy Consumption, Balancing Data Transfer and Energy Harvesting Intervals, Minimize Energy Harvesting Lost.
• Parameters:
Association, Hovering Time, Operation Schedule, Multi-workflow, Computing Resources, Trajectory, Subcarrier, Slot Allocation, Charging Time, Offloading Capacity, Transmit Power.
• Optimization Alogrithms:
MDPs and K-means Clustering, Improved K-means Clustering, Mean Field Game, Decomposition of Non-convex Optimization, Sub-problem Solution using LP, SCA with Water-filling Alogrithm, LP, RL, DQN, DDPG, Q-learning, Double Q-Learning.

Figure 7.1 An overview of factors impacting EH in UAV-assisted WSNs/IoTs

communication links. A detailed diagram of a UAV-assisted network is presented in Figure 7.2, consisting of N SNs/IoTDs, U UAVs multiple base stations and a satellite.

The UAV-assisted WSNs/IoT network can be a combination of various configurations and transmission links. Depending on the application needs and network nature, UAV-assisted WSNs/IoT networks can follow different configurations. In the literature, UAVs are deployed mainly in three configurations, multiple UAVs network, single mobile UAV network, and single UAV network

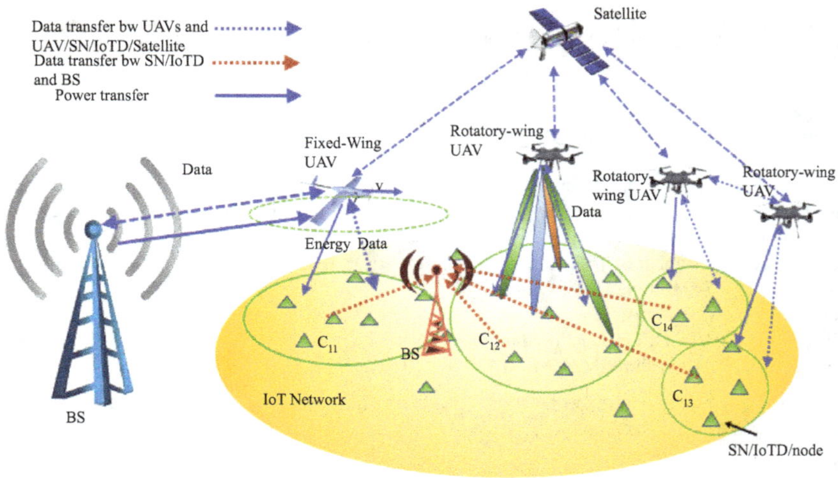

Figure 7.2 UAV-assisted WSNs/IoT network

with beamforming. All three of these configurations are shown in Figure 7.2 while separately these configurations are explained further in their respective sections with elaborated figures. The transmission links include between SNs/IoTDs and UAVs, from UAVs to BS, BS to SNs/IoTDs, and among multiple UAVs. The nodes could form clusters or transmit independently. Clustering is the division of nodes into clusters, clustered nodes are either served at the same time or by the same UAV in the case of multiple UAV-assisted WSNs/IoT networks. The clusters are depicted by green circles. The energy transfer links are shown as solid lines and data links are depicted as dashed lines. The links involving UAVs are blue, while the transmission among BS and SNs/IoTDs is shown with red links.

The BS present in the network ranges from macro, which is using UAV for offloading, to micro that is just deployed for assistance in larger networks. There are mainly two kinds of UAVs, fixed-wing, which can fly only and rotatory-wing UAVs, which can hover at a position in addition to flying [14]. UAVs receive control or other data from a terrestrial BS or satellite BS and power from some terrestrial charging point or mounted battery. UAVs can receive energy from a BS or charging station through wireless/light power transfer or can also have solar panels mounted on them for harvesting solar energy. In UAV-assisted WSNs/IoT networks, SNs/IoTD nodes get energy from UAVs, in case they get deprived of energy and charging resources; in addition, they can transmit data to UAVs for offloading or processing. Certain factors are considered while designing efficient UAV-assisted networks. These include trajectory, clustering, and association. The parameters further depend on various factors like UAV power, energy harvesting capabilities, the relative

Table 7.1 List of parameters

Parameter	Description
n	IoTD/sensors/ground nodes/ users
N	Maximum number of SN/IoTD in a network
N_0	Number of nodes that do not offload their task to UAV
u	UAV
C_u	The computing capacity of u-th UAV
C_{un}	The computational capacity of UAV u dedicated for executing the task of SN/ IoTD n
C_n	Total number of cycles required for task of node n
$C_{n,max}$	The maximum computing capacity of SN/IoTD n
A_s	Association matrix
$E_{c,n}$	Energy consumed by a node n
E_H	The maximum harvested energy
$E_{h\ n,w}[s]$	Harvested energy by node n in slot s on subcarrier w
E_n	The maximum energy left on sensor n
E_P	UAV's propulsion energy
E_{th}	Minimum energy threshold
ε_n	Power conversion efficiency
g_{nu}	Total channel loss for node n when served by UAV u
P_U	UAV's transmission power
P_{Uu}	Energy transferred by a UAV u
P_{UL}	Uplink power
q	UAV's trajectory
$v_{um}(t)$	UAV's velocity at time "t"
$a_u(t)$	UAV's acceleration at time "t"
s	Slots
T_{on}	The offloaded task of SN/IoTD n
t_{nu}^{Tr}	Time required to offload the task of node n to UAV u
τ_F	UAV's flight time
τ_{EH}	Energy harvesting time
τ_n	Hovering time over node n
$R_n^u[s]$	The data rate of user n in the uplink direction
γ_{th}	Minimum data threshold
ξ_n	Hovering point over node n
$d_{\xi_n,\xi_{n+1}}$	Distance between consecutive nodes
w	Subcarrier

placement of network components, channel, time allocation, etc. A detailed list of parameters related to the UAV-assisted WSNs/IoT networks is presented in Table 7.1.

Trajectory decision concerns the decision regarding speed, hovering position, and path planning, these factors are discussed in detail in Section 7.3. Similarly, association, discussed in Section 7.4, deals with the decision of which cluster/node is attached to the UAV in which time slot in the case of a single UAV and the decision regarding which SN/IoTD is attached to which UAV in the case of a multiple UAV-assisted IoT network. Another important design decision is to allocate resources of UAV, such as time, channels/subcarriers, and transmit power. Schemes relevant to resource allocation are discussed in Section 7.5.

A larger and dense IoT network has higher demands for power, data, and computing. A single UAV might not be able to fulfill the needs of all the devices due to limited battery. Even with a large battery size or energy harvesting capabilities, meeting the energy requirements of the entire network before nodes are completely run out of battery is difficult. Therefore, in such scenarios, multiple UAVs are deployed to meet the energy and data requirements of SNs/IoTDs [15]. The important design parameters for this scenario are clustering and association, resource allocation, and selecting appropriate hovering positions. In a single mobile UAV-assisted WSNs/IoT network, a single UAV is used to assist the WSNs/IoT network. To cover the whole region, UAVs fly over the entire network area. With hovering capacities of the latest UAVs, UAVs can also hover at a certain position to satisfy the increased needs of a node/nodes. In such a network, it is important to design a trajectory in addition to deciding association and resource allocation, as discussed in Section 7.3. In the case of a single UAV, another approach to simultaneously assist multiple SNs/IoTDs is to utilize beamforming. Beamforming results in concentrated energy in a particular direction, which results in better energy harvesting and information transfer.

7.3 Impact of UAV trajectory on EH

The placement and trajectory of UAV play a deterministic role in overall network efficiency. The trajectory planning of UAV is constrained by multiple factors such as line-of-sight/non-line-of-sight (LoS/NLoS) links, node distribution and behavior, data and power transmission strategies.

In UAV-assisted WSNs or IoTs, the path planning of UAV is a complex procedure as the collaboration between WSNs and UAVs requires matching transmission schedules and flight patterns. While considering energy harvesting (EH), the basic aim is to reduce the distance between the EH sensor node and UAVs hover point either by applying certain limitations over paths like a spiral, circular, etc., to suit the requirements or by formulating an optimized problem considering various operational parameters such as flight time, hover time, EH time, and WSNs/IoT node distribution [16]. The optimization problems could be lengthy and complex like brute force/exhaustive search or heuristically designed fast sub-optimal solutions with effective hover points and flying patterns. Constrained optimization problems are usually non-convex and require sub-optimal solutions. Even the simplest hover point optimization problems result in mixed-integer linear problems. The aim of introducing various optimization problems and algorithms is to reduce computational complexity and costs while providing maximum EH capabilities.

7.3.1 UAV deployment/placements

The operational constraints like the UAV altitude, its 2D/3D placement, flight time, hovering time, and velocity greatly impact the energy harvesting capabilities of UAV-assisted WSNs/IoT and need to be optimized to get desired results.

7.3.1.1 Static deployment/placement

Single or multiple UAVs are deployed in the 2D/3D layout to provide information and power transfer over a specified WSN/IoT network, as shown in Figure 7.3(a). The EH in this scenario depends on the hovering location and distribution of nodes in the WSNs/IoT networks. The UAV is placed to provide coverage to the maximum number of nodes with a minimum distance to harvest energy. The association scheme of nodes also plays a vital role in efficiently harvesting energy. As the UAV is not moving in this scenario, the critical nodes are also taken into account while deciding the static placement to maximize coverage for them. Multiple UAVs are deployed over larger networks to enhance coverage capabilities.

7.3.1.2 Hover-and-fly model

In this model, the number of waypoints is determined following the node distribution in the associated WSNs/IoT. The points are chosen to minimize the distance between UAV and served node to maximize the harvested energy. The UAV hovers over these points for specified intervals depending upon slot duration and flies to the next location with specified velocity. The hovering and flying modes continue until the UAV completes the flight path and moves back to its initial location. The initial point in most of the cases is over the control center and similar to the final point for periodic UAV paths. As the UAV hovers over a certain node or set of nodes, it transmits power and collects information from them. The hover-and-fly model is depicted in Figure 7.3(b). It is deployed in the literature with various iterative algorithms providing optimal or sub-optimal solutions for trajectory planning.

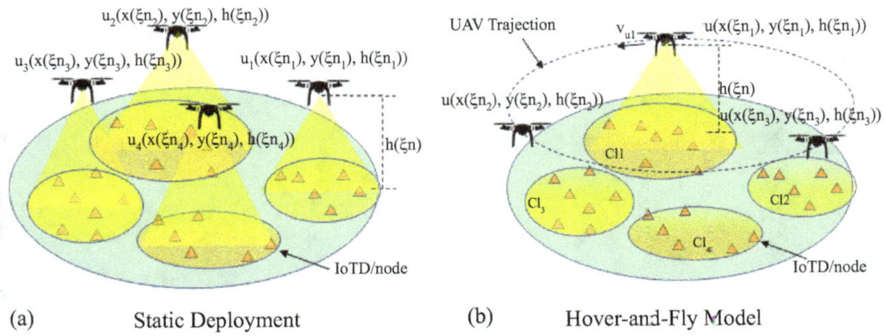

Figure 7.3 UAV Deployment Techniques in UAV-assisted WSNs/IoTs: (a) static deployment and (b) hover-and-fly model

7.3.2 Optimization problems and algorithms

Several optimal problems are formulated in the literature to enhance the energy harvesting capabilities of UAV-assisted WSNs/IoT networks based on trajectory planning. In general, a UAV-assisted EH-enabled network is considered where the UAV first transfers the wireless power transmit (WPT) signal to SNs/IoTDs, and then the SNs/IoTDs send wireless information transfer (WIT) signal to the UAVs. The objective in most of the EH-enabled networks is to maximize the harvested energy E_H and maintain fairness among the nodes accessing this harvested energy, i.e., maximizing the minimum harvested energy among nodes. In order to achieve these objectives the trajectory planning of UAV is done which is constrained by various parameters like UAV's path q, hovering points ξ, hover time τ, flying time τ_F, power transfer time τ_P, energy harvesting time τ_{EH}, the distance between hovering points $d_{\xi_n,\xi_{n+1}}$, the distance between UAV and SNs/IoTDs $d_{UAV,n}$, the distance between hovering point and SNs/IoTDs $d_{\xi_n,n}$, UAV's velocity v_{um} and acceleration a_u, harvested power P_H, uplink transmit power P_{UL}, and the number of transmission slots S. A general objective function can be constrained to anyone or a set of the parameters discussed above. A simple objective function to maximize the minimum harvested energy by nodes subject to certain parameters, say, p_1, p_2, p_n is described in (7.1). This objective function is further subject to several parameters or bounds. Some of the frequently used constraints in trajectory planning for EH-enabled WSNs/IoT networks are defined in (7.1b–7.1l).

$$\max_{p_1, p_2, \dots, p_n} \quad \min E_H \tag{7.1a}$$

$$q_n(0) = 0, q_n\left(\tau_n^F\right) = d_{\xi n, \xi n+1}, \tag{7.1b}$$

$$||q[s] - q[s-1]|| \le SV_{\max}, \tag{7.1c}$$

$$(x(\xi_n) - x_n(\xi_n))^2 + (y(\xi_n) - y_n(\xi_n))^2 \le R_n^{\max}, n = 1, 2, .., N, \tag{7.1d}$$

$$d_{Xn} \le R_n^{\max}, n = 1, \dots, N, \tag{7.1e}$$

$$|v_{um}(t)| \le V_{\max}, |a_u(t)| \le A_{\max}, \tag{7.1f}$$

$$v_{um}(0) = v_{um}\left(\tau_n^F\right) = 0, \tag{7.1g}$$

$$\tau_n^p \ge 0, n = 1, \dots, N, \tag{7.1h}$$

$$\sum_{n=1}^{N} \sum_{s=1}^{s} P_U[s] \le P_0, \tag{7.1i}$$

$$0 \le P_{UL}[s] \le P_{UL}^{\max}, \tag{7.1j}$$

$$\int_0^T P_{H,nl}\left(P_{R,n}\right) dt \ge E, given, P_{R,n} = \frac{\beta_0 P_U}{d_{UAV,n}}, \tag{7.1k}$$

$$E_0(v) \sum_{n=0}^{N} d_{\xi n, \xi n+1} + (P_H + P_P)||\tau_p|| \le Eu, \tag{7.1l}$$

An objective function can be constrained to a single above-defined parameter or a set of parameters and bounds depending on the problem formulation. Equation (7.1b) defines that the initial point of the path at time $t = 0$ is 0 and at any time during the UAV's flight depends on the distance between its hovering points. Equation (7.1c) restricts the maximum allowable trajectory and bounds the path between two consecutive slots. Equation (7.1d) describes the distance between node n with coordinates $(x_n(\xi_n), y_n(\xi_n))$ and UAV hovering over point ξ_n at location $(x(\xi_n), y(\xi_n), H)$ to be within maximum horizontal transmission range (MHTR) R_n^{max}. Equation (7.1e) also specifies the distance between any transmission pair to be within MHTR. 7.1f puts a bound to UAV's velocity and acceleration in any time t to be less than specified permissible values. Equation (7.1g) states that the UAV's velocity at the initial point and over specific hovering point is zero. Equations (7.1h), (7.1i), and (7.1j) specify the power constraints. The time interval over which WPT occurs should be greater than 0. Power transfer does not occur if the hover time is zero. The power UAV provides is bound to its maximum power and the uplink power in any slot could not exceed the maximum uplink power. Equation (7.1k) bounds the harvested power P_H as a function of the received power. $P_{R,n}$ is the received power on node n, and the harvested power is a non-linear function of it. Equation (7.1l) states that overall power consumption is less than the total power available for UAV for flight and power transfer. Some of the existing literature on optimization problem formulations for optimal trajectory for EH-enabled networks is defined below.

7.3.2.1 Successive convex approximation

In successive convex approximation (SCA), a sequence of more compliant/manageable sub-problems is formulated as a substitute of the original non-convex problem, reducing the original function into simpler possibly convex alternatives.

In [17], the remote secondary users (RSUs) harvest energy from the UAV and transfer their data through the UAV. A SCA-based iterative algorithm (SCAIA) is proposed to optimize the UAV trajectory. The aim is to provide the maximum coverage so that the EH by RSUs is maximized. A simple hover-and-fly model is considered initially where hover points equal the number of RSUs. The lowest possible distance is maintained between RSUs and hover point to ensure effective charging of nodes and to make sure that no RSU is left with insufficient energy. The optimal UAV trajectory where maximum energy can be harvested by RSUs is obtained by

$$\max_{x(\xi_n),y(\xi_n)} E_H(x(\xi_n), y(\xi_n)) = \sum_{n=1}^{N} \frac{\tau_{EH} P_U g}{(x(\xi_n) - x_n(\xi_n))^2 + (y(\xi_n) - y_n(\xi_n))^2 + H^2} \tag{7.2}$$

where E_H is the maximum harvested energy of the RSU n with coordinates $(x_n(\xi_n), y_n(\xi_n))$ harvested in time τ_{EH} from UAV having hover point coordinates as $(x(\xi_n), y(\xi_n), H)$. P_U is UAV's transmission power and g specifies the channel gain. The distance between the node and the UAV is restricted to be within MHTR subject to (7.1d). The objective function is changed to an affine function after the first-order Taylor expansion at hover points and can be solved by CVX. The SCA algorithm converges to give optimal hover points where EH is maximum. The resulting optimized hover points are as close to RSUs as possible.

WSN nodes are said to harvest energy from UAV in half-duplex or full-duplex modes in [18]. The energy consumption of UAV is kept to a minimum while accomplishing the minimum data requirement of users. The basic problem deals with minimizing energy and planning path. The objective function for the path planning sub-problem of a UAV flying with propulsion energy E_P from user n to $n+1$ over path q in time τ_F is given by

$$\min_{n,q,\tau_F} \sum_{n=1}^{n+1} E_P(q_n(t), \tau_n^F) \tag{7.3}$$

The trajectory is a function of the distance between consecutive hover points. The velocity ($v_{um}(t)$) of UAV is zero at initial and final points and has some value, up to V_{max}, over the interval t with an acceleration of $a_u(t)$. The objective function is constrained by (7.1b), (7.1g), and (7.1f). The trajectory is optimized using SCA and the visiting order is optimized via the dual method. The complexity of the proposed algorithm is $O(IK^3)$ with I being the number of iterations.

The trajectory design to maximize the minimum harvested energy among multiple nodes in a UAV-assisted WPT network is considered in [19]. A successive hover-and-fly (SHF) model is considered where UAV hovers for a specified time and then flies to the next location with maximum velocity. The UAV does not adopt a straight path but follows curves and arcs over the turning points. The curved paths are simplified to multiple straight line segments. The UAV path is defined as a function of these line segments and hover points. The series of convex problems is optimally solved by convex approximation. The main objective function subject to (7.1k) is defined in (7.4) with E being an auxiliary variable. The optimization is defined for non-linear EH

$$\max_{x(t),y(t)} \min E_H \tag{7.4}$$

The harvested energy is effectively delivered to sensor nodes when the UAV adopts a specified path. However, this path planning greatly depends on the scheduling of sensor nodes so a joint optimization problem for path planning and scheduling is formulated in [20]. The aim is to maximize the energy transfer from UAV along with maintaining fairness among the harvesting nodes by maximizing the minimum harvested energy. The original problem is convex and is decomposed into two sub-problems and solved using SCA. The second sub-problem deals with the optimal UAV trajectory considering fixed wake-up scheduling. The objective is to maximize the energy subject to (7.1b) and (7.1c).

A UAV-assisted OFDM based scheme is used to provide EH to an IoT network in [21]. A joint trajectory and resource allocation optimization is used to maximize the minimum achievable rate for information transfer from SN/IoTD to the UAV.

The authors jointly optimize trajectory, power, subcarrier, and slot allocation. An alternative iterative algorithm is proposed to find the optimal sum rate which gives improved results compared to average and water-filling algorithms. The basic problem is divided into two sub-problems. At first, the joint power, subcarrier, and slot allocation are optimized for the fixed UAV trajectory, whereas the trajectory is optimized in second for fixed resource allocation. The objective function bound to (7.1c) is defined as follows:

$$\max R_{\min}^{n} \qquad given, \;\; R_{\min} = \min\left(\sum_{s=1}^{S} R_n[s]\right) \tag{7.5}$$

The problem is solved using SCA. The lower bounds are obtained by the first-order Taylor series. The resulting affine function is convex and solved using CVX.

Practical non-linear EH is considered in [22] along with typical linear EH for wireless powered communication networks (WPCN). Two scenarios are discussed for both linear and non-linear schemes, namely, integrated UAV and separate UAV. In an integrated UAV, the UAV acts as a hybrid access point (AP) transmitting energy and then receiving the information from WPCN nodes, whereas in the separate UAV scheme, one UAV performs WET and the other receives WIT signals. The linear EH-based system uses the CCCP procedure and is discussed in the next section. In the non-linear EH model, the trajectory and resources are jointly optimized by the SCA approach. The objective function for jointly optimizing trajectory q and uplink power P_{UL} for given time resource allocation in slot s is given by

$$\max_{R_{\min}, P_{UL}, q} \; R_{\min} \tag{7.6}$$

The objective function is constrained by (7.1c) and (7.1j). Two concave surrogate functions are derived by applying the Taylor expansion to the objective function. Using these, the problem can be approximated to a convex one. The convexity can be solved by existing convex solvers and the solution can be calculated iteratively, found out at i-th iteration, based on SCA.

7.3.2.2 Concave–convex procedure

The concave–convex procedure (CCCP) is a way to construct discrete-time iterative dynamical systems that are guaranteed to decrease global optimization and energy functions monotonically.

The linear EH-based systems for both UAV scenarios, i.e., integrated UAV and separate UAV, in [22] are solved using the CCCP framework. The objective function is the same as in (7.6) for linear EH. The given function is non-convex so an auxiliary variable is introduced to transform it into a convex problem. Once formulated the first-order Taylor approximation is applied and a tight lower bound is obtained over the variable. The resulting equations can be solved using existing convex optimization tools like CVX. Similarly, for separate UAV two auxiliary variables are introduced. The resulting problem contains non-convex constraints as a difference of two convex problems and can be handled by using CCCP. Iterations are performed till convergence.

An IoT network is powered by the UAV-assisted power splitting simultaneous wireless information and power transfer (SWIPT) mechanism in [23]. The efficiency of information and power transfer is enhanced by optimizing UAV's trajectory and power transfer. Rather than a winner-take-all methodology, a fair harvesting opportunity is provided where minimum harvested energy is maximally shared among all the nodes by optimal trajectory design and power allocation to UAV. The objective function to obtain the maximum of minimum harvested energy E_H over path q and UAV power P_U is given by:

$$\max_{P_U[s],q[s]} \min E_H \tag{7.7}$$

The power constraints and maximum allowable trajectory are restricted by (7.1i) and (7.1c). The non-convex problem is approximated using auxiliary variables and utilizing CCCP into series of convex problems. The computation complexity is given by $O((NS)^{3.5})\log(1/\varepsilon)$, where N is the number of nodes, S is the number of slots, and ε is the solution accuracy.

7.3.2.3 Geometry-based algorithm

While designing an energy harvesting mechanism for wireless sensor networks, the energy constraints of the UAV are also crucial. The energy consumption of a UAV to complete its flight time is also taken into consideration in [24]. SWIPT is considered for EH and data transmission along with energy consumed during flight and hovering of UAV. An objective function is formulated as in (7.8) to maximize energy harvesting subject to hovering time τ_n in which power is delivered to node n at distance d_{ξ_n} between the sensor node and hovering point ξ_n.

$$\max_{\tau_p,\xi} \min_{n=1,..,N} E_n(\tau_n^p, \xi_n) \tag{7.8}$$

where E_n is the maximum energy left on sensor n after data transfer and energy harvesting. The problem is subject to (7.1e), (7.1h), and (7.1l). The optimization problem is non-convex and cannot be solved for both variables, i.e., hovering time and hovering location, simultaneously. At first, the maximization problem is formed to solve hovering time which results in a convex problem solved by Lagrange multipliers. Then, a geometry-based algorithm is proposed to determine hovering location.

7.3.2.4 Modified approximations and variants

Modifications in existing iterative algorithms and utilization of their variants are carried out in certain optimization problems to provide more reliable results and close approximations.

A multi-UAV network to provide coverage and power to large-scale IoT is discussed in [25]. As the UAV has its energy and power constraints so a single UAV's energy is not enough to support information and power transfer for a large-scale network. Some other parameters like load balancing, effective path planning, and interference among UAVs arise which need to be tackled effectively. An energy-consuming model for multi-UAV is designed as a function of their respective speed. The number of UAVs is expected to be minimum by maximizing their energy utilizing capability. A relationship is formulated between UAV trajectory length and WPT, WIT model, and the trajectory optimization problem is formulated. A heuristic algorithm termed MAVNS is proposed which is a combination of a global search strategy, memetic algorithm (MA), and a local search method, variable neighborhood search (VNS). The objective function for the trajectory of M UAVs having length L is defined by (7.9) where G is a three-dimensional variable depicting the arcs belonging to UAVs.

$$\min_{G} \; L_{\max} + \sqrt{\frac{1}{M-1}\sum_{i=1}^{N}(L_i - L)^2} \tag{7.9}$$

The formulated problem depicts the maximum distance traversed by a UAV is to be minimized which seems to be a variant of the travelling salesman problem (TSP). For multiple UAVs, it becomes MTSP and the problem is min–max mTSP which also takes into account the load balancing among UAVs. To improve the algorithm computation capability a combination of local and global search methods is applied. A summary of trajectory optimization schemes is presented in Table 7.2.

7.3.3 UAV's energy constraints

As the UAV hovers over the WSN to provide WPT, its energy levels are also crucial. The energy levels should be enough to complete a round-trip flight time and return to the control center. While designing a system, this constraint is also considered and studies are being done on energy saving and harvesting for UAV's flight.

7.3.3.1 Solar-powered UAVs

In the case of large WSNs/IoT, the UAV's energy is crucial. It needs to be enough to support the efficient hovering of UAV for longer durations to provide sufficient coverage to all the SNs/IoTDs. The charging of the UAV battery from specified ground stations or control center requires extra time and hinders efficient working. Solar-powered UAVs eliminate this issue by harvesting solar energy and charging the UAVs during flight. The adjustment in the trajectory of UAV in accordance with solar energy density and spatial distribution maximizes the harvested energy.

An efficient energy and radio resource management framework based on intelligent power cognition at SUAVs is proposed in [26]. It differs from other machine learning (ML) methods as it improves learning through interactions of agents with the uncertain environment including the spatial distributions of solar energy density, the channel state evolution, and the traffic patterns of wireless communication applications in adaption to the environment changes. The EH, flight trajectory, and information transmission are intelligently adjusted by SUAVs

Table 7.2 Summary of trajectory optimization schemes for UAV-assisted EH in WSNs/IoT networks

Paper	Objective	Parameters	Method to optimize trajectory
[17]	Maximize coverage and EH	Distance between RSUs and hovering points, UAV's transmission power	SCAIA, CVX, SCA
[18]	Minimizing energy consumption of UAV, maintaining minimum data requirement of users	Distance between consecutive hover points, UAV's velocity and acceleration	SCA, dual method
[19]	Maximize the minimum harvested energy	UAV path as a function of line segments and hover points	SCA
[20]	Maximize energy transfer from UAV, maintaining fairness among harvesting nodes	Path planning, scheduling of sensor nodes	SCA
[21]	Information transfer in IoT at maximized minimum achievable rate	Trajectory, power, subcarrier and slot allocation	SCA, CVX, Taylor series
[22]	Joint trajectory and resource optimization for linear and non-linear EH	UAV's trajectory, uplink power	SCA, CVX, Taylor expansion, CCCP
[23]	Fair distribution of minimum harvested among nodes	UAV's trajectory, power transfer	CCCP
[24]	Maximize energy harvesting	Hovering time, hovering location	Lagrange multipliers, geometry-based algorithm
[25]	Multi-UAV coverage and power transfer to large-scale IoT	Load balancing, path planning, interference among UAVs	MAVNS, MA, MTSP, TSP, VNS

to enhance solar energy utilization for achieving high performance resulting in longer hovering and flying periods and improved lifetime. The maximization of the total system throughput within the lifetime of the SUAV is optimally decided by reinforcement learning (RL). Simulation results show that the proposed power cognition scheme can simultaneously improve the communication throughput and the harvested energy for SUAVs.

The path planning of SUAVs considering the terrain and obstacles is described in [27] and [28]. The former formulate an optimized trajectory algorithm to maximize solar energy harvesting by UAV for tracking applications over the obstacle environment. The latter designs a software known as the meteorology aware trajectory planning and analysis software for solar-powered UAVs (METPASS) taking into account various environmental and geographical factors, such aas winds, cloud coverage, solar radiation, terrain, and obstacles, combined with flight time, and parameters.

7.3.3.2 Magnetic resonance coupling

The efficient hovering of the UAV requires an unhindered supply of energy for its flight and power delivery. The energy capacity of battery-powered UAVs could limit their working. A large battery would increase UAV's load and the small one needs frequent replacement or recharging so it is replaced by an alternate technique known as MRC. The technique is effective in powering the UAV within the near-field range. It consists of two coils, one being placed on the ground, either static or mobile, and the other one is placed on a UAV. The coils are called transmit and receive coils, respectively.

An inductive resonance coupling system is proposed in [29] to power the UAV for hovering. The model suggests the orientation and geometries of the coil-over which power transfer is maximized. The relative location of the UAV is also sensed relative to the placement of the transmit coil over the ground and the UAV adjusts according to maximize the power transfer. It is also demonstrated that the hovering time of UAV is improved based on the model.

The MRC-WPT network to power uninterrupted working of UAV is proposed in [30]. Two geometrically different coils are utilized as a transmit coil and receiver coil. The ground transmit coil is a flat spiral coil (FSC), whereas the UAV-mounted receiver coil is a multi-turn coil (MTC). Results are observed for both aligned and misaligned setup of coils over various distance ranges. In the case of an aligned coil setup, the power transfer is improved to $20.46W$ with an 85.25% efficiency. For a misaligned coil setup, the power transfer is improved to $17.1W$ with a 71% efficiency. Moreover, an extended UAV's battery life of 851 min is also observed with a power-saving of 96.9% resulting in an improved hovering time.

7.4 Impact of association on EH

The problem of user association deals with finding the optimum pairing between an SN/IoTD and a UAV. A straightforward criterion for selecting an association is to associate an SN/IoTD with a UAV with the best channel. However, this simple method cannot cater to the needs of a whole network. Particularly, in the case of UAV, in which channel is varying for mobile UAV and adaptable in the case of fly and hovering UAV.

In the case of multiple UAVs, this problem is extended to design an association scheme involving the association of nodes to the most suitable UAV from a group of UAVs during the complete operation time of UAVs, as shown in Figure 7.4. An efficient user association scheme can benefit IoT operations in several ways such as improving the efficiency of energy harvesting, decreasing packet loss and transmission power required for data transmission to UAV, decrease in time delay, and an extended lifetime of SN/IoTDs and UAVs.

Most of the schemes in the literature that improve device association also improve other factors such as a sequence of workflow and resource (time, power, trajectory, and computing resources) allocation. Depending on the objectives and control parameters, the association problem takes different mathematical forms and in the literature, various solutions are presented. In this section, we present a brief overview of schemes that find an optimum association in UAV-assisted IoT

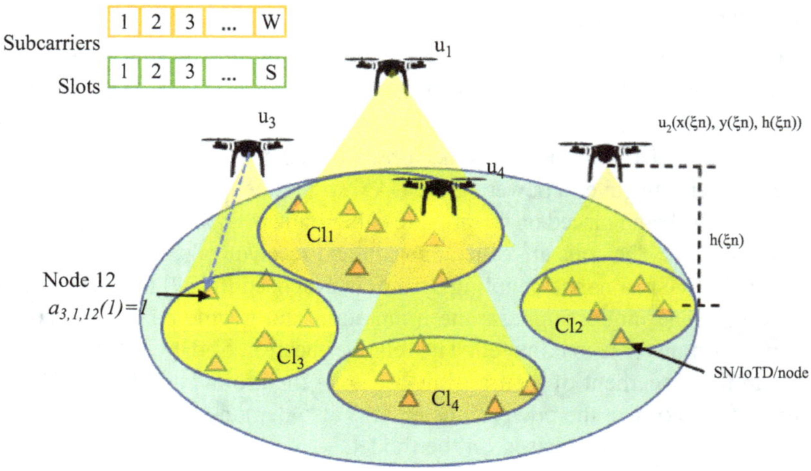

Figure 7.4 Multi-UAV-assisted WSNs/IoT network, $a_{3,1,12}[s] = 1$ indicates that Node 12 is associated with U_3 in slot 1 over subcarrier 1

networks. We start by presenting a general optimization problem and then discuss specific schemes related to the optimization of association for improving energy harvesting directly or indirectly.

7.4.1 General optimization problem

In the literature, the optimization problem to find optimum association and resource allocation involves both direct and indirect maximization of energy harvesting. In direct problems, energy harvesting is considered as an objective function, while in indirect approaches, the considered objective is some other function such as data rate/network capacity that depends on improved energy harvesting. A general optimization problem for direct improvement of energy harvesting is as follows:

$$\max_{a_{w,n}[s], p^d_{w,n}[s], W[s], Ts[s]} \sum_{n=1}^{N} min\ E_H\ \forall s \in S \tag{7.10a}$$

$$\text{s.t.} \quad \sum_{w \in WI} a_{w,n}[s]\ Da_n[s] \geq \gamma_{th}, \tag{7.10b}$$

$$\sum_{w \in W_E} a_{w,n}[s] E_{hn,w}[s] \geq E_{th}, \tag{7.10c}$$

$$\sum_{w \in W} a_{w,n}[s]\ p^d_{w,n}[s] \leq P_U, \tag{7.10d}$$

$$\sum_{n=1}^{N} a_{w,n}[s] = 1\ a_{w,n}[s] \in \{0, 1\}, \tag{7.10e}$$

$$W^I[s] + W^E[s] = \mathbf{W} \tag{7.10f}$$

$a_{w,n}[s]$ represents association variable, $a_{w,n}[s] = 1$ if node n is assigned w subcarrier in slot s. This term can have another variable, u, i.e. $a_{u,w,n}[s]$ to show whether a user is connected to UAV u using w subcarrier in slot s. Equation (7.10b) ensures minimum data rate requirements of an SN/IoTD n, (7.10c) ensures minimum energy requirements. Equation (7.10d) presents the upper limit on the sum of powers of all beams transmitted by a UAV, while (7.10e) restricts that only one user is attached to one set (w) of subcarriers and (7.10f) restricts the upper bound on the sum of subcarriers. W^E and W^I represent subcarriers dedicated to energy harvesting and information transfer, respectively.

7.4.1.1 K-means clustering

A K-means clustering is a method to divide data set into k clusters such that clusters have minimum distance with the data sets within the clusters and as separate from other clusters as possible. The cost function is the distance between cluster objects and the distance from the center of clusters. Initially, K random objects are selected from the data set as the center of the cluster, and nodes are associated with the nearest cluster based on the cost function. The cost function is re-evaluated for the given clustering, this process continues until the convergence of the cost optimization function [31].

Three different configurations of UAV-assisted IoT networks are considered in [15]. The first scenario considered a dense IoT network that is served by a single UAV with a multi-beam antenna array. The beam pattern optimization is discussed in Section 4.1.3. The second scenario is considered where multiple UAVs with a single transmission beam are deployed to cover a large area. In this work, the authors have focused to improve the QoS of SNs/IoTDs by maximizing the minimum data rate. First, SNs/IoTDs are divided into clusters using k-means clustering and each UAV serves one cluster. Further, for finding optimum values of time, power, and trajectory, deep Q learning is used. In the third configuration presented in [15], a single flying UAV is serving to satisfy the dynamic needs of SNs/IoTDs. For this case, the trajectory of UAV is optimized using the deep deterministic policy gradient (DDPG) for satisfying the dynamic needs of SNs/IoTs.

In [32], the authors aimed to minimize packet loss by providing sufficient energy to needy devices. They have considered joint optimization of cluster association, hovering time, and flight schedule. IoT devices are divided into clusters using an improved K-means clustering. To formulate clusters that improve charging efficiency, nodes are ordered according to the prediction of their energy consumption. Top k nodes are selected as a mean vector $\{M_v = \mu_1, \mu_2, ..., \mu_k\}$, the distance of each node is calculated from this vector of nodes and a node j is selected in i cluster if $d_{ji} < L$, wherein the simulation, authors have considered $L = 5m$. The hovering position of the UAV is selected above the cluster mean. This improved k-means clustering and hovering position results in improved efficiency of energy harvesting.

7.4.1.2 Non-convex joint optimization problems

In [33], the proposed solution aimed to minimize UAV energy consumption which in turn results in improved energy efficiency in a multi-UAV network. UAVs are assisting SNs/IoTs by providing energy harvesting and computation capabilities. Association, charging time, and offloading capacity are optimized jointly to achieve minimum energy consumption. The complex joint optimization problem is divided into three subproblems; SNs/IoT association and capacity offloading optimization, charging time optimization, and transmit power optimization.

To solve the association problem, the joint problem of association and capacity offloading is converted to association problem by using the following value of capacity offloading for an SNs/IoT n, if it is associated with UAV u

$$C_{un} = \frac{C_n}{T - \tau_{EH} - t_{nu}^{Tr}}, \tag{7.11}$$

where C_{un} represents the computational capacity of UAV u dedicated for executing the task of SNs/IoT n and C_n represents a total number of cycles required to execute a task of node n. The T represents latency constraint for the offloaded task, while t_{nu}^{Tr} represents the time required to offload the task of node n to UAV u and τ_{EH} represents the time that is required for transmitting RF energy. For this value of computational capacity, an association problem is divided into $N - N_0$ subproblems, where N_0 represents the number of nodes that are not offloading their task to UAV, all possible binary combinations for these problems are evaluated to find the best combination of values that result in minimum energy consumption.

The authors have analyzed the energy performance of the proposed optimization-based solution for varying values of energy harvesting rate, number of IoT devices, and bandwidth. Network performance is not compared to any other approach or with base case i.e when no optimization is applied.

In [12], the network model consists of a single UAV assisting SNs/IoTs by providing power and computing resources. In this work, the authors have considered multi-workflow optimization. In the multi-workflow, service operation for each SN/IoTD consists of three stages: energy harvesting stage, data transfer stage, and task computing stage. Instead of waiting for the completion of this whole operation and then moving to service the next node, in the proposed scheme, the sequence of these tasks is optimized for a concurrent operation of multiple SNs/IoTs. The objective is to minimize the energy consumption of UAV, authors have included energy consumed in UAV mobility in addition to energy consumed in power transfer and computing tasks. The proposed scheme jointly optimizes UAV hovering time, duration of power transfer, and allocation of computing resources in addition to optimizing node association. The joint optimization problem is divided into four subproblems for finding optimum values of the above-mentioned variables. After finding optimum values for allocated computing resources and services sequence, the SNs/IoTDs association is optimized with the objective of minimizing energy consumption. In the given system model, each node can upload data to a UAV for one hovering position. Therefore, the association problem is to select an optimum hovering position for each node. For the given values of computing resource and sequence, the SNs/IoTDs association problem is converted to a linear programming problem (LP), which is solved using convex optimization tool CVX in MATLAB.

7.4.1.3 Beamforming

A single UAV with multiple antennas can serve multiple SNs/IoTDs through beamforming as shown in Figure. 7.5. In the case of multi-beam antennas, the association problem converts to association and adaptation of beam phases to improve harvested energy or energy efficiency. In classical time-domain filtering, the received signal is linearly combined with different filtering weights to achieve low, high, or bandpass filtering. Thus, conventional beamforming can be considered as an extension of matched filtering [34]. A beamformer is made of a processor along with an array of antennas. In addition to time/frequency filtering, a beamformer enhances the signal received from the desired direction and suppresses the effects of signal/signals received from other direction/directions [35]. This is achieved by selecting the right delay and phases for each antenna and adding these signals. There are numerous schemes proposed in the literature for beamforming that can differ on the amount of information they require, the desired objective, and robustness [36].

In the literature, many schemes are proposed to improve the efficiency of energy harvesting in IoT through beamforming. However, most of these schemes [37–40] consider a static transmitter with an ample source of energy and require channel state information (CSI). To estimate CSI, all the devices have to send orthogonal pilot sequences [41]. Due to a large number of IoT devices, finding unique orthogonal sequences for all the devices in the network results in the transmission of pilot signals that take a longer period. Longer duration-based pilot sequences result in decrease time left for power signal transfer while consuming more energy in the transmission of these signals. Furthermore, the complexity of channel estimation using these signals increases linearly with the number of SNs/IoTDss involved [42]. Thus, this approach is expensive from both energy and

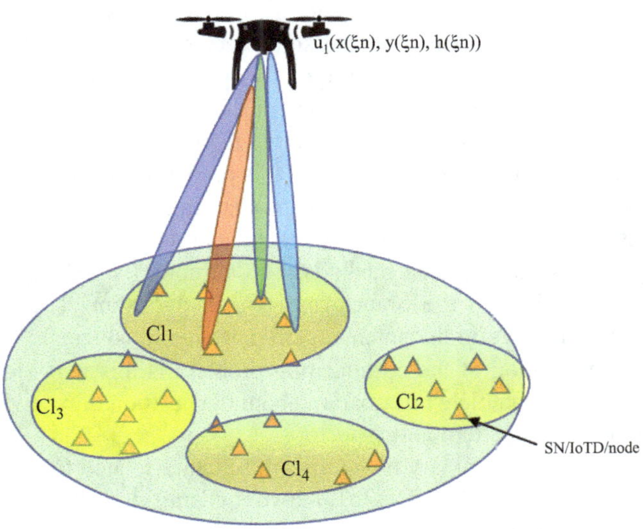

Figure 7.5 Single UAV-assisted WSNs/IoT network with beamforming

computation perspectives. Switched beamforming has been used to relax demand for complete information of CSI [14,43,44]. In this scheme, a combination of predefined beams is used to generate a beam pattern and this combination is determined based on the average channel gain of these predefined beams. This method cannot achieve multiplexing gain in the case of data transmission to multiple independent users due to the limitation of predefined beams [44]; however, this is not an issue in WPT which is in nature a broadcast transmission [45].

In [46], authors have proposed a switched beam scheduling scheme for dynamic IoT devices. They have considered a common pilot signal transmitted simultaneously by all the requesting users. The common pilot signal can decrease both the required time to transmit the pilot signal and the complexity involved in channel estimation. Moreover, the pilot signal needs not be a sequence, instead, a single tone can be used as a pilot signal [47]. All these pilot signals are summed at the receiver and the receiver has to estimate channel state and schedule beam. The main task for adaptive beam scheduling considering the common pilot signal is to estimate IoTD's context and channel information and design patterns based on the estimated results. To decrease possible error in the estimation and beam scheduling, the authors have proposed to use deep neural networks (DNNs) to estimate channel state and use RL, deep Q network (DQN), and deep deterministic policy gradient (DDPG) to train DNN for determining optimal beam scheduling policy.

The proposed scheme has been evaluated over time and in comparison to the Omni beam pattern scheme for the value of satisfaction rate, where the satisfaction rate is the ratio of served power requests and total power requests. Results exhibit a significant improvement in the performance. To apply this scheme in UAV-assisted energy harvesting, one must need to evaluate the performance of this scheme with the constraint on the total energy of the UAV.

Another work that focuses on beamforming in UAV-assisted WSNs/IoT networks is presented in [15]. A single UAV with multiple beams is serving IoTDs in a dense IoT network. For maximizing the sum received power, beam pattern, the trajectory of UAV, and charging time are jointly optimized. Due to the non-convex nature of the problem, the optimization problem is divided into three sub-problems. Given the optimum values of position and altitude, a multi-objective evolutionary algorithm based on decomposition (MOEA/D) is proposed to find optimum values of phases of antenna arrays for a fixed value of charging time. Further, to deal with the fairness issue, UAV charging time is formulated as a max–min problem and solved using convex optimization methods. Furthermore, the 3D trajectory of the UAV is optimized through the branch and bound algorithm. The result shows the effect of optimization on the antenna gain pattern and coverage; however, the mathematical formulation of the optimization problem is missing. Moreover, this optimization requires complete information about the position of receiving nodes in order to design a beam pattern and trajectory.

Table 7.3, depicts a summary of significant schemes related to optimization of association in UAV-assisted WSNs/IoT network identifying different objectives that are focused in those schemes. It also presents specific parameters that are optimized to achieve desired objective. The optimization methods that are used to

Table 7.3 Summary of association optimization schemes for UAV-assisted EH in WSNs/IoT networks

Paper	Objective	Parameters	Method to optimize association
[12]	Minimize energy consumption	Multi-workflow, time, computing resources	Decomposition of non-convex optimization, sub-problem solved using LP
[15]	Maximizing sum received power for single UAV with a multi-beam antenna Maximizing the minimum average throughput for multiple UAVs serving IoT network	Association (beam pattern design), charging time, power, and trajectory Association, charging time, trajectory	MOEA/D , Deep Q learning K-means clustering
[32]	Minimizing packet loss	Association, hovering time and operation schedule	Improved K-means clustering
[33]	Minimizing energy consumption of UAV	User association, charging time and off-loading capacity, transmit power	Decomposing non-convex optimization and Brute-force method
[46]	Satisfaction rate for energy harvesting	Design of beam pattern	RL, DQN, and DDPG

solve designed optimization problems are also highlighted. This table provides a guideline for selecting an appropriate scheme according to application-specific objectives and available control variables.

7.5 Impact of resource allocation on EH

Important resources of UAV in UAV-assisted EH for SNs/IoTDs are set of subcarriers, time/slots, and transmit power. The large set of subcarriers with better channel conditions dedicated to energy harvesting increases the energy harvesting at SN/IoTDs. Similarly, an increase in the duration for which a UAV is transmitting radio waves to charge an SN/IoTD directly increases the harvested energy at an SN/IoTD [21]. Furthermore, received power at SNs/IoTDs is directly proportional to the transmit power of the UAV [23]. Allocation of the above-mentioned resources in such a way that a UAV can serve all the SNs/IoTDs meeting their energy and QoS requirements while having limited energy is a challenging task. Most of the literature related to optimizing resource allocation considers the optimization of multiple resources simultaneously.

7.5.1 Optimization problems and algorithms

A general optimization problem for maximizing energy harvesting in UAV-assisted EH in IoT networks by optimizing resource allocation is presented in Section 4.1.

7.5.1.1 Successive convex approximation

In [33], discussed in Section 4.1.2, the joint problem of association, computing resources, charging time and power is decomposed into four optimization problems. The decomposed charging time optimization problem can easily be solved based on the condition of least energy harvesting requirements using the following formula:

$$\tau_{EH} = \max\left\{\frac{E_{c,n}}{\sum_{u\in U}\varepsilon_n g_{n,u} P_{Uu}}, \forall n \in N\right\} \tag{7.12}$$

Given the values of association matrix, computing resource, and charging time, the optimization problem of finding the optimum value of transmission power of UAV to minimize energy consumption is a convex optimization problem that can be solved using LP methods such as Dantzig's Simplex method, interior-point method or using the MATLAB optimization tool such as CVX.

In [23], mentioned in Section 7.3.2.2, joint optimization of power and trajectory is proposed to maximize the minimum energy harvesting at each node. The joint problem is divided into two subproblems; one subproblem deals with transmit power optimization and the other deals with trajectory optimization

$$\max_{P_U} E_H \,\forall\, n \in N \tag{7.13a}$$

$$\text{s.t.} E_{cn} \leq E_H, \tag{7.13b}$$

$$(7.10\text{b}),\ (7.10\text{d}), \tag{7.13c}$$

$$P^d_{w,n}[s] \geq 0 \,\forall\, n \in N \text{ and } s \in S \tag{7.13d}$$

Equation (7.13b) imposes a constraint on the minimum energy harvesting at every node, harvested energy E_H must be greater than the energy consumption requirements E_{cn}, $P^d_{w,n}[s]$ represents power transmitted to node n by a UAV in slot s and on channel w. Equations (7.10b), (7.10d), and (7.13d) constraints ensure the minimum data rate, the maximum transmitted power limit, and the positive nature of $P^d_{w,n}[s]$. All the given constraints and objective functions are convex with respect to the optimization variable P_U. Therefore, any convex optimization method, such as the interior point method can be used.

7.5.1.2 Lagrange multiplier and water-filling

In [21], discussed in Section 7.3.2.1, authors have considered sub-carrier and slot allocation in addition to the transmit power and trajectory optimization, to maximize minimum achievable data rate in the uplink direction. The objective function of subproblem for subcarrier, power, and slot allocation is formulated as follows:

$$\max_{a_{w,n}[s],\, P^d_{w,n}[s], W[s], T_s[s]} \sum_{n=1}^{N} R^u_n[s] \,\forall s \in S \tag{7.14}$$

where $p^d_{w,n}[s]$ represents power allocated to SN/IoTD n for sub-carrier w in slot s, $T_s[s]$ is a part of the slot which is dedicated for the uplink transmission while the remaining portion of slot $(1 - T_s[s])$ is dedicated for downlink transmission. $R^u_n[s]$ represents the data rate of user n in the uplink direction. While maximizing the data rate, bounds on the maximum and minimum values of the number of subcarriers, slots, power, and the minimum acceptable data rate are considered. The problem is converted to maximizing harvested energy as the uplink data rate is directly affected by the amount of harvested energy.

$$\max_{p^d_{w,n}[s]} \sum_{n=1}^{N} \sum_{w \in W^E} a^n_w \, [s] \, E_W \, , ns \, \forall s \in S \tag{7.15}$$

Furthermore, only $p^d_{w,n}[s]$ is considered as a control variable. The resulting problem is a convex optimization problem, which is solved using the Lagrange multiplier and water-filling algorithm to find the best value for power. $p^d_{w,n}[s]$ is then used to find optimum values of W^{E*} and W^{I*}, which represent subcarriers dedicated for energy harvesting and information transfer, respectively. The sub-gradient method is used to update these values until convergence. After calculating optimum values for $p^d_{w,n}[s]$ and uplink power $p^{U*}_n[s]$, the $T^*_s[s]$ is determined. The result shows a significant improvement in the uplink data rate as compared to non-optimized solution. This scheme considers the known value of CSI and a single UAV is considered that are serving different users during its flight. In [48], authors have optimized hovering and flying time of the UAV for three distinct optimization problems: maximizing sum throughput, minimizing total time, and minimizing total energy consumption, given the constraint on the maximum speed and minimum data that is to be transmitted from sensor nodes to UAV. All three optimization problems are convex functions of the control variable; however, instead of solving them using linear programming methods, coupling of constraints is used to define lemmas and to find optimal values of the hovering and flying time. The resulting algorithms are far less complex than linear programming-based solutions. The solution of the optimization problem exhibit that for different objectives; optimal values of hovering and flying time follow different trends.

7.5.1.3 Machine learning-based schemes

Time dedicated to transmitting power by UAV directly impacts the amount of harvested energy in SNs/IoTDs. This time must be allocated efficiently to satisfy the needs of all the SNs/IoTDs in the network.

In [49], authors have investigated how network performance is affected by the time allocated for energy harvesting and data collection. They have analyzed the trade-off between minimizing time delay in data collection and the amount of energy harvested by IoT. A variable α is used to divide time, where $(1 - \alpha)T_{\text{cover}}$ time is dedicated for information collection and $\alpha \times T_{\text{cover}}$ is dedicated to energy harvesting. Battery state is modeled using a Markov chain model while energy harvesting process is exponential. The required time that must be allocated for energy harvesting depends on the number of devices whose battery is below a certain safe threshold and rate of energy harvesting. Furthermore, the time requires to transfer payload data at a certain rate is calculated. α is calculated by considering both the above-mentioned factors. For scheduling energy harvesting, three policies have been considered; lowest battery first, highest battery first, and uniform charging schedule. The average battery level curves show the best performance when a low battery first approach is followed.

In [50], the authors have proposed a double Q-learning-based solution for optimizing transmission schedules for both power and data. The objective is to minimize energy harvesting loss while considering the data queue and energy status of the battery. Double Q learning reduces the approximation error which results due

Table 7.4 Summary of resource optimization schemes for UAV-assisted EH in WSNs/IoT networks

Paper	Objective	Parameters	Method to optimize resource allocation
[12]	Minimize energy consumption	Multi-workflow, time, and computing resources	Decomposition of non-convex optimization, Lagrange dual method for computing resources, flow-shop scheduling for service sequence
[15]	Maximizing sum received power	Association, charging time, power, and trajectory	Sequential unconstrained optimization, convex optimization techniques
	Maximizing the minimum average throughput	Association, charging time, and trajectory	DQN trained through DQL
[21]	Maximize the minimum achievable information transfer rate from SN/IoTD to the UAV	Trajectory, power, subcarrier, and slot allocation	SCA with the water-filling algorithm
[32]	Minimizing packet loss	Association, hovering time, and operation schedule	Mean-field game theory
[33]	Minimizing energy consumption of UAV	User association, charging time and offloading capacity, and transmit power	Decomposing non-convex optimization, Constrained optimization
[48]	Maximize sum throughput, minimizing total time, minimizing energy consumption	Hovering and flying time, and UAV transmit Power	Convex optimization and constraint coupling
[49]	Balancing time for data transfer and energy harvesting	Time resources	Birth–death process
[50]	Minimize energy lost	Transmission time schedule	Q learning
[51]	Minimize energy harvesting lost	Time	Double Q-learning

to unknown battery and data queue status of every node at the UAV. To deal with the divergence of the Q-learning algorithm in a larger IoT network, a double Q-learning-based solution is proposed in [51]. Due to suppressing the estimation error, the proposed method results in an improved solution. The effect of varying values of network size and data queue length on the packet loss rate is evaluated. The result shows better performance as compared to a greedy approach.

Although power optimization of an SN/IoTD will not improve harvested energy, it can significantly improve the utilization of available energy. In [32], after optimizing user association, the transmit power of SNs/IoTDs is optimized for decreasing interference. Association is optimized through clustering as mentioned in Section 7.4. As clusters are calculated and the flight of the UAV is scheduled, the node's transmit power is optimized to decrease interference in peer-to-peer communication. The optimization problem is designed as a mean-field game and solved by each node. The power optimization problem requires each node to know the mean-field distribution to solve the mean-field game. This information is disseminated by a central controller. The result depicts a decrease in packet loss as compared to an optimized charging operation. Furthermore, the obtained SINR is improved as compared to a case when all nodes transmit the same power.

Table 7.4 presents a summary of significant schemes which focus on optimizing resource allocation in UAV-assisted WSNs/IoT networks identifying different objectives that are focused in those schemes.

7.6 Conclusion

Due to several attractive features of UAV-assisted energy harvesting for WSNs/IoT networks, extensive research is happening in this area. The research focuses on optimizing trajectory, association, and resource allocation to improve energy harvesting directly and through other application-specific objectives. In this chapter, we have presented an overview of significant schemes in this regard. This overview can help select a suitable scheme that meets specific application needs by optimizing desired control variables. Furthermore, this chapter identifies that most of these schemes consider the limited energy of UAV and there is a need of exploring different proposed solutions in the presence of energy harvesting capabilities. Moreover, it is observed that most of the analytical optimization schemes have considered perfect information of channel state, which is not the case, especially for a mobile UAV. Limited literature exists that employ statistical approaches to cater to unknown channel state and estimates CSI instead of considering its known value.

A recent approach in UAV-assisted communication is to use reflecting surfaces along with UAV. These surfaces can significantly improve the directionality of energy transfer and thus can improve energy harvesting at SNs/IoTDs. However, analysis of UAV-assisted WSNs/IoT networks in the presence of reflecting surfaces pertaining to application-specific energy needs is missing in the literature. Exploring this dimension can result in more efficient energy harvesting schemes for UAV-assisted WSNs/IoT networks.

List of acronyms

AP	Access point
CCCP	Concave-convex procedure
CSI	Channel state information
CTS	Clear to send
DDPG	Deep deterministic policy gradient
DNN	Deep neural network
DQL	Deep Q learning
DQN	Deep Q networks
EH	Energy harvesting
EPEM	Energy prediction-based energy management
FSC	Flat spiral coil
IoTD	IoT device
IoTs	Internet-of-things
LoS	Line-of-sight
MA	Memetic algorithm
METPASS	Meteorology aware trajectory planning and analysis software for solar-powered UAVs
MHTR	Maximum horizontal transmission range
MILP	Mixed-integer linear problems
ML	Machine learning
MOEA/D	Multi-objective evolutionary algorithm based on decomposition
MRC	Magnetic resonance coupling
MTC	Multi-turn coil
NLoS	Non-line-of-sight
OFDM	Orthogonal frequency-division multiplexing
QoS	Quality of service
RL	Reinforcement learning
RSU	Remote secondary user
SCA	Successive convex approximation
SCAIA	Successive convex approximation based iterative algorithm
SHF	Successive Hover-and-fly
SN	Sensor node
SUAV	Solar-powered UAV
SWIPT	Simultaneous wireless information and power transfer
TSP	Travelling salesman problem
UAV	Unmanned aerial vehicle
VNS	Variable neighborhood search

WIT	Wireless information transfer
WPCN	Wireless powered communication networks
WPT	Wireless power transfer
WSNs	Wireless sensor networks

References

[1] European Drones Outlook Study, Luxembourg City, Luxembourg, Nov. 2016. Accessed: 2020-10-29. http://www.sesarju.eu/sites/default/files/documents/reports/European_Drones_Outlook_Study_2016.pdf.

[2] Ouahouah S, Taleb T, Song J, and Benzaid C. Efficient offloading mechanism for UAVs-based value added services. In: *2017 IEEE International Conference on Communications (ICC)*. IEEE; 2017. p. 1–6.

[3] Ueyama J, Freitas H, Faical BS, *et al.* Exploiting the use of unmanned aerial vehicles to provide resilience in wireless sensor networks. *IEEE Communications Magazine.* 2014;52(12):81–87.

[4] Motlagh NH, Bagaa M, and Taleb T. UAV-based IoT platform: a crowd surveillance use case. *IEEE Communications Magazine.* 2017;55(2):128–134.

[5] Xu J, Ota K, and Dong M. Big data on the fly: UAV-mounted mobile edge computing for disaster management. *IEEE Transactions on Network Science and Engineering*. 2020.

[6] Hassan A, Ahmad R, Ahmed W, Magarini M, and Alam MM. Managing critical nodes in UAV assisted disaster networks. In: *2020 17th Biennial Baltic Electronics Conference (BEC)*; 2020. p. 1–5.

[7] Griffin B, Detweiler C. Resonant wireless power transfer to ground sensors from a UAV. In: *2012 IEEE International Conference on Robotics and Automation*. IEEE; 2012. p. 2660–2665.

[8] Kurs A, Karalis A, Moffatt R, Joannopoulos JD, Fisher P, and Soljačić M. Wireless power transfer via strongly coupled magnetic resonances. *Science.* 2007;317(5834):83–86.

[9] Wang H, Wang J, Ding G, Wang L, Tsiftsis TA, and Sharma PK. Resource allocation for energy harvesting-powered D2D communication underlaying UAV-assisted networks. *IEEE Transactions on Green Communications and Networking.* 2018;2(1):14–24.

[10] Li Q, Gao J, Liang H, Zhao L, and Tang X. Optimal power allocation for wireless sensor powered by dedicated RF energy source. *IEEE Transactions on Vehicular Technology.* 2019;68(3):2791–2801.

[11] Zhou F, Wu Y, Hu RQ, and Qian Y. Computation rate maximization in UAV-enabled wireless-powered mobile-edge computing systems. *IEEE Journal on Selected Areas in Communications.* 2018;36(9):1927–1941.

[12] Du Y, Yang K, Wang K, Zhang G, Zhao Y, and Chen D. Joint resources and workflow scheduling in UAV-enabled wirelessly-powered mec for IoT systems. *IEEE Transactions on Vehicular Technology.* 2019;68(10):10187–10200.

[13] Su C, Ye F, Wang LC, Wang L, Tian Y, and Han Z. UAV-assisted wireless charging for energy-constrained IoT devices using dynamic matching. *IEEE Internet of Things Journal*. 2020;7(6):4789–4800.

[14] Zhang H, Jiang Y, Sundaresan K, Rangarajan S, and Zhao B. Wireless multicast scheduling with switched beamforming antennas. *IEEE/ACM Transactions on Networking*. 2012;20(5):1595–1607.

[15] Feng W, Tang J, Yu Y, *et al.* UAV-enabled SWIPT in IoT networks for emergency communications. *IEEE Wireless Communications*. 2020;27(5):140–147.

[16] Popescu D, Stoican F, Stamatescu G, Chenaru O, and Ichim L. A survey of collaborative UAV WSN systems for efficient monitoring. *Sensors*. 2019; 19(21):4690. Available from: http://dx.doi.org/10.3390/s19214690.

[17] Shi S, Li Y, Gu S, Huang T, and Gu X. Time allocation optimization and trajectory design in UAV-assisted energy and spectrum harvesting network. *IEEE Access*. 2020;8:160537–160548.

[18] Yang Z, Xu W, Shikh-Bahaei M. Energy efficient UAV communication with energy harvesting. *IEEE Transactions on Vehicular Technology*. 2020;69(2): 1913–1927.

[19] Yuan X, Yang T, Hu Y, Xu J, and Schmeink A. Trajectory design for UAV-enabled multiuser wireless power transfer with nonlinear energy harvesting. *IEEE Transactions on Wireless Communications*. 2020.

[20] Wang Y, Hua M, Liu Z, Zhang D, Ji B, and Dai H. UAV-based mobile wireless power transfer systems with joint optimization of user scheduling and trajectory. *Mobile Networks and Applications*. 2019;p. 1–15.

[21] Na Z, Zhang M, Wang J, and Gao Z. UAV-assisted wireless powered internet of things: joint trajectory optimization and resource allocation. *Ad Hoc Networks*. 2019 11;98:102052.

[22] Park J, Lee H, Eom S, and Lee I. UAV-aided wireless powered communication networks: trajectory optimization and resource allocation for minimum throughput maximization. *IEEE Access*. 2019;7:134978–134991.

[23] Huang F, Chen J, Wang H, *et al.* UAV-assisted SWIPT in Internet of Things with power splitting: trajectory design and power allocation. *IEEE Access*. 2019;7:68260–68270.

[24] Baek J, Han SI, Han Y. Optimal UAV route in wireless charging sensor networks. *IEEE Internet of Things Journal*. 2020;7(2):1327–1335.

[25] Wu P, Xiao F, Huang H, and Wang R. Load balance and trajectory design in multi-UAV aided large-scale wireless rechargeable networks. *IEEE Transactions on Vehicular Technology*. 2020;69(11):13756–13767.

[26] Zhang J, Lou M, Xiang L, and Hu L. Power cognition: enabling intelligent energy harvesting and resource allocation for solar-powered UAVs. *Future Generation Computer Systems*. 2020;110:658–664.

[27] Wei X, Yao P, and Xie Z. Comprehensive optimization of energy storage and standoff tracking for solar-powered UAV. *IEEE Systems Journal*. 2020;14(4):5133–5143.

[28] Wirth L, Oettershagen P, Ambuhl J, and Siegwart R. Meteorological path planning using dynamic programming for a solar-powered UAV. In: *2015 IEEE Aerospace Conference*; 2015. p. 1–11.

[29] Plaizier GM, Andersen E, Truong B, He X, Roundy S, and Leang KK. Design, modeling, and analysis of inductive resonant coupling wireless power transfer for micro aerial vehicles (MAVs). In: *2018 IEEE International Conference on Robotics and Automation (ICRA)*. IEEE; 2018. p. 1–6.

[30] Jawad AM, Jawad HM, Nordin R, Gharghan SK, Abdullah NF, and Abu-Alshaeer MJ. Wireless power transfer with magnetic resonator coupling and sleep/active strategy for a drone charging station in smart agriculture. *IEEE Access*. 2019;7:139839–139851.

[31] MacQueen J. Some methods for classification and analysis of multivariate observations. In: *Proceedings of the fifth Berkeley symposium on mathematical statistics and probability*, vol. 1. Oakland, CA, USA; 1967. p. 281–297.

[32] Li L, Xu Y, Zhang Z, Yin J, Chen W, and Han Z. A prediction-based charging policy and interference mitigation approach in the wireless powered Internet of Things. *IEEE Journal on Selected Areas in Communications*. 2018;37(2):439–451.

[33] Xu H, Pan C, Wang K, Chen M, and Nallanathan A. Resource allocation for UAV-assisted IoT networks with energy harvesting and computation offloading. In: *2019 11th International Conference on Wireless Communications and Signal Processing (WCSP)*; 2019. p. 1–7.

[34] Krim H and Viberg M. Two decades of array signal processing research: the parametric approach. *IEEE Signal Processing Magazine*. 1996;13(4):67–94.

[35] Chen JC, Kung Y, and Hudson RE. Source localization and beamforming. *IEEE Signal Processing Magazine*. 2002;19(2):30–39.

[36] Sharma SK, Chatzinotas S, and Ottersten B. Cognitive beamforming for spectral coexistence of hybrid satellite systems. In: *Cooperative and Cognitive Satellite Systems*. Amsterdam: Elsevier; 2015. p. 415–452.

[37] Chu Z, Zhou F, Zhu Z, Hu RQ, and Xiao P. Wireless powered sensor networks for internet of Things: maximum throughput and optimal power allocation. *IEEE Internet of Things Journal*. 2018;5(1):310–321.

[38] Yang G, Ho CK, and Guan YL. Multi-antenna wireless energy transfer for backscatter communication systems. *IEEE Journal on Selected Areas in Communications*. 2015;33(12):2974–2987.

[39] Zeng Y and Zhang R. Optimized training design for wireless energy transfer. *IEEE Transactions on Communications*. 2015;63(2):536–550.

[40] Zeng Y, Clerckx B, and Zhang R. Communications and signals design for wireless power transmission. *IEEE Transactions on Communications*. 2017;65(5):2264–2290.

[41] Jose J, Ashikhmin A, Marzetta TL, and Vishwanath S. Pilot contamination and precoding in multi-cell TDD systems. *IEEE Transactions on Wireless Communications*. 2011;10(8):2640–2651.

[42] Huang H, Yang J, Huang H, Song Y, and Gui G. Deep learning for super-resolution channel estimation and DOA estimation based massive MIMO system. *IEEE Transactions on Vehicular Technology*. 2018;67(9):8549–8560.

[43] Lakshmanan S, Sundaresan K, Rangarajan S, and Sivakumar R. Practical beamforming based on RSSI measurements using off-the-shelf wireless clients. In: *Proceedings of the 9th ACM SIGCOMM conference on Internet measurement*; 2009. p. 410–416.

[44] Sundaresan K, Ramachandran K, and Rangarajan S. Optimal beam scheduling for multicasting in wireless networks. In: *Proceedings of the 15th annual international conference on Mobile computing and networking*; 2009. p. 205–216.

[45] Lee HS and Lee JW. EHLinQ: distributed scheduler for D2D communication with RF energy harvesting. *IEEE Systems Journal*. 2019.

[46] Lee HS and Lee JW. Adaptive wireless power transfer beam scheduling for non-static IoT devices using deep reinforcement learning. *IEEE Access*. 2020;8:206659–206673.

[47] Lee S, Zeng Y, and Zhang R. Retrodirective multi-user wireless power transfer with massive MIMO. *IEEE Wireless Communications Letters*. 2017;7(1):54–57.

[48] Ye HT, Kang X, Joung J, and Liang YC. Optimization for full-duplex rotary-wing UAV-enabled wireless-powered IoT networks. *IEEE Transactions on Wireless Communications*. 2020;19(7):5057–5072.

[49] Arabi S, Elbiaze H, Sabir E, and Sadik M. Tradeoffs for data collection and wireless energy transfer dilemma in IoT environments. In: *2018 IEEE International Conference on Communications (ICC)*. IEEE; 2018. p. 1–6.

[50] Li K, Ni W, Abolhasan M, and Tovar E. Reinforcement learning for scheduling wireless powered sensor communications. *IEEE Transactions on Green Communications and Networking*. 2018;3(2):264–274.

[51] Li K, Ni W, Wei B, and Tovar E. Onboard double Q-learning for airborne data capture in wireless powered IoT networks. *IEEE Networking Letters*. 2020;2(2):71–75.

Chapter 8

A review on resonant beam communication with simultaneous wireless data transmission and energy harvesting techniques

Ehab Ali[1], Abdelhamid Younis,[1] and Raed Mesleh[2]

Abstract

Limited power sources in wireless sensor network and Internet of Things (IoT) devices are major obstacles that prevent extensive growth. Thereafter, wireless energy harvesting methods were developed to provide energy for these devices and prolong their lifetime. In this chapter, an in-depth review of the essential and resonant beam charging (RBC) energy harvesting method is described. Besides, a detailed explanation of the system model, working mechanism, and mathematical analysis of the RBC system is furnished and discussed. In addition, improving the overall system efficiency using an adaptive control system is introduced. The recent RBC-simultaneous wireless information and power transfer (SWIPT) system, which is implemented to support data and power transfer, is studied and named resonant beam communication (RBCom). An analytical model and mathematical analysis of RBCom are provided. Finally, the scheduling algorithm for multiple users charging is presented while considering other latent alternatives.

8.1 Motivations and applications

Wireless communication has seen enormous development in recent decades. The data rate of smartphones and Internet of Things (IOT) devices has increased exponentially from one generation to another to cope up with the increased demand and applications [1]. Smart devices in new generations become more sophisticated, and more power is consumed by active and standby systems. Accordingly, power management for portable wireless devices imposes new challenges in the areas of hardware interface, energy management, and battery lifetime. Improvements in these

[1]Electrical and Electronics Engineering Department, University of Benghazi, Benghazi, Libya
[2]Electrical and Communication Engineering Department, School of Electrical Engineering and Information Technology, German Jordanian University, Amman, Jordan

challenges are still moderate and additional enhancement is mandatory in these areas [2,3]. Among the very promising techniques in that direction is the wireless power transfer (WPT) method, which enables remote charging for mobile devices [4].

The latest and most promising WPT technology is resonant beam charging (RBC), which offers a wide range of advantages and overcomes several challenges that are encountered in other alternative methods. In addition, the RBC system is capable of simultaneously transmitting data through free space optical communication (FSOC) along with providing energy harvesting. Such benefit is very appealing beyond fifth-generation (5G) wireless communication systems. As such, this chapter provides a detailed discussion and analysis of recent technologies in the areas of WPT and RBC systems.

In terms of possible applications, it should be noted that RBC wireless charging method is a rather new approach, and very limited literature is available on its use in some potential applications. Yet, theoretical studies anticipate that the RBC structure can be used in a wide range of scenarios. For instance, RBC can be used to charge environmental sensors such as water, temperature, and humidity sensors. In addition, home and personal smart devices such as smartwatches, headphones, smartphones, and healthcare wearable devices are other possible consumers. Moreover, some recent research employed RBC to charge unmanned aerial vehicles (UAVs), which are deployed to collect data reliably from a group of buoy sensors far off the coast in a three-dimensional (3D) coordinate system [5].

8.2 Introduction

IoT, which links billions of smart devices together with nearly wire-free connection, provides various features and leads for a smart and easy lifestyle in foreseeable future [6–8]. Mobile devices, such as sensors, actuators, smartphones, portable computers (laptops), and other smart gadgets, can be powered using batteries or a power outlet [9–11]. On one hand, providing charging plugs limits mobility and requires wired infrastructure. On the other hand, remote charging requirement and battery life of electronic devices is a major obstruction, which prevents the implementation of wireless charging networks [2,3]. However, the demand for unconventional approaches to conquer these challenges led to the emergence of a new exciting research field, called WPT [1,4] or wireless energy transfer [12].

The inspiration of WPT was first presented at the end of the 1890s by ac inventor Nikola Tesla [13]. Experimentally, Tesla managed to light up 200 bulbs and run an electric motor over 25 mi distance. In 1901, he invented his Wardencliffe Tower to remotely transfer electrical power through air [14]. After Tesla's achievement, no remarkable progress was conducted due to the shortage of the necessary technology for WPT development.

Nowadays, after more than a century, major advances in electronics, magnetrons, and microwaves have risen, which motivates many researchers including Massachusetts Institute of Technology (MIT) scientists to indulge in WPT studies [15,16]. These research studies conduct various WPT technologies such as inductive coupling, magnetic resonance, RF, and laser [17]. Furthermore, many leading

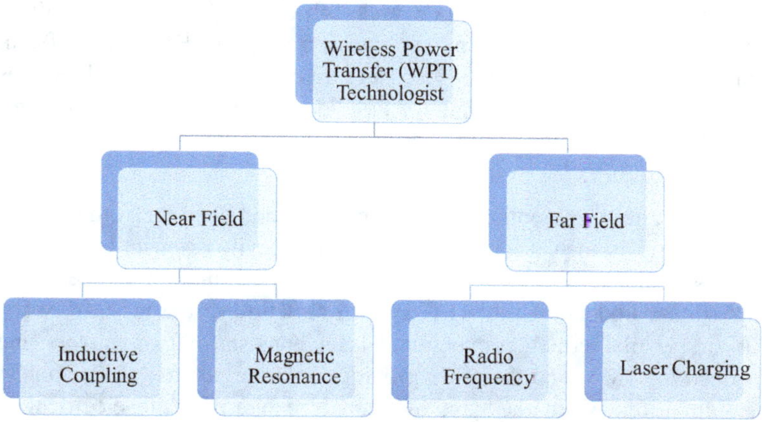

Figure 8.1 Wireless power transfer technologist

enterprises, such as Sony, Fujitsu, Wamper, Energizer, and Intel, exploit the advantage of WPT features and embedded them in their products [18].

WPT technologies are generally classified according to their effective charging range into two types, as illustrated in Figure 8.1. In particular, the near-field charging method with an effective distance limited to few meters and a far-field charging method with relatively a longer charging range are identified [14]. Near-field techniques are not suitable for mobile device charging due to their bulky coil size requirement and short working range and are only used in applications such toothbrush or electrical car charging [19,20]. In contrast, RF waves have the capabilities of charging remote devices several meters away [21]. Nevertheless, it faces challenges of safety issue, los requirement, and low charging efficiency [22].

Accommodating the aforementioned challenges inspires the search for other modern technologies. A promising solution is foreseen in the laser-based charging method named DLC or RBC [23,24]. Recent studies reported that RBC can charge devices that are 10 m away with 2 W of power [25].

Due to the laser nature and the distinguishing structure of the RBC system, RBC is utilized with some novel features that set it apart from other WPT technologies such as self-aligning, safe charging, multi-user charging, high power, compact size, and free electromagnetic interference. In addition, it supports swipt, which guarantees its compatibility for mobile applications [4].

8.3 Literature review

RBC is a new research area offering a wide range of research opportunities with many open issues for investigation and improvement. For instance, RBC channel modeling and system efficiency analysis plays an important role in RBC development [4,17,26]. Charging efficiency optimization and power tracking using the adaptive system is an open research problem that has attracted substantial research lately [5,24,27–29]. Scheduling algorithm and access techniques designed for RBC multi-user scenarios is another open area for investigations [12,23,30–32]. In addition, simultaneous data and power transfer adopting RBC swipt feature are investigated in a recent article [33–35]. Hereunder, some of RBC's latest research topics that provide the fundamentals for RBC development are discussed in detail.

8.3.1 RBC principles

RBC is a laser system with optical components detached from each other at relatively a larger distance. Distance is considered as the intra-cavity resonant beam that allows wireless power transmission and separates the RBC transmitter and receiver [17,27].

To facilitate understanding RBC system structure, a traditional laser system model is first explained. A conventional laser consists of two mirrors one with 100% reflectivity (M1) and the other provides 95% reflectivity (M2) along with a gain medium [36]. When the gain medium is excited by the electrical source, photons released from the gain medium begin to travel back and forth between the two reflector mirrors and repeatedly amplified by the gain medium [4]. Furthermore, only photons that travel perfectly perpendicular between M1 and M2 will undergo this process and create the laser beam, as illustrated in Figure 8.2 [27].

In the RBC system, a retro-reflector is used instead of mirrors to reflect light back to its direction of entry regardless of its angle of incidence. Retroreflectors are commonly used in safety clothing, road signs, bicycle reflectors, and others [4]. An RBC system structure is depicted in Figure 8.3, where both 100% theoretical reflectivity retro-reflector mirror ($\Re_1$) and a gain medium are placed on the RBC transmitter. On the other side of the cavity, the receiver contains a 95% reflectivity retro-reflector mirror ($\Re_2$) and a PV panel [4,17]. The RBC resonating cavity consists of the retro-reflector mirrors, $\Re_1$ and $\Re_2$, and the gain medium [28]. The unique RBC structure provides both safety and self-alignment, where the photons that travel within the intra-cavity in the LOS direction are the ones that create the laser beam. Yet, any obstacle between the receiver and the transmitter will break the intra-cavity resonating laser and interrupt the whole process [27].

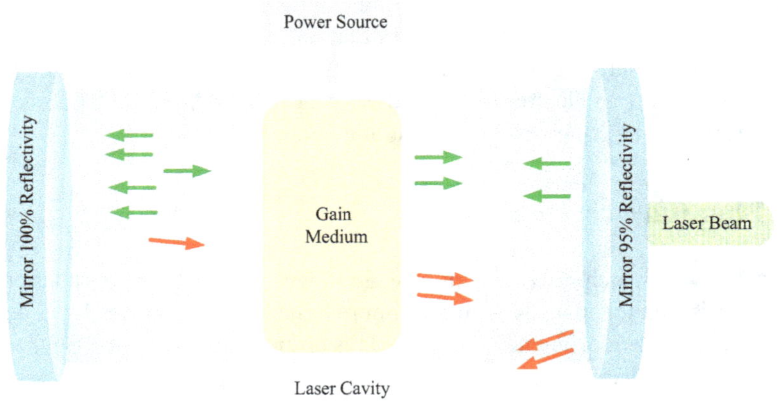

Figure 8.2 Laser structure

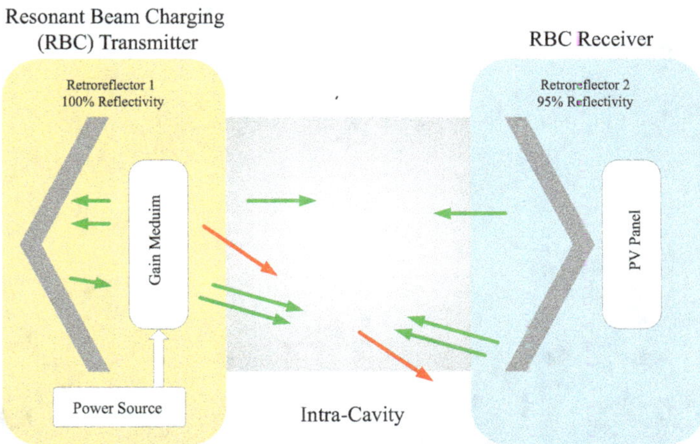

Figure 8.3 Resonant beam charging system structure

8.3.1.1 RBC mechanism

Parallel to a normal laser, the population inversion reached by the gain medium in the RBC after being provoked electrically by the power sources will result in storing energy in the gain medium [26]. The stored energy represents the gain medium photons by which the intra-cavity resonating laser is formed due to photons reflection between $\Re_1$ and $\Re_2$ [26]. Furthermore, a limited amount of the formed intra-cavity beam is permitted to pass through toward the PV panel and form the external-cavity laser due to the imperfect reflectivity of $\Re_2$. This external beam is then converted into electrical power using the PV panel at the receiver [23,28].

As described later in the chapter, the system efficiency of RBC is affected by many factors including laser wavelength, electrical to laser conversion efficiency, transmission attenuation, PV panel conversion efficiency, etc. [17]. In order to improve total system efficiency, a power control system is built on top of the traditional RBC system to optimize the charging efficiency and develop an ARBC system [24,28].

8.3.2 ARBC principles

The ARBC system is equipped with power control at the transmitter side and power monitoring at the receiver side. Both represent the feedback control system, which offers many advantages in terms of improving the overall charging efficiency, increasing battery lifetime, avoiding unnecessary heat, and securing charging [27,28].

For standard RBC schemes, massive capacity reduction takes place in an undercharging situation, especially for the Li-ion battery, which is commonly used in IoT devices [37,38]. However, battery overcharging has its drawbacks in terms of the possibility of battery ruining or even causing danger [39]. Moreover, additional energy converted by the receiver PV panel will cause the thermal effect and lead to a significant drop in system efficiency [27].

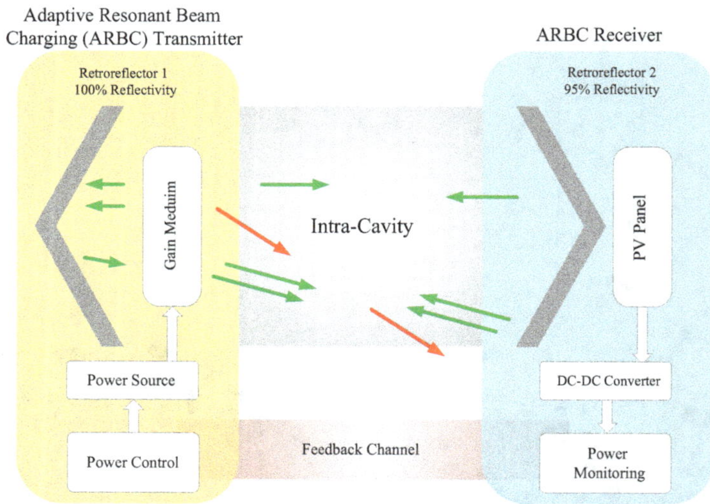

Figure 8.4 Adaptive resonant beam charging system structure

The adaptive control in ARBC aims to get rid of RBC charging issues and optimize the charging efficiency by tracking the battery SOC. According to battery SOC, the preferred charging current and charging voltage are fed to the battery [24,28]. The ARBC system structure, which consists of a power supply and gain medium at the transmitter; PV panel, DCDC converter, and battery at the receiver is illustrated in Figure 8.4. The feedback channel, which is composed of power monitor and power controller, is across both ends [24,28].

8.3.2.1 ARBC charging mechanism

Identifying the battery charging profile is an essential step in the optimization process. Based on different manufacturing components, charge profiles may vary from one battery to another [40]. The charge profile of a single cell Li-ion battery is depicted in Figure 8.5, which is widely used in IoT devices. In addition, the Li-ion battery charge profile consists of four stages as follows [39,41]:

Stage 1: tc – at the start of the charging process when the voltage applied to the battery is under 3 V, charging current increases gradually to 200 mA and the voltage will rise to 3 V.

Stage 2: cc charge – when the battery voltage exceeds 3 V, the tc stage ends and cc charge begins with a constant charging current between 200 mA and 1,000 mA feeding the battery. The voltage rises toward 4.2 V.

Stage 3: cv charge – at 4.2 V, the stage switch from cc to cv and charging current will drop to 20 mA. However, in order to maximize the capacity, the voltage variation should be kept under 1%.

Stage 4: ct – this stage takes place under one of the following approaches. *I*: When charging current drops below 20 mA, the charging process terminates. This

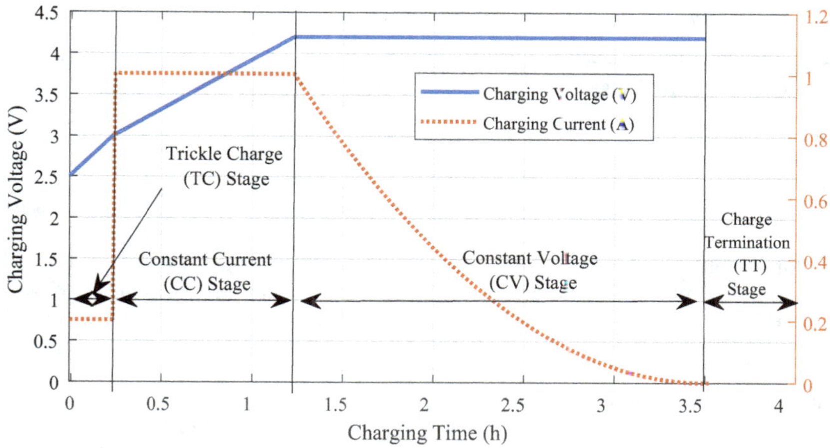

Figure 8.5 Li-ion battery charging profile

method is known as the minimum current charge. *II*: The time cutoff method, where the charging time period is set up in order to terminate the charging process. For example, a 1-hour charging period approach terminates the charging procedure after 1 hour since the cv stage is invoked. However, a mix of the two methods is an applicable solution.

At the start of the charging process, the pv panel output current/voltage could be different from the optimal current/voltage required for charging. This happens due to the changeable received laser power and the pv panel characteristics [24,27]. Besides, the power monitor aims to debrief the SOC of the battery and calculate the required laser charging power and subsequently inform the power control through the feedback channel [28]. The feedback channel can be set using any wireless connection approach such as infrared laser communication, Wi-Fi, Bluetooth, etc. [24].

In the ARBC receiver, a DCDC converter is used to adjust the output current I_o and output voltage V_o into their optimal values. The optimization mechanism can take different tracking approaches based on battery SOC [24,27,28], charging profile and distance [5], or transmitter coverage [29]. However, an algorithm should be designed in order to facilitate the charging process.

To summarize the differences between the normal laser system and the RBC system, Table 8.1 illustrates a comparison in terms of system structure, cavity size, used reflectors, system function, and feedback.

8.4 System model

8.4.1 RBC/ARBC analytical modeling and transmission efficiency

The ARBC/RBC system model is presented in this section along with analytical derivation of the total system transmission efficiency [26,30].

Table 8.1 Comparison between normal laser and RBC system

Aspect of comparison	Normal laser	RBC system
System structure	Laser components are combined with each other	Transmitter and receiver are separated from each other
Cavity size	The intra-cavity is related to laser size	Intra-cavity is the distance between the transmitter and the receiver
Reflectors	Mirror is used to reflect the photons within the cavity	Retroreflectors are used to reflect light back to the original source
Objective	Generate the laser beam	To charge electrical devices by converting beam into electrical power through the photovoltaic
Feedback control	No feedback control is used	Feedback control could be used to improve system efficiency

8.4.1.1 Input electrical power to stored energy

The input electrical power P_E is generated based on the stimulating current I_{in} and voltage V_{in} of the power source as [17,24]

$$P_E = I_{in} V_{in}. \tag{8.1}$$

At the RBC transmitter, the electrical power source P_E drives the laser pump power P_α. The pump power created when the input applied current I_{in} exceeds certain threshold I_{th} as [42]

$$P_\alpha = \gamma \frac{hf_\ell}{q} (I_{in} - I_{th}), \tag{8.2}$$

where γ is the modified coefficient, h is the Planck constant, f_ℓ is the laser frequency, and q is the electronic charge constant. The generated pump power is transferred within the RBC cavity and excites the gain medium to realize the population inversion [26]. The pump power is then stored in the gain medium. When continuous input of electrical power is applied to the gain medium, the direct relationship between input electric power and stored power relation can be written in terms of conversion efficiency as [43]

$$P_S = \eta_\alpha^S P_\alpha, \tag{8.3}$$

where η_α^S is the electrical power to stored energy conversion efficiency. Note that part of the electrical power is dissipated as heat and not converted to P_S [26].

8.4.1.2 Stored energy to beam power transmission

Excited by spontaneous radiation [30,43], the stored energy is converted to beam laser power which travels through the cavity as an intra-cavity beam. Furthermore and due to the imperfect reflectivity of the receiver $\Re_2$ retro-reflector mirror, a small amount of beam power is allowed to break through and form the external cavity beam, which is collected by the PV panel. Factors such as transmission distance and efficiency may have an impact on the stored energy to beam power conversion process [17].

In addition, two types of losses determine the transmitted beam strength and should be considered. First, the loss occurring due to system components and beam transverse mode, which can be molded as a loss from the partially transparent mirror $\Re_2$ [26]. Second, the diffraction loss, which mainly influences the transmission efficiency. Diffraction loss is caused due to the geometry of $\Re_1$ and $\Re_2$ and the transmission distance. In the resonant cavity, the single transmission diffraction loss of the resonant beam is a function of the transmission distance presented by [26,30] and formulated as

$$\Psi(d) = Ne^{-2\pi(\mathfrak{r}^2/\lambda(z+d))}, \tag{8.4}$$

where z represents the distance between the gain medium and $\Re_1$, $\mathfrak{r}$ is the cavity aperture radiance, N is an integral related to the cavity mode, λ is the beam wavelength, and d represents the distance between the medium gain and $\Re_2$. From (4), the impact of diffraction loss on the transmitted beam is calculated as [43]

$$F(d) = \frac{2(1 - \Gamma_2)\mu}{(1 + \Gamma_2)\Psi(d) - (1 + \Gamma_2)\ln\Gamma_2}, \tag{8.5}$$

where Γ_2 is the effective reflectivity of the partially transparent mirror $\Re_2$, μ is the overlap efficiency, and $\Psi(d)$ is the total diffraction loss. Finally, the external cavity beam power can be determined by [34]

$$P_\beta = P_S F(d) + \beta, \tag{8.6}$$

where β is a constant determined by the system's internal parameters [26].

8.4.1.3 Beam power to electrical power conversion

In order to charge the device connected to the RBC receiver and finish the charging process, the beam power should be converted into electrical power using the PV panel [44,45]. In RBC, the PV panel is modeled as a single-diode equivalent circuit, as illustrated in Figure 8.6 [46,47]. The PV panel single-diode model consists of a photo-current source, I_φ, connected in parallel to a diode whose forward current is

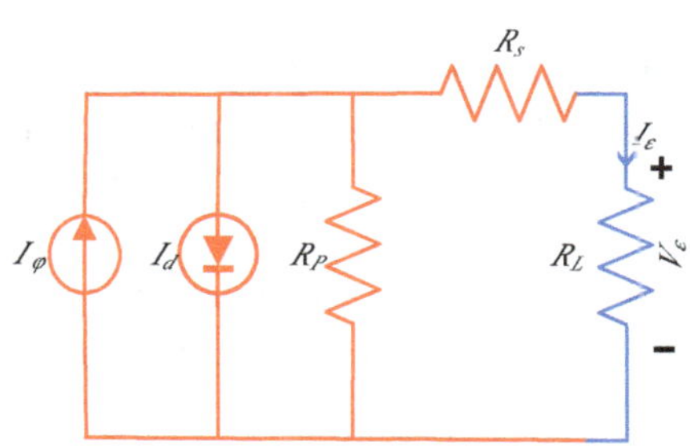

Figure 8.6 Photovoltaic single-diode model

I_d. In addition and for practical reasons, two resistors R_p and R_s are added in parallel and series, respectively.

The PV panel current–voltage $(I-V)$ characteristics are described by [46,47]

$$I_\varepsilon = I_\varphi - I_d - \frac{V_d}{R_p}, \tag{8.7}$$

$$I_d = I_\delta\left(e^{((V_d/n_s\Lambda V_J))} - 1\right), \tag{8.8}$$

$$V_d = V_\varepsilon + I_\varepsilon R_s, \tag{8.9}$$

where I_δ is the reverse saturation current, n_s denotes the number of series connected cells in the PV panel, and Λ is the diode ideality factor. $V_J = \kappa T/q$ is the junction thermal voltage of the diode, with κ being the Boltzmann's constant, and T is the PV cell temperature in Kelvin [34]. The PV panel output can be obtained by multiplying the PV output voltage V_ε with the current I_ε as [17]

$$P_\varepsilon = V_\varepsilon I_\varepsilon. \tag{8.10}$$

In order to find the direct relationship between P_ε and P_β, a number of curve-fitting approximation techniques, e.g., linear-fitting method and square-root fitting method, were conducted in [24,28] based on the RBC design parameter to yield

$$P_\varepsilon = a_\varepsilon P_\beta + b_\varepsilon, \tag{8.11}$$

where a_ε and b_ε are curve-fitting coefficients. Furthermore, the direct relation based on beam power to electrical power conversion η_ε can be estimated by [28]

$$P_\varepsilon = \eta_\varepsilon P_\beta, \tag{8.12}$$

Yet, using fixed beam power, the electrical output power may vary based on the applied load [26]. In ARBC case, an MPPT technique is used in order to maximize the PV power [48]. However, the efficiency of DCDC should be taken into consideration, where the output charging power is given by [28]

$$P_{dc} = \eta_{dc} P_\varepsilon. \tag{8.13}$$

Although some studies consider an ideal DCDC converter in their calculation [24,27,49,50], others count factors such as battery charging efficiency for more accurate results [28].

8.4.1.4 Total transmission power and efficiency

The received charging power for the ARBC system can be calculated as

$$P_{dc} = \eta_{dc}\eta_\varepsilon\left[\eta_\alpha^S \gamma \frac{hf_\ell}{q}(I_{\text{in}} - I_{\text{th}})F(d) + \beta\right]. \tag{8.14}$$

Also, the total transmission efficiency is obtained as

$$\eta_o = \frac{P_{dc}}{P_a}. \tag{8.15}$$

Table 8.2 System parameters

Parameter	Symbol	Value	Unit
Stimulating current	I_{in}	9.2	A
Generated laser power	P_a	20	W
Current threshold	I_{th}	0.5	A
Modified coefficient	γ	1.5	–
Planck constant	h	$6.62606957 \times 10^{-34}$	*J.s*
Electronic charge constant	q	1.6×10^{-19}	*C*
Laser wavelength	λ	810	nm
Laser frequency	f_ℓ	370	*THz*
Electrical power to stored energy conversion efficiency	η_a^S	0.432	–
Distance	d	5	m
Cavity aperture radiance	$\mathfrak{r}$	1.5	mm
Internal parameters constant	β	−5.64	–
Overlap efficiency	μ	0.8	–
Cavity mode integral	N	1	–
Effective reflectivity of the partially transparent mirror $\Re_2$	Γ_2	0.88	–
Beam power to electrical power conversion	η_ε	0.5	–

Note, in the case of the RBC system, an ideal DCDC converter is assumed with $P_{dc} = 1$.

To illustrate the RBC received charging power and the total system transmission efficiency, Monte-Carlo simulation results were obtained according to the parameters listed in Table. 8.2 and results are depicted in Figure 8.7*.

8.4.2 Resonant beam communication

The resonant beam communication (RBCom) system is a recent RBC swipt system, which can support data transmission along with charging capability [33,34]. By adopting RBC features and FSOC, several advantages are incurred including large bandwidth, free license of the optical spectrum, electromagnetic free interference, short communication distance, and low power consumption [35,51,52]. It is anticipated that RBC swipt can fulfill the criteria and specifications of future 6G communication systems [53].

An RBCom system structure is illustrated in Figure 8.8. A modulator and current regulator were linked to the transmitter in addition to a typical RBC system. Also and at the receiver side, the power harvesting branch and the information decoding branch (CEH) were connected [54]. Supported by the conducted modification current regulator, the system combines the information signal that travels with the charging power using the RBC laser beam [33,34]. At the receiver, the information signal and charging power are detached and, respectively, guided into the communication decoder branch and power harvesting branch [33,34].

*Distance between the gain medium and $\Re_1$ is neglected.

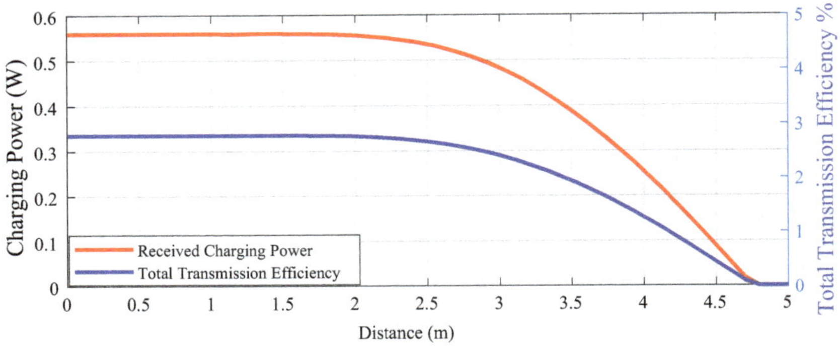

Figure 8.7 Received charging power and transmission efficiency

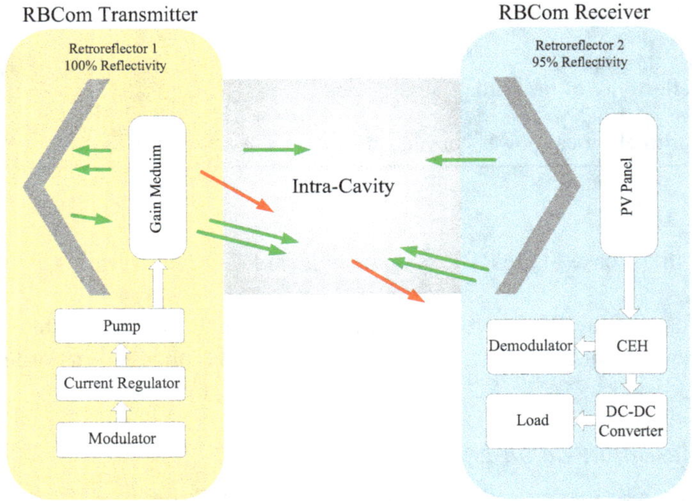

Figure 8.8 Resonant beam communications system structure

8.4.2.1 RBCom analytical model

RBCom modulator: In RBCom, laser pump input I_{in} consists of a charging dc I_{chg} and a communication information ac current i_{com}. Charging current I_{chg} ensures that $I_{in} > I_{th}$ and provides charging power, whereas i_{com} contains the information data

$$I_{in} = I_{chg} + i_{com} \tag{8.16}$$

$$i_{com} = I_s \cos(\omega_s t + \theta_s) \tag{8.17}$$

where I_s is the signal amplitude, ω_s is the frequency in rad · Hz, and θ_s denotes the signal phase.

RBCom receiver circuit: The conducted modification on the PV panel model is shown in Figure 8.9. To support ac signal analysis, the cell diode is changed into

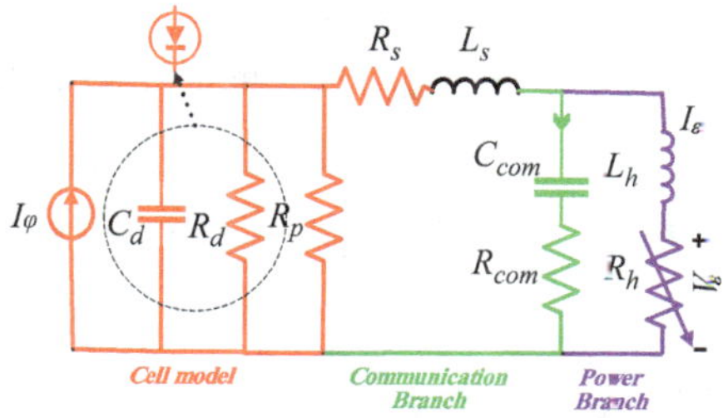

Figure 8.9 Resonant beam charging simultaneous wireless information and power transfer (RBC-swipt) receiver

a shunt dynamic resistor R_d and a shunt capacitor C_d and the connection wires are modeled as a serial inductor L_s. The shunt diode resistance R_d is a function of voltage V_d determined by [55]

$$R_d(V_d) = \frac{dI_d}{dV_d} = \frac{nV_J}{I_\delta e^{\frac{V_d}{nV_J}}}, \tag{8.18}$$

when $I_d \gg I_\delta$ the diode resistance is approximated by $R_a = n(V_J/I_d)$. Furthermore, the shunt capacitor is a sum of a transition capacitance C_{tran} and a diffusion capacitance C_{diff}, $C = C_{\text{tran}} + C_{\text{diff}}$ that can be obtained from [56,57].

The two branches attached to the PV panel model are described as [33,34]

1. Power harvesting branch, which consists of a choking inductor L_h to remove the ac component and a changeable load resistant R_h.
2. Communication decoder branch responsible for the information absorption, which contains a dc preventable capacity C_{com} and an information capture resistance R_{com}.

In the RBCom receiver, the load resistance is adjusted to guarantee that the PV is working in the linear region. In this region, the PV panel works in a manner similar to the constant-current source, where for a given PV input laser power P_ε, the resulting photo-current I_φ is almost constant with the increase of the output voltage, i.e., $I_{pc} \approx I_\varphi$. Also, the received photo-current can be formulated as [58]

$$I_{pc} = \rho P_\varepsilon \tag{8.19}$$

$$= \rho\eta_\varepsilon \left[\eta_\alpha^S \gamma \frac{hf_\ell}{q} \left(I_{\text{chg}} + i_{\text{com}} - I_{\text{th}} \right) F(d) + \beta \right], \tag{8.20}$$

where ρ is the optical-to-electrical conversion in A/W and can be measured under the standard test condition (stc, 25 °C temperature and 1,000 W/m^2 irradiance). The linear region enlargement can be achieved by decreasing the series resistance R_s. In addition, decreasing the parallel resistance R_p will adversely affect the constant-current property in the linear region [46].

Receiver circuit ac response: The transfer function of the received ac signal i_{pc} at the communication resistance R_{com} is formulated as [54]

$$\left|h_{\text{pv}}(\omega)\right|^2 = \left|\frac{v_{\text{com}}(\omega)}{i_{pc(\omega)}}\right|^2$$

$$= \left|\frac{\frac{R_{LC}}{R_s+j\omega L_s+R_{LC}}\frac{R_{\text{com}}}{\frac{1}{j\omega C_{\text{com}}}+R_{\text{com}}}}{\frac{1}{R_d}+\frac{1}{\frac{1}{j\omega C_d}}+\frac{1}{R_{sh}}+\frac{1}{R_s+j\omega L_s+R_{LC}}}\right|^2, \tag{8.21}$$

where ω is the angular frequency, $j=\sqrt{-1}$, and R_{LC} is the equivalent parallel impedance of the two branches, which can be calculated by

$$R_{LC} = \frac{1}{\frac{1}{j\omega L_h+R_h}+\frac{1}{\frac{1}{j\omega C_{\text{com}}}+R_{\text{com}}}}. \tag{8.22}$$

Based on the above analysis, the average power of the signal at R_{com} is estimated as

$$S = \frac{\left|h_{\text{pv}}(\omega)\right|^2\frac{i_{pv}^2}{2}}{R_{\text{com}}} \tag{8.23}$$

To demonstrate the SNR of the RBCom system, shot noise and thermal noise have to be considered. Shot noise is related to the PV panel converted electrical power, including the converted laser charging power and the background radiance such as sunlight and lamplight. The shot noise can be modeled as the shunt current source parallel to the input of the PV panel as

$$I_{shot} = \rho(P_{\text{BG}}+P_\varepsilon), \tag{8.24}$$

where P_{BKG} is the background radiance power, which can be approximated as [39]

$$P_{\text{BG}} = \eta_{op}\Psi_{\text{BG}}B_{\text{op}}A_{pv}\Theta_{\text{pv}}T, \tag{8.25}$$

where η_{op} is the optical efficiency, Ψ_{BG} is the strength of background radiance, B_{op} is the optical bandwidth of the $\Re_2$ attached filter, A_{pv} is the receiving area of the PV panel, Θ_{pv} is the solid angle of the PV panel field of view, and T is the transmittance of $\Re_2$. Moreover, shot noise is a Gaussian distributed noise with a one-sided PSD in A^2/Hz given by

$$\begin{aligned} N_{\text{SH}}^{\text{pv}} &= 2qI_{shot} \\ &= 2q\rho(P_{\text{BG}}+P_\varepsilon). \end{aligned} \tag{8.26}$$

Therefore, the shot noise effect on R_{com} is described by

$$N_{\text{sh}}(\omega) = \left|h_{\text{pv}}(\omega)\right|^2 2q\rho(P_{\text{BG}}+P_\varepsilon). \tag{8.27}$$

Thermal noise is linked to the heat generated by the receiver circuit resistances. An appropriate representation of the resistances' noise contribution is a voltage source serially connected to each resistance. The PSD in A^2/Hz for a resistance with R ohms can be written as [59]

$$N_{\mathrm{TH}}^{R} = 4kTR, \tag{8.28}$$

where k is the Boltzmann constant, T is the temperature in Kelvin, and R is the resistor in ohms. Applying the superposition theorem, the total thermal noise effect at the receiver load R_{com} can be calculated as the sum of each resistance thermal noise contribution as [59]

$$\begin{aligned} N_{\mathrm{TH}}(\omega) &= |h_{R_{com}}(\omega)|^2 N_{\mathrm{TH}}^{R_{com}} + |h_{R_{\mathrm{p}}}(\omega)|^2 N_{\mathrm{TH}}^{R_{\mathrm{p}}} \\ &+ |h_{R_h}(\omega)|^2 N_{\mathrm{TH}}^{R_h} + |h_{R_d}(\omega)|^2 N_{\mathrm{TH}}^{R_d} \\ &+ |h_{R_s}(\omega)|^2 N_{\mathrm{TH}}^{R_s}, \end{aligned} \tag{8.29}$$

where $|h_{R_{com}}(\omega)|^2$, $|h_{R_{\mathrm{p}}}(\omega)|^2$, $|h_{R_h}(\omega)|^2$, $|h_{R_d}(\omega)|^2$, and $|h_{R_s}(\omega)|^2$ represent the transfer function of the related network elements described in [54] and $N_{\mathrm{TH}}^{R_{com}}$, $N_{\mathrm{TH}}^{R_{\mathrm{p}}}$, $N_{\mathrm{TH}}^{R_h}$, $N_{\mathrm{TH}}^{R_d}$, and $N_{\mathrm{TH}}^{R_s}$ are the thermal noise PSD of each of the circuit's resistance. Based on (23), (27), and (29), the SNR of the RBCom system can be written as

$$SNR = \frac{S}{B(N_{\mathrm{TH}}(\omega) + N_{\mathrm{SH}}^{\mathrm{pv}})} \tag{8.30}$$

where B is the bandwidth of the designed receiver circuit. Furthermore, the RBCom system may suffer from Echo-interference occurs due to the RBCom system structure and working mechanism [35,60]. An optical filtering and frequency shifting method was presented in [35,60] to eliminate the Echo-interference. The exemplary design proposed in the mentioned research introduces an interference free RBCom system.

8.4.3 RBC/ARBC multiple user charging

The RBC/ARBC multi-user system utilizes the signal transmitter to charge multiple receivers. LOS connection is established between the transmitter and each receiver separately. Similar to the ARBC single receiver scenario discussed previously in Section 8.4.2, the power monitor tracks the battery charging status and informs the transmitter power control [23]. At the transmitter and after the power controller is updated with battery SOC, it drives the electrical source to feed the gain medium with the required power to create the preferred laser beam power for the individual receiver [23]. To support multi-user charging, various scheduling algorithms such as first access first charge (FAFC), high priority charge (HPC), time-division multiple access, channel-dependent charge, and coverage model are proposed and designed [12,23,30–32]. Hereinafter, FAFC and HPC RBC scheduling algorithms are discussed, which provide charging organization and improvement.

8.4.3.1 FAFC scheduling algorithm

The FAFC scheduling algorithm is a fair charging procedure that aims at maintaining the maximum time period for all connected receivers [23]. The charging concept of FAFC is described as follows:

1. Charging time slot is made through breaking down the charging time into equivalent sections.
2. First connected receiver charges first and consecutive receivers follow in order.
3. Preferred charging power is sent to multiple connected receivers within the charging time slot while other receivers wait their turn.
4. Discharging power varies from one receiver to another based on its working status.
5. Each cycle the time slot ends, the charged receiver goes to the bottom of the queue and update its SOC.

The FAFC scheduling procedure for N_r receivers and T_s time slots is depicted in Figure 8.10. The charging process is divided into time slots $T_1, T_2, \ldots, T_n$. In each time slot, a number of receivers are selected to be charged. Receivers illustrated by colored rectangles and stacked on the vertical axes represent the order of the queued receivers. The left and the right side of the vertical axes demonstrate the required charging power P_r and the receiver discharging power P_{dis}, respectively. At the end of the time slot, the charged receiver is sent to the end of the queue and the following receiver will be selected to be charged. A detailed working algorithm for FAFC is described hereunder.

1. Connectivity judgment: The charging process takes place if the connection between the transmitter and the receiver is established; otherwise, the scheduling process is terminated.

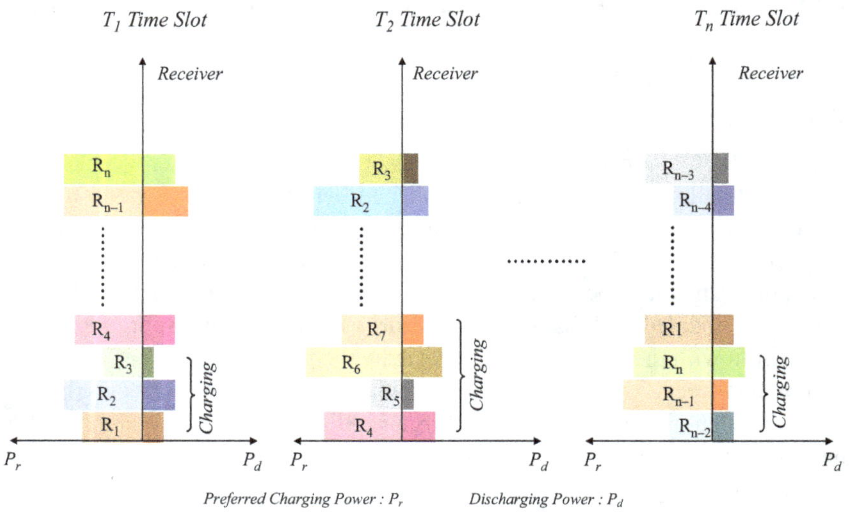

Figure 8.10 First access first charge schematic diagram

2. Time segmentation: Divide the overall charging time into equal time slots.
3. Lining up: Receivers are lined in a queue based on their access time to the transmitter.
4. Feedback: Through a feedback channel, the transmitter collects the receiver's information including device type, battery type, and required charging power.
5. Receiver picking: A certain number of receivers are selected to be charged based on the transmitter's available charging power and receivers' required power.
6. Charging/discharging: Each receiver is discharged based on its working status and charged with its preferred charging power.
7. State update: After the charging time slot ends, a new SOC is sent to the transmitter.
8. Organization: Each charged receiver is sent to the end of the line and the procedure continues with the next receiver.
9. Finish procedure: The charging procedure is terminated if one of the following conditions are met: (1) all receivers' batteries are fulfilled, (2) one or more of the receivers is turned off, and (3) the defined time charging threshold is reached.

During each charging time slot, the power is allocated for each receiver using a simple approach. At the start of the charging time slot, the available charging power is P_A, the first receiver required power is P^1_{req}, and the transmitter provides power P^1_t. The transmitter provides power to the receiver according to its requirement $P^1_t = P^1_{req}$. The new available transmitter power is $P^{new}_A = P_A - P^1_t$. Repeatedly, the same steps are repeated on the following receiver until the requirement of the current receiver is higher than the available power $P^n_{req} > P^{new}_A$. Finally, the remaining power to the next receiver $P^n_{req} = P^n_t = P^{new}_A$ is updated and proceed to the next charging slot.

8.4.3.2 HPC scheduling algorithm

The HPC scheduling algorithm aims at optimizing the power allocation, the control charging order, and maintaining the quality of charging service (QoCS) for each receiver [32]. The charging priority is used to arrange the charging order based on the required charging power and the battery SOC.

First, the evaluation rules of the cps, QoCS, and the Earning are discussed. The cps factor is related to the power level and the duration of charging. The higher the number, the higher the cost. QoCS relies on battery SOC and charging characteristics such as time, rate, and duration of charging. Finally, the earning value is based on charging duration and pricing. The details of the HPC scheduling algorithm is discussed in what follows:

1. Time segmentation: Divide charging time into an equal time slot, each with duration determined based on charging status.
2. Feedback: Transmitter is informed through the feedback channel with battery SOC and charging info.

3. Lining up: According to priority determined using former feedback information in the previous step, receivers are lined up in queue from the highest to lowest priority.
4. Receiver picking: Charging process begins with the highest priority receiver.
5. Charging/discharging: Selected receiver is charged based on the predetermined charging strategy [32] and discharged corresponding to its consumption.
6. State update: After charging and discharging are completed, SOC information is sent to the transmitter.
7. Finish procedure: Charging procedure is terminated if one of the following conditions are met (1) all receiver batteries are filled, (2) all receivers are turned off, or (3) specified time charging threshold is reached.

Power allocation starts by comparing the available charging power P_A with the required charging power of the highest priority receiver P_r^{p1}. If $P_A > P_r^{p1}$, the first receiver will be charged with its required power and the charging process will resume after calculating the newly available charging power P_A^{new} using $P_A^{new} = P_A - P_r^{p1}$. If $P_A < P_r^{p1}$, the following two cases are considered: (1) $P_A^{new} = 0$ and process is finished and (2) $P_A^{new} \neq 0$, check the next receiver and if the receiver battery is full, proceed to the next receiver.

To summarize the previously mentioned scheduling algorithms, Table 8.3 illustrates a comparison between FAFC and HPC in terms of their objective, charging order, charging receiver picking, charging process, and charging termination.

Table 8.3 Comparison between FAFC and HPC schedule algorithm

Aspect of comparison	First access first charge	High priority charge
Objective	Maintain maximum time period for all connected receivers	Optimizing the power allocation, the control charging order and maintaining the quality of charging service for each receiver
Charging order	Receiver are charged based on their connection time	Receiver is charged based on their charging priority
Charging receiver picking	A group of receivers are selected to be charged based on the transmitter's available charging power and receivers required power	Charging process begins with the highest priority receiver
Charging process	Each charged receiver within a time slot is sent to the end of the line and the procedure continues with the next receiver	The highest priority receiver will be repeatedly charged until its battery is fulfilled
Charging termination	All batteries are fulfilled, charging time threshold is reached and one or more of the receivers is turned off	All batteries are fulfilled, charging time threshold is reached and all receivers are turned off

8.5 Open research issues and opportunities

As discussed earlier, RBC is a new WPT trend with numerous research opportunities. However, the field is still far from being fully explored and several open research questions, listed below, are yet to be answered.

- Network architecture design and protocols development for power charging and data communications are essential to build a complete RBC WPT network [4].
- Existing PV panel efficiency is around 50% [17], and several improvements are needed to enhance this number.
- Analysis for RBCom charging and data transfer performance including system efficiency, capacity, bit error rate, and energyefficiency need several enhancements. In addition, a fresh perspective to RBC swipt design should be considered.
- Various system parameters and different operating conditions should be considered in future research. For example, using several battery types, different laser wavelengths and working temperatures. This would allow compatibility with a diverse number of wireless network devices.
- Develop an RBC MIMO system to support multi-transmitter and multi-receiver charging and data transfer.

8.6 Conclusion

In this chapter, recent advances, trends, and open research issues for a laser-based far-field WPT named RBC are reviewed. This new WPT method outweigh other WPT technologies in term of self-alignment, safety, high power, and charging distance along with other advantages. In addition, the mechanism and mathematical formulation of the RBC/ARBC system model and efficiency are described. In addition, the RBCom system and the RBC scheduling algorithm are discussed. Besides, preliminary results regarding the received charging power and the transmission efficiency were presented.

List of acronyms

AC	Alternating current
ARBC	Adaptive resonant beam charging
BER	Bit error rate
CC	Constant current
CDC	Channel-dependent charge
CPS	Charging pricing strategy
CT	Charge termination
CV	Constant voltage
DC	Direct current

DC-DC	Direct current to direct current
DLC	Distributed laser charging
EE	Energy efficiency
FAFC	First access first charge
FSOC	Free-space optical communication
HPC	High priority charge
IoT	Internet of Things
LOS	Line of sight
MIMO	Multi-input multi-output
MPPT	Maximum power point tracking
PSD	Power spectral density
PV	Photovoltaic
QoCS	Quality of charging service
RBC	Resonant beam charging
RBCom	Resonant beam communications
RF	Radio frequency
SNR	Signal-to-noise ratio
SOC	State of charge
STC	Standard test conditions
SWIPT	Simultaneous wireless information and power transfer
TC	Trickle charge
TDMA	Time-division multiple access
UAVs	Unmanned aerial vehicles
WET	Wireless energy transfer
WPC	Wireless powered communications
WPT	Wireless power transfer
WSN	Wireless sensor network

References

[1] Huang K and Zhou X. Cutting the last wires for mobile communications by microwave power transfer. *IEEE Communications Magazine.* 2015;53 (6):86–93.

[2] Fadhel YB, Ktata S, Rahmani S, and Al-Haddad K. Used methods to wirelessly powered implantable medical devices. In: 2016 17th International Conference on Sciences and Techniques of Automatic Control and Computer Engineering (STA). IEEE; 2016. p. 475–480.

[3] Zhou Y, Huang C, Jiang T, and Cui S. Wireless sensor networks and the internet of things: optimal estimation with nonuniform quantization and bandwidth allocation. *IEEE Sensors Journal.* 2013;13(10):3568–3574.

[4] Liu Q, Wu J, Xia P, *et al.* Charging unplugged: will distributed laser charging for mobile wireless power transfer work. *IEEE Vehicular Technology Magazine*. 2016;11(4):36–45.

[5] Chen W, Zhao S, Shi Q, and Zhang R. Resonant beam charging-powered UAV-assisted sensing data collection. *IEEE Transactions on Vehicular Technology*. 2019;69(1):1086–1090.

[6] Wu Q, Ding G, Xu Y, *et al.* Cognitive internet of things: a new paradigm beyond connection. *IEEE Internet of Things Journal*. 2014;1(2):129–143.

[7] Feng B, Zhang H, Zhou H, and Yu S. Locator/identifier split networking: a promising future Internet architecture. *IEEE Communications Surveys and Tutorials*. 2017;19(4):2927–2948.

[8] Xiao R, Laput G, Zhang Y, and Harrison C. Deus EM Machina: on-touch contextual functionality for smart IoT appliances. In: Proceedings of the 2017 CHI Conference on Human Factors in Computing Systems; 2017. p. 4000–4008.

[9] Aloi G, Caliciuri G, Fortino G, *et al.* Enabling IoT interoperability through opportunistic smartphone-based mobile gateways. *Journal of Network and Computer Applications*. 2017;81:74–84.

[10] Lazarescu MT. Design of a WSN platform for long-term environmental monitoring for IoT applications. *IEEE Journal on emerging and selected topics in circuits and systems*. 2013;3(1):45–54.

[11] Deng F, Yue X, Fan X, Guan S, Xu Y and Chen J. Multisource energy harvesting system for a wireless sensor network node in the field environment. *IEEE Internet of Things Journal*. 2018;6(1):918–927.

[12] Liu M, Wang G, Giannakis GB, Xiong M, Liu Q and Deng H. Wireless power transmitter deployment for balancing fairness and charging service quality. *IEEE Internet of Things Journal*. 2020;7(3):2223–2234.

[13] Tesla N. The transmission of electrical energy without wires as a means for furthering peace. *Electrical World and Engineer*. 1905;1:21–21.

[14] Lu X, Niyato D, Wang P, Kim DI and Han Z. Wireless charger networking for mobile devices: Fundamentals, standards, and applications. *IEEE Wireless Communications*. 2015;22(2):126–135.

[15] Jadidian J and Katabi D. Magnetic MIMO: How to charge your phone in your pocket. In: Proceedings of the 20th Annual International Conference on Mobile Computing and Networking; 2014. p. 495–506.

[16] Kurs A, Karalis A, Moffatt R, Joannopoulos JD, Fisher P, and Soljačić M. Wireless power transfer via strongly coupled magnetic resonances. *Science*. 2007;317(5834):83–86.

[17] Zhang Q, Fang W, Liu Q, *et al.* Distributed laser charging: a wireless power transfer approach. *IEEE Internet of Things Journal*. 2018;5(5):3853–3864.

[18] Valtchev SS, Baikova EN, and Jorge LR. Electromagnetic field as the wireless transporter of energy. *Facta Universitatis-Series: Electronics and Energetics*. 2012;25(3):171–181.

[19] Garnica J, Chinga RA, and Lin J. Wireless power transmission: from far field to near field. *Proceedings of the IEEE*. 2013;101(6):1321–1331.

[20] Assawaworrarit S, Yu X, and Fan S. Robust wireless power transfer using a nonlinear parity–time-symmetric circuit. *Nature*. 2017;546(7658):387–390.
[21] Lu X, Wang P, Niyato D, Kim DI and Han Z. Wireless networks with RF energy harvesting: a contemporary survey. *IEEE Communications Surveys & Tutorials*. 2014;17(2):757–789.
[22] Shinohara N. *Wireless power transfer via radiowaves*. Wiley Online Library; 2014.
[23] Fang W, Zhang Q, Liu Q, Wu J and Xia P. Fair scheduling in resonant beam charging for IoT devices. *IEEE Internet of Things Journal*. 2018;6(1):641–653.
[24] Zhang Q, Fang W, Xiong M, Liu Q, Wu J and Xia P. Adaptive resonant beam charging for intelligent wireless power transfer. *IEEE Internet of Things Journal*. 2018;6(1):1160–1172.
[25] WCL Technology. Available from: https://www.wi-charge.com/.
[26] Wang W, Zhang Q, Lin H, Liu M, Liang X and Liu Q. Wireless energy transmission channel modeling in resonant beam charging for IoT devices. *IEEE Internet of Things Journal*. 2019;6(2):3976–3986.
[27] Zhang Q, Shi X, Liu Q, Wu J, Xia P and Liao Y. Adaptive distributed la ser charging for efficient wireless power transfer. In: 2017 IEEE 86th Vehicular Technology Conference (VTC-Fall). IEEE; 2017. p. 1–5.
[28] Zhang Q, Liu M, Lin X, Liu Q, Wu J and Xia P. Optimal resonant beam charging for electronic vehicles in internet of intelligent vehicles. *IEEE Internet of Things Journal*. 2018;6(1):6–14.
[29] Zhang Q, Wang G, Chen J, Giannakis GB and Liu Q. Mobile energy transfer in internet of things. *IEEE Internet of Things Journal*. 2019;6(5):9012–9019.
[30] Fang W, Wang G, Giannakis GB, Liu Q, Wang X and Deng H. Channel-dependent scheduling in wireless energy transfer for mobile devices. *IEEE Transactions on Vehicular Technology*. 2020;69(3):3330–3340.
[31] Xiong M, Liu M, Zhang Q, Liu Q, Wu J and Xia P. TDMA in adaptive resonant beam charging for IoT devices. *IEEE Internet of Things Journal*. 2018;6(1):867–877.
[32] Fang W, Zhang Q, Liu M, Liu Q and Xia P. Earning maximization with quality of charging service guarantee for IoT devices. *IEEE Internet of Things Journal*. 2018;6(1):1114–1124.
[33] Xiong M, Liu Q, Liu M, and Xia P. Resonant beam communications. In: ICC 2019 – 2019 IEEE International Conference on Communications (ICC). IEEE; 2019. p. 1–6.
[34] Xiong M, Liu Q, Liu M, Wang X and Deng H. Resonant beam communications with photovoltaic receiver for optical data and power transfer. *IEEE Transactions on Communications*. 2020;68(5):3033–3041.
[35] Xiong M, Liu Q, Wang G, Giannakis GB and Huang C. Resonant beam communications: principles and designs. *IEEE Communications Magazine*. 2019;57(10):34–39.
[36] Silfvast WT. *Laser fundamentals*. Cambridge University Press; 2004.
[37] Goodenough JB and Park KS. The Li-ion rechargeable battery: a perspective. *Journal of the American Chemical Society*. 2013;135(4):1167–1176.

[38] Shen W, Vo TT, and Kapoor A. Charging algorithms of lithium-ion batteries: an overview. In: 2012 7th IEEE Conference on Industrial Electronics and Applications (ICIEA). IEEE; 2012. p. 1567–1572.

[39] Dearborn S, *et al.* Charging Li-ion batteries for maximum run times. *Power Electronics Technology*. 2005;31(4):40–49.

[40] Cleveland T. Battery charger adapts to multiple chemistries. *Power Electronics Technology*. 2008;34(7):26–31.

[41] Hwang YS, Wang SC, Yang FC, and Chen JJ. New compact CMOS Li-ion battery charger using charge-pump technique for portable applications. *IEEE Transactions on Circuits and Systems I: Regular Papers*. 2007;54(4):705–712.

[42] Van Zeghbroeck B. *Principles of semiconductor devices*. Colarado University. 2004;34.

[43] Koechner W. *Solid-state laser engineering*. Vol. 1. Springer; 2013.

[44] Edouard M and Njomo D. Mathematical modeling and digital simulation of PV solar panel using Matlab software. *International Journal of Emerging Technology and Advanced Engineering*. 2013;3(9):24–32.

[45] Aziz MS, Ahmad S, Husnain I, Hassan A and Saleem U. Simulation and experimental investigation of the characteristics of a PV-harvester under different conditions. In: 2014 International Conference on Energy Systems and Policies (ICESP). IEEE; 2014. p. 1–8.

[46] Gray JL. The physics of the solar cell. *Handbook of photovoltaic science and engineering*. 2011;2:82–128.

[47] Sera D, Teodorescu R, and Rodriguez P. PV panel model based on datasheet values. In: 2007 IEEE International Symposium on Industrial Electronics. IEEE; 2007. p. 2392–2396.

[48] de Brito MAG, Galotto L, Sampaio LP, e Melo GdA and Canesin CA. Evaluation of the main MPPT techniques for photovoltaic applications. *IEEE Transactions on Industrial Electronics*. 2012;60(3):1156–1167.

[49] Peng FZ, Zhang F, and Qian Z. A magnetic-less DC-DC converter for dual voltage automotive systems. In: Conference Record of the 2002 IEEE Industry Applications Conference. 37th IAS Annual Meeting (Cat. No. 02CH37344). Vol. 2. IEEE; 2002. p. 1303–1310.

[50] Salem LG, Louie JG, and Mercier PP. 12.9 A flying-domain DC–DC converter powering a Cortex-M0 processor with 90.8% efficiency. In: 2016 IEEE International Solid-State Circuits Conference (ISSCC). IEEE; 2016. p. 234–236.

[51] Haas H, Chen C, and O'Brien D. A guide to wireless networking by light. *Progress in Quantum Electronics*. 2017;55:88–111.

[52] Haas H. LiFi is a paradigm-shifting 5G technology. *Reviews in Physics*. 2018;3:26–31.

[53] David K and Berndt H. 6G vision and requirements: Is there any need for beyond 5G? *IEEE Vehicular Technology Magazine*. 2018;13(3): 72–80.

[54] Wang Z, Tsonev D, Videv S, and Haas H. On the design of a solar-panel receiver for optical wireless communications with simultaneous energy

harvesting. *IEEE Journal on Selected Areas in Communications.* 2015; 33(8):1612–1623.

[55] Boylestad RL and Nashelsky L. *Electronic devices and circuit theory.* Pearson; 2013.

[56] Schlosser V and Ghitas A. Measurement of silicon solar cells ac parameters. In: Proceedings of the Arab Regional Solar Energy Conference; 2006.

[57] Mandal H and Nagaraju J. GaAs/Ge and silicon solar cell capacitance measurement using triangular wave method. *Solar Energy Materials and Solar Cells.* 2007;91(8):696–700.

[58] Zhou W and Jin K. Optimal photovoltaic array configuration under Gaussian laser beam condition for wireless power transmission. *IEEE Transactions on Power Electronics.* 2016;32(5):3662–3672.

[59] Kahn JM and Barry JR. Wireless infrared communications. *Proceedings of the IEEE.* 1997;85(2):265–298.

[60] Xiong M, Liu Q, Wang G, Giannakis GB, Zhang S and Huang C. Analytical models for resonant beam communications. In: 2019 11th International Conference on Wireless Communications and Signal Processing (WCSP); 2019. p. 1–6.

Chapter 9

Simultaneous wireless information and power transfer in Internet of Things

Faisal Karim Shaikh[1], Mohsin Memon[1], and Sherali Zeadally[2]

Abstract

Recently, we have seen massive integration of devices with the Internet which helped in envisioning the concept of Internet of Things (IoT). IoT comprises a distributed network of low powered, low storage, low processing, and low wireless communication range devices. Typically, these devices are battery operated with limited lifespan and replacing the batteries is a cumbersome process. Therefore, effective energy management and energy harvesting schemes need to be exploited. Generally, the major source of power drain is associated with the wireless communication between the devices. On the contrary, information gathering and transfer are attributed as the main functions of the IoT network Thus, simultaneous wireless information and power transfer (SWIPT) is a promising choice for IoT. We describe various SWIPT architectures and enabling technologies in IoT. We review the various approaches and techniques proposed in the literature for SWIPT-enabled IoT. Finally, we highlight future design challenges of SWIPT that must be addressed to deliver information and energy continuously and reliably to IoT devices.

9.1 Introduction

The Internet of Things (IoT) consists of a large network of things that include a wide range of devices with varying networking, computing, and storage capabilities. IoT enables things to interact with each other wirelessly and exchange various types of information (e.g., sensor data, multimedia data). According to [1],

[1]Institute of Information and Communication Technologies, Mehran University of Engineering and Technology, Jamshoro, Pakistan
[2]College of Communication and Information, University of Kentucky, Lexington, USA

by 2022, there will be 29.7 billion devices connected to the Internet, out of which 18.1 billion will be related to the IoT. IoT enables smart communication capabilities for things using ubiquitous computing, embedded networks, communication technologies, and Internet protocols [2]. This ecosystem continues to grow and includes various application domains such as smart cities [3,4], smart transportation, and smart grids [5,6]. IoT has already started to drastically shape the quality of our daily lives in terms of improved convenience, ubiquitous access to a range of services anytime, anywhere, and from any device. According to [7], the most important challenge for IoT is smart energy management. Every active component in the IoT network consumes a certain amount of energy to perform its functionality.

Recently, we have witnessed a significant increase in the amount of data produced by IoT [8,9] despite the use of scarce energy resources [10]. This results in rapid depletion of batteries, requiring battery replacement frequently. Once a network has been deployed, changing batteries can also be risky for scenarios that require continuous monitoring. In such cases, battery replacement can be an expensive and laborious process. Thus, energy harvesting (EH) is the only likely option to provide unlimited energy resources to such low-powered devices operating in an IoT environment [11]. There are various ways through which energy can be harvested using the ambient environment for IoT networks [12]. However, it will be more beneficial if the information and power can simultaneously be transferred to the destination. Therefore, at information receivers, data is restored, and electromagnetic energy is collected and transformed to electrical power for EH and is called wireless power transfer (WPT). Furthermore, in IoT, the data is received by all nearby nodes due to the broadcast nature of the signals, and along with the receiver, the other unintentional receivers may also use the same signal using WPT. Generally, the WPT can be divided into two categories [13]: (a) EH from ambient signals and (b) EH from the dedicated fully controlled power source. The major concern when dealing with WPT is the distance of energy transmissions. With the advances of communication networks such as IoT and 5G cellular communication, the major focus of WPT is on near-field energy transmissions. The near-field WPT has many drawbacks including (a) maintaining field strength as per safety requirement, (b) tuning of resonant inductance, (c) use of high frequency at the source, (d) lack of air ionization techniques, and (e) high initial cost [14–16]. Alongside, there are far-field WPT techniques available too such as satellite, television, and radio towers, but they are not much useful for IoT applications.

Integrating WPT with information-centric networks such as IoT requires a technology, which can efficiently transfer both power and information to the end devices. Thus, the concept of simultaneous wireless information and power transfer (SWIPT) was first coined in [17] theoretically. Recently, we have seen a lot of progress in SWIPT in wireless communication networks [13,18,19]. SWIPT can be beneficial in terms of power consumption, low interference, spectral efficiency, and latency by enabling simultaneous transmission of both information and power. A proper design of SWIPT can provide energy-efficient wireless communication and lead to state-of-the-art applications and services for next-generation sustainable societies.

We organize the rest of the chapter as follows. Section 9.2 describes various aspects of SWIPT and its usage in IoT. As energy consumption is such a crucial issue, Section 9.3 surveys the state-of-the-art results in the area of SWIPT enabled IoT and explores its feasibility for various IoT applications. We discuss future directions and challenges in SWIPT-enabled IoT in Section 9.4. Section 9.5 concludes the chapter.

9.2 Simultaneous wireless information and power transfer

SWIPT has recently gained a lot of attention and it has been developed out of various WPT technologies to enable SWIPT. Figure 9.1 shows a typical SWIPT scenario where static and mobile units are used to transmit the information and power to end devices. To provide an efficient tradeoff between the information delivery rate and the amount of energy harvested, fundamental design changes are required in IoT networks.

To enable efficient SWIPT, the end device should be able to perform information decoding and EH capabilities simultaneously. Generally, when the RF signal arrives at the device, the EH process destroys the information contained in the signal. Thus, both operations cannot be performed simultaneously on the received signal in SWIPT. Hence, in practice, to enable SWIPT the received signal may be divided into two signals or separate antennas will be required for

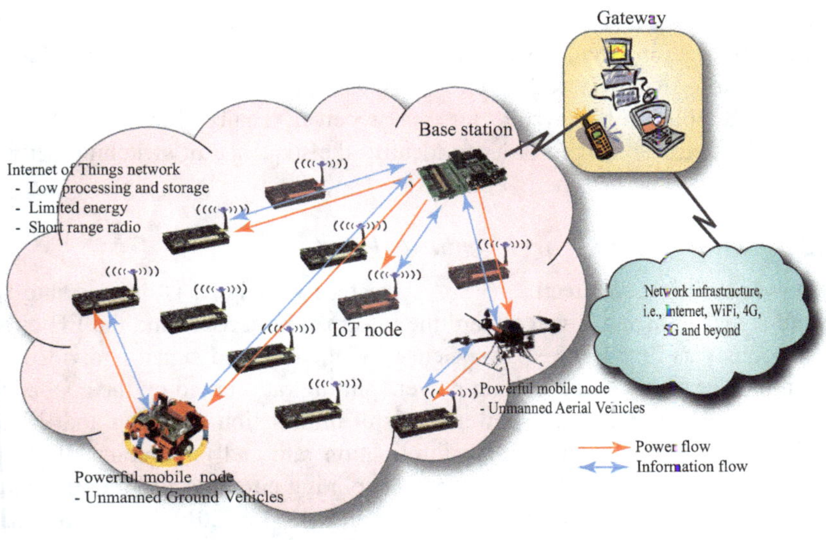

Figure 9.1 Generic SWIPT model for IoT

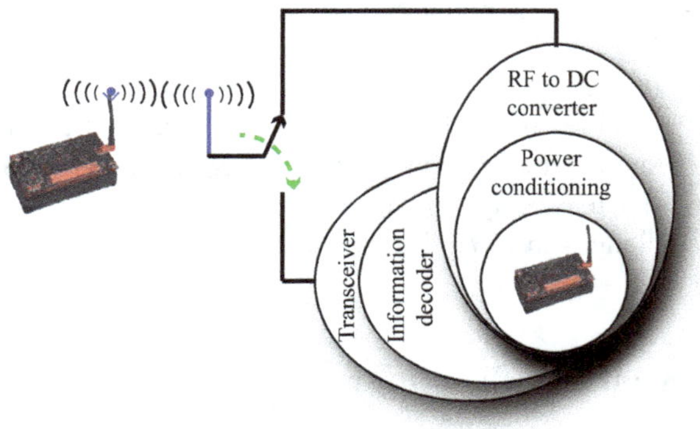

Figure 9.2 Periodically switching SWIPT architecture

information reception and EH. Furthermore, for a reliable and continuous energy supply, a single-antenna receiver may not be sufficient due to a limited collection of energy. Therefore, a distributed array deployment of antennas such as multiple input multiple output (MIMO) may be required for sufficient energy collection. Accordingly, the SWIPT architecture classification can be done as follows: (a) periodically switching architecture, (b) the use of a power splitting architecture, (c) the use of a separate receiver architecture, (d) the use of an antenna switching architecture, (e) using a relay-enabled architecture, (f) the use of an interference optimization architecture, and (g) the use of a mobility-enabled architecture.

9.2.1 Periodically switching architecture (PSA)

In a periodically switching architecture (PSA) (Figure 9.2), the antenna is switched between information decoding and EH circuit periodically. The receiver device includes an information decoding circuit, an EH circuit, and a switch to perform periodic switching. In this architecture, the receiver requires accurate synchronization with the transmitter in order to identify the sequence of switching between information decoding and the EH circuit.

9.2.2 Power splitting architecture (PSpA)

In a power splitting architecture (PSpA), the received signal is divided into two separate signals which are sent toward the information decoder and the EH circuit simultaneously as Figure 9.3 shows. Generally, the received signal is divided into different power levels with a certain power splitting ratio based on the application requirements. Having different power levels of information decoding and EH can provide a good balance between the information rate and the amount of energy harvested. In the receiving device, the power splitting circuit is also required along with the information decoder and the EH circuit. In [20], the authors have demonstrated that power splitting achieves the best tradeoff between the information rate and the amount of energy harvested.

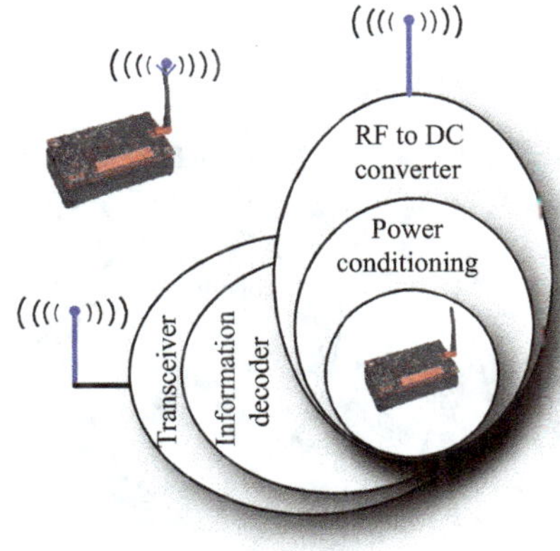

Figure 9.4 Separate receiver SWIPT architecture

9.2.3 Separate receiver architecture (SRA)

In the separate receiver architecture (SRA) (as Figure 9.4 shows), two separate antennas are used for information decoding and EH purposes at the receiver. The transmitter uses multiple antennas to transmit the information and energy signals toward the receiver. Generally, different channels are occupied by the different antennas. Thus, EH and information transfer can be done simultaneously. A SRA can be easily built by using off-the-shelf components for EH and information reception.

9.2.4 Antenna switching architecture (ASA)

For the antenna switching architecture (ASA), the receiver device must have two or more antennas available for reception. Generally, the antenna switching is used for SWIPT and it has low complexity [21,22]. In the antenna switching SWIPT architecture, the subset of antennas can be used for information reception and other antennas can be used for EH purposes as Figure 9.5 shows. By adopting the appropriate number of antenna switchings, a good tradeoff between EH and information rate can be achieved.

9.2.5 Relay-enabled architecture (RA)

When the network deployment is large, to reach end devices, we need to traverse multiple hops, and in this case, the relay-enabled architecture (RA) is useful. The bigger the distance between the source and the destination, the higher the power needed for information transmission. This will quickly drain the node battery and lower the network lifetime. The relay nodes can be used for both information relay and power transfer to other nodes in the network. Figure 9.6 shows a relay-enabled SWIPT architecture for IoT.

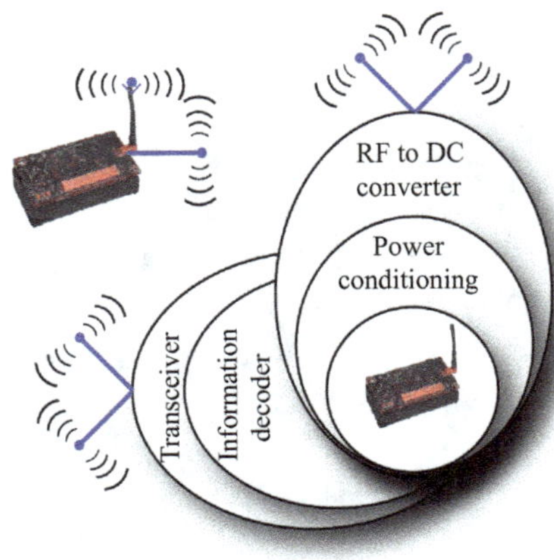

Figure 9.5 Antenna switching SWIPT architecture

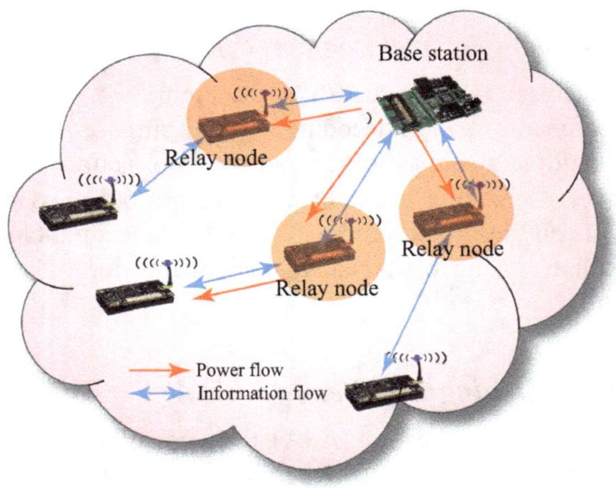

Figure 9.6 Relay enabled SWIPT architecture

9.2.6 Interference optimization architecture (IOA)

For large networks and especially in IoT, the simultaneous transmission of data across multiple sources and multiple or single destination nodes is very common. This also introduces interference among the nodes. Generally, interference has been a limiting factor in traditional wireless communication networks as the variance is introduced in transmitted signal and the reciever node receives the deterioted signal. Thus, traditional wireless communication relies on either no or very low interference. However, interference avoidance and mitigation techniques may cause inefficient use of wireless channels [23]. Recently, the focus has been moved to the potential use of interference instead of avoiding and mitigating the interference [24]. The interference has been considered harmful for information transfer

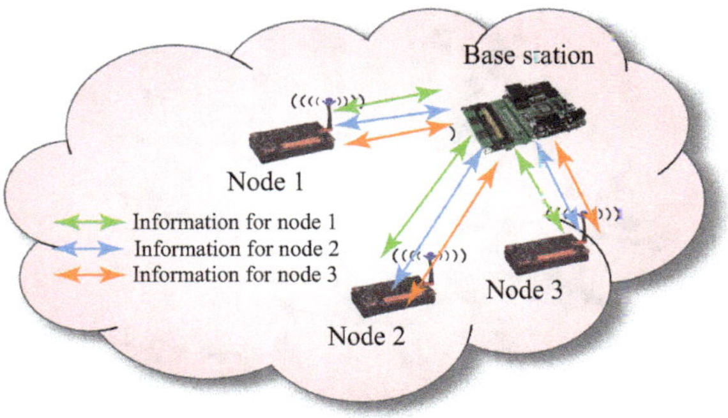

Figure 9.7 Interference optimization architecture

but can be useful for EH. Therefore, a proper trade-off between information transfer and EH is required to facilitate efficient SWIPT in IoT. Accordingly, the idea of beamforming and precoding in IoT-enabled SWIPT has got more attention. In interference optimization architecture (IOA), for efficient interference management, the nodes in the networks are enabled to decode the interference, i.e., receive all signals and process whether they are related to the node or not. For SWIPT interference, decoding allows each transmitter's beam to cross (interfere with) other beams in order to increase the power transfer without loss of information as Figure 9.7 shows.

9.2.7 UAV-enabled Architecture (UA)

Recently, we have seen widespread use of unmanned aerial vehicles (UAVs) acting as a communication bridge in IoT networks [25,26]. In a UAV-enabled architecture (UA), the UAVs can fly in an IoT network to transmit information and energy simultaneously. For large network deployments, multiple UAVs can be more feasible. Thus, multi-UAV-enabled SWIPT in the IoT network will play an important role in the future. Figure 9.8 depicts a generic scenario of a multi-UAV-enabled SWIPT architecture.

9.3 SWIPT-enabled Internet of Things

Current and future trends in the popularity of IoT devices for personal and commercial uses are motivating researchers to develop techniques that embed SWIPT into devices. Next, we review the various techniques of SWIPT used in IoT networks.

The authors of [27] improved the capacity of the non-orthogonal multiple access (NOMA) SWIPT-enabled IoT with RA and Rayleigh fading channels. They have considered PSA for NOMA-SWIPT with direct links wherein a base station sends information to the IoT devices via direct links or using the relay node with the EH circuit. The throughput capacity enhancement of NOMA-SWIPT was high

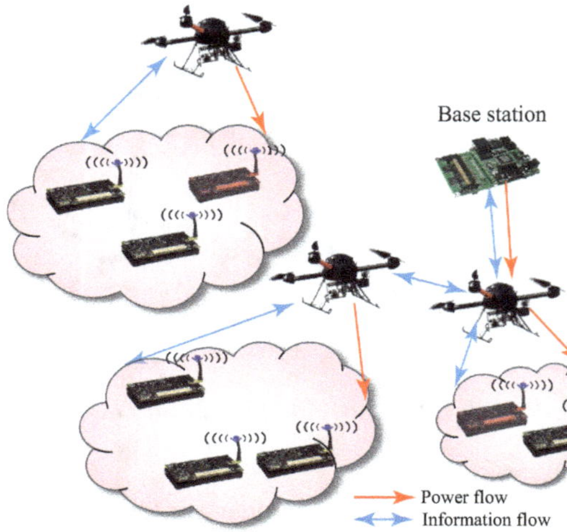

Figure 9.8 UAV-enabled SWIPT architecture

compared to systems without direct links and orthogonal multiple access (OMA) schemes.

Mao *et al.* proposed a technique to minimize power consumption of a SWIPT system using the transmit beamformer [28]. The authors used PSpA for cognitive multiple-input single-output NOMA-SWIPT. They considered a nonlinear EH model to mimic the practical energy conversion from RF. This approach was shown to be more effective than the linear EH model.

In [29], the authors used the SWIPT-enable relay which uses the RF energy supplied by two bi-directional relaying NOMA user groups for recharging. First, the relay node recharges itself using PSA through the information signals from the NOMA user group. Second, the relay node exchanges the information among the NOMA user groups. The proposed scheme leads to a significant improvement of throughput and capacity gain compared to other bi-directional multiple access schemes.

In [30], the authors proposed the IOA-aided SWIPT scheme for cellular IoT networks over Nakagami-m fading channels. The information signal and interference signals are considered for EH. The relay nodes are selected based on the maximum energy harvested from the signals. The EH-enable relay node forwards its information and source information on the same channel using NOMA.

In [31], the authors proposed a PSA- and IOA-based full duplex SWIPT system for device-to-device (D2D) communication to minimize energy consumption. Additionally, the authors have introduced the self-energy recycling phase in which the node harvests the energy from the self-interference. The overall transmission time and power of the nodes were optimized using the proposed scheme.

The authors of [32] studied the achievable information rate of the half-duplex quantize-and-forward (QF) cooperation in SWIPT. In QF, the node first quantizes the received signal, i.e., removes the noise, and then forwards the signal to the destination. The QF scheme was compared with the existing cooperation schemes such as amplify-and-forward (AF) and decode-and-forward (DF) using numerical simulations. The PSpA and RA are used for the proposed scheme over additive white Gaussian noise and slow fading channels. The source and destination are provided by the external power supply and the relay node utilizes the SWIPT technology.

In [33], the authors proposed an optimal solution in the IoT distributed antenna system using ASA, SRA, and PSpA. For a single IoT node, the optimal closed-form solution was explored by analyzing the Karush–Kuhn–Tucker (KKT) conditions. The KKT conditions are first-order necessary conditions for a solution in nonlinear programming to be optimal. For multiple nodes, the authors proposed an efficient sub-optimal algorithm to achieve energy efficiency.

In [34], the authors proposed a self-powered adaptive PSpA-based SWIPT architecture that relies on RF EH in low-power IoT devices. They extended the operational range of IoT devices using the dual-mode operation, i.e., single or multi-tone communication. The proposed mechanism achieves battery-free operation for received signals above 20 dBm. For low-received power, the authors proposed a peak-to-average power ratio scheme for the demodulation of information signals.

In [35], the authors developed a resource allocation mechanism for secure communications using IOA by enabling artificial noise. The proposed scheme also takes into account the nonlinear EH and quality of service required for secure communication. In order to optimize the rank constrained for resource allocation, a semidefinite programming relaxation approach is utilized. The proposed scheme showed promising results compared to linear EH models.

In [36], Wang *et al.* proposed a solution to the non-convex problem of power allocation and power splitting at the receiver nodes using zero-forcing dirty paper coding (ZF-DPC) in the IoT for multiple-input-single-output (MISO) SWIPT based on PSpA. Before applying ZF-DPC, semidefinite relaxation, zero-forcing, and maximum ratio transmission techniques are used to solve the non-convex problem. Later, the ZF-DPC technique is applied to mitigate the multiuser interference and obtain the optimal closed-form solution.

In [37], the authors designed a PSA-based beamforming SWIPT protocol for massive MIMO (mMIMO) SRA with a fixed half-duplex (HD) non-linear EH terminal. The proposed model is composed of two phases, i.e., (a) downlink training and (b) two-layer beamformers for the reception of signals. To increase the spectral efficiency, the base station sends the periodic switching schedules to the receivers. The simulation results show the dominance of the proposed protocol on spectral efficiency performance compared with the conventional mMIMO SWIPT protocols.

The author of [38] proposed an mMIMO two-way RA using PSpA. For high information rate applications, two classical and linear relay precodings, i.e., zero-forcing reception/zero-forcing transmission and maximum-ratio combining/maximum-ratio transmission are used under large-scale mMIMO facing channels. The authors derived numerical results to evaluate the asymptotic performance of the achievable sum rates when the number of relay antennas grows to infinity and proved them using simulations. Further, the authors also discussed the impact of the power factor and the number of antennas on the achievable information rate.

In [39], the authors investigated the max–min EH fairness problem in MISO SWIPT multicasting IoT based on PSpA. They proposed an efficient algorithm for optimal precoding and power splitting ratios. They compared the proposed scheme with benchmark schemes and outperformed them for MIMO RF energy transfer both with and without information decoding.

In [40], the authors studied mMIMO energy efficiency optimization using ASA. The authors proposed an iterative algorithm based on the Dinkelbach method [41] for energy efficiency by exploiting coordinated multi-point (CoMP) transmission and reception. The proposed algorithm has a faster convergence compared with the maximum signal-to-interference-plus-noise ratio (MaxSINR) algorithm.

In [42], the authors developed an emergency communication system based on UA SWIPT that can be deployed in disaster areas. In the proposed scheme, UAV operates as a wireless charger to deliver energy to receivers. For a multi-UAV scenario, the authors proposed a technique that allows optimal resource allocation at UAVs and accordingly produces a UAVs trajectory to traverse the whole field. To achieve energy efficiency, the authors developed a predictive approach for EH and information rate between source and destination nodes which then optimize the trajectory of the UAV.

In [43], the authors proposed a joint resource allocation UA-based PSpA approach for a multicast scenario that involves UAV to send information and energy to multiple nodes in the region. The proposed approach optimizes the trajectory and transmits the power of the UAV to maximize the achievable information rate between source and destination nodes and the rate at which energy is harvested. The authors used the quadratic transform technique to convert the non-convex problem to a simple and tractable problem. For joint optimization of information rate and EH, they proposed an efficient iterative algorithm.

Table 9.1 compares the various approaches for IoT-enabled SWIPT. Based on our literature review above, we found that the power splitting architecture along with the relaying architecture is the most popular architectures to enable SWIPT in IoT. However, interference optimization and UAV-enabled architectures have been receiving a lot of attention recently and show promise in enabling SWIPT in IoT.

9.4 Future research challenges for SWIPT in Internet of Things

Battery-operated IoT remains a significant energy management issue. Information transfer and EH in an IoT system brings about several challenges. Based on the above review and observations, this section discusses some open challenges and major issues for an SWIPT-enabled IoT which must be addressed in the future.

9.4.1 Security

Increasing the transmission power of the signal opens up security concerns in SWIPT because by doing so we can enhance the information rate and EH between the source and the destination but this also increases the chances of eavesdropping. There is some research on the physical layer security of SWIPT-enabled IoT

Table 9.1. SWIPT enabled IoT

Ref #	Architecture	System model	Technique	Results
[27]	PSA, RA	Improvement in IoT relay systems	Maximum ratio combining (MRC) shows network capacity enhancement	Ergodic sum capacity better than orthogonal multiple access
[28]	PSpA	Non-linear model for efficient resource allocation	A penalty function-based algorithm to address non-convex problem	Consumes lower power as compared to the linear model
[29]	PSA, RA	Bidirectional relaying SWIPT	Perfect successive interference cancellation (SIC) and imperfect SIC	Provides significant network throughput and capacity gain
[30]	IOA, RA	Cellular IoT network under multi-relay scenario	Dual-hop interference aided SWIPT	Delay-tolerant throughput, outage probability, delay-limited capacity, and energy efficiency
[31]	PSA, IOA	A time switching-based ful-duplex SWIPT	An iterative algorithm to enhance transmission time and power	Provides self-energy recycling compared to conventional time switching systems
[32]	PSpA, RA	Power-splitting (PS) factor is optimized	Half-duplex quantize-and-forward (QF) cooperation	Information rate of QF cooperation is better than amplify-and-forward and decode and-forward cooperation
[33]	ASA, PSpA, SRA	Distributed antenna (DA) systems	Power splitting is used by varying transmit power	Energy efficiency of the proposed scheme is better than benchmark schemes in single and multiple IoT devices scenario
[34]	Adaptive PSpA	Dual-mode SWIPT	Uses a high-efficiency RF-DC converter	Sensitivity is −7 dBm and the measured peak power efficiency of the RF energy harvester is 69% at 20 dBm input power level
[35]	IOA	Communication security SWIPT	Resource allocation algorithm design is formulated as a non-convex optimization problem	Significant performance gain in terms of the harvested energy compared to existing designs

(Continues)

Table 9.1. (*Continued*)

Ref #	Architecture	System model	Technique	Results
[36]	PSpA	Multiple-input single output (MISO) SWIPT	ZF-DPC technique assists with eliminating multiuser interference and derives the optimal solution	Zero-forcing dirty paper coding (ZF-DPC) provides minimum signal-to-interference-plus-noise ratio (SINR) than semidefinite relaxation (SDR) and ZF
[37]	PSA, SRA	Massive MIMO system with non-linear EH terminals	Base station (BS) transmit power and time-switching ratios are optimized	Beam-domain SWIPT protocol's spectral efficiency (SE) is better than the conventional massive MIMO
[38]	PSpA, RA	Massive MIMO two-way relaying systems	Large-scale fading effects of MIMO channels are used for asymptotic sum-rates	Asymptotic performance of four classical power scaling cases has been studied and evaluated using simulation
[39]	PSpA	Power fairness optimization in MISO	Efficient algorithm for the optimal precoding and power splitting ratio	Optimizes power and achievable transmission rate trade-off
[40]	ASA	Coordinated user-base station association algorithm	Coordinated multipoint (CoMP) transmission and reception	Faster convergence to attain maximum energy efficiency
[42]	UA	Predictive energy efficiency scheme	Trajectory optimization and beam pattern, a dynamic path planning scheme	Simulation results demonstrate the effectiveness of the proposed schemes
[43]	UA, PSpA	UAV-enabled SWIPT with power splitting	Efficient iterative algorithm for the joint optimization using quadratic transform	Proposed scheme demonstrated good performance and efficiency through numerical simulation results

networks [44]. However, there are many other issues that need to be addressed such as the tradeoff between latency and security, cross-layer security, and authentication of nodes. More research is needed on dynamic and adaptive security mechanisms to be implemented in SWIPT-enabled IoT networks.

9.4.2 Mobility

Mobility in the SWIPT-enabled IoT network is appealing and provides many advantages. However, mobility, especially the high-speed mobility of nodes, transmitters, and bridging nodes (e.g., UAVs), is associated with many challenges [45]. Mobility affects the accurate availability of channel state information, thus requiring robust beamforming techniques to handle the mobility. Furthermore, the time-varying nature of information delivery and EH require dynamic, robust, and real-time mechanisms for resource allocation for mobile nodes [46]. With mobility, the node's energy decreases which increases the importance of the robust path planning and efficient relay selection mechanisms for the mobile node.

9.4.3 Hardware and software limitation

Hardware impairment is one of the main challenges in practical implementations of SWIPT-enabled IoT [47]. Embedding antennas for information reception along with rectifiers and matching networks for EH into a small device is challenging. Furthermore, the use of MIMO and massive MIMO by integrating an array of antennas on the nodes in an IoT network opens up other issues which must be addressed efficiently. Other hardware issues include impedance mismatch, oscillator phase noise, quadrature imbalance, phase imbalance, and high-power amplifier non-linearity, which may degrade the performance of the SWIPT-enabled wireless network.

Software designs of SWIPT-enabled IoT need to be robust and smart enough to utilize energy availability and information dissemination appropriately. The operating systems should be able to handle various tasks in an energy-efficient manner. One approach is to design an operating system capable of extending its functionality according to the variable nature of EH. The unavailability of simulation software is also one of the future challenges in the area of SWIPT-enabled IoT.

9.4.4 Network coding techniques

Wireless signals suffer from interference, noise, and fading, which result in the corruption of the received signal. Thus, proper coding schemes need to be adopted. Generally, polar codes and low-density parity-check codes perform well for SWIPT-enabled wireless networks [19,48]. Low-power and adaptive network coding techniques for SWIPT-enabled IoT should be investigated in the future, as they show promise in this research area.

9.4.5 Artificial intelligence and machine learning

Artificial intelligence (AI) and machine learning (ML) algorithms can play an important role to efficiently predict the amount of EH and information rate based on various factors of SWIPT [49]. AI and ML can improve the overall performance of the SWIPT-enabled IoT network. ML can also be utilized for enhancing the security of SWIPT systems by learning the attackers' patterns. AI can be used to estimate the channel state information and interference which will help in the optimal transmission of the signal. Additionally, ML can be used in optimal switching between information delivery and EH, selection of antennas, resource allocation, and so on [50,51].

9.5 Conclusion

In this chapter, we have discussed the use of SWIPT in IoT networks together with various enabling architectures for SWIPT. We have divided SWIPT architectures into seven categories, i.e., periodically switching architecture, power splitting architecture, separate receiver architecture, antenna switching architecture, relay-enabled architecture, interference optimization architecture, and UAV-enabled architecture. We then reviewed the recently proposed mechanisms for SWIPT-enabled IoT highlighting their research contributions. Finally, we discussed a few challenges that must be addressed and we have also identified future research directions in the area of SWIPT-enabled IoT networks.

List of acronyms

Acronyms	Definitions
AF	Amplify-and-forward
ASA	Antenna switching architecture
AI	Artificial intelligence
BS	Base station
CoMP	Coordinated multi-point
DF	Decode-and-forward
D2D	Device-to-device
DA	Distributed antenna
EH	Energy harvesting
HD	Half-duplex
IOA	Interference optimization architecture
IoT	Internet of Things
KKT	Karush–Kuhn–Tucker
LDPC	Low-density parity-check
ML	Machine learning
mMIMO	Massive MIMO
MIMO	Multiple input multiple output
MISO	Multiple input single output
MRC	Maximum ratio combining
MaxSINR	Maximum signal-to-interference-plus-noise ratio
NOMA	Non-orthogonal multiple access
OMA	Orthogonal multiple access
PSA	Periodically switching architecture
PS	Power-splitting
PSpA	Power splitting architecture
QF	Quantize-and-forward
RA	Relay-enabled architecture
SDR	Semi definite relaxation
SRA	Separate receiver architecture
SINR	Signal-to-interference-plus-noise ratio
SWIPT	Simultaneous wireless information and power transfer
SE	Spectral efficiency
SIC	Successive interference cancellation

(Continues)

(*Continued*)

Acronyms	Definitions
UAVs	Unmanned aerial vehicles
UA	UAV-enabled architecture
WPT	Wireless power transfer
ZF-DPC	Zero-forcing dirty paper coding

References

[1] Ericsson. Mobility Report; 2016. [Online; accessed 13-12-2019]. https://www.ericsson.com/assets/local/mobility-report/documents/ 2016/ericsson-mobility-report-november-2016.pdf.

[2] Bello O, and Zeadally S. Toward efficient smartification of the Internet of Things (IoT) services. *Future Generation Computer Systems*. 2019;92:663–673.

[3] Chowdhry BS, Shaikh FK, and Mahoto NA, editors. *IoT Architectures, Models, and Platforms for Smart City Applications*. IGI Global; 2020.

[4] Khatoun R, and Zeadally S. Smart cities: Concepts, architectures, research opportunities. *Communications of the ACM*. 2016;59(8):46–57.

[5] Shah SH, and Yaqoob I. A survey: Internet of Things (IOT) technologies, applications and challenges. In: *2016 IEEE Smart Energy Grid Engineering (SEGE)*; 2016. p. 381–385.

[6] Al-Fuqaha A, Guizani M, Mohammadi M, *et al.* Internet of Things: A survey on enabling technologies, protocols, and applications. *IEEE Communications Surveys Tutorials*. 2015;17(4):2347–2376.

[7] Lin J, Yu W, Zhang N, *et al.* A survey on internet of things: Architecture, enabling technologies, security and privacy, and applications. *IEEE Internet of Things Journal*. 2017;4(5):1125–1142.

[8] Tsai CW, Lai CF, Chiang MC, *et al.* Data mining for Internet of Things: A survey. *IEEE Communications Surveys Tutorials*. 2014;16(1):77–97.

[9] Mukherjee A, Paul HS, Dey S, *et al.* ANGELS for distributed analytics in IoT. In: 2014 IEEE World Forum on Internet of Things (WF-IoT)*; 2014*. p. 565–570.

[10] Arshad R, Zahoor S, Shah MA, *et al.* Green IoT: An investigation on energy saving practices for 2020 and beyond. *IEEE Access*. 2017;5:15667–15681.

[11] Shaikh FK, and Zeadally S. Energy harvesting in wireless sensor networks: A comprehensive review. *Renewable and Sustainable Energy Reviews*. 2016;55:1041–1054.

[12] Zeadally S, Shaikh FK, Talpur A, *et al.* Design architectures for energy harvesting in the Internet of Things. *Renewable and Sustainable Energy Reviews*. 2020;128:109901.

[13] Perera TDP, Jayakody DNK, Sharma SK, *et al.* Simultaneous wireless information and power transfer (SWIPT): Recent advances and future challenges. *IEEE Communications Surveys & Tutorials*. 2017;20(1):264–302.

[14] Yoon IJ. Wireless power transfer in the radiating near-field region. In: *2015 USNC-URSI Radio Science Meeting (Joint with AP-S Symposium)*. IEEE; 2015. p. 344.

[15] Jawad AM, Nordin R, Gharghan SK, *et al.* Opportunities and challenges for near-field wireless power transfer: A review. *Energies.* 2017;10(7):1022.

[16] Le AM, Truong LH, Quyen TV, *et al.* Wireless power transfer near-field technologies for unmanned aerial vehicles (UAVs): A review. *EAI Endorsed Transactions on Industrial Networks and Intelligent Systems.* 2020;7 (22):162831.

[17] Varshney LR. Transporting information and energy simultaneously. In: *2008 IEEE International Symposium on Information Theory*. IEEE; 2008. p. 1612–1616.

[18] Chae SH, Jeong C, and Lim SH. Simultaneous wireless information and power transfer for Internet of Things sensor networks. *IEEE Internet of Things Journal.* 2018;5(4):2829–2843.

[19] Hossain MA, Noor RM, Yau KLA, *et al.* A survey on simultaneous wireless information and power transfer with cooperative relay and future challenges. *IEEE Access.* 2019;7:19166–19198.

[20] Zhou X, Zhang R, and Ho CK. Wireless information and power transfer: Architecture design and rate-energy tradeoff. *IEEE Transactions on Communications.* 2013;61(11):4754–4767.

[21] Krikidis I, Sasaki S, Timotheou S, *et al.* A low complexity antenna switching for joint wireless information and energy transfer in MIMO relay channels. *IEEE Transactions on Communications.* 2014;62(5):1577–1587.

[22] Golipour S, Andargoli SM, and Ghazalian R. A low complexity power allocations algorithm for DF MIMO SWIPT relaying systems. *Internet Technology Letters.* 2021;4(2):e213.

[23] Gu Y, Aissa S. Interference aided energy harvesting in decode-and-forward relaying systems. In: *2014 IEEE International Conference on Communications (ICC)*. IEEE; 2014. p. 5378–5382.

[24] Li A, Spano D, Krivochiza J, *et al.* A tutorial on interference exploitation via symbol-level precoding: Overview, state-of-the-art and future directions. *IEEE Communications Surveys & Tutorials.* 2020;22(2):796–839.

[25] Alsamhi SH, Afghah F, Sahal R, *et al.* Green internet of things using UAVs in B5G networks: A review of applications and strategies. *Ad Hoc Networks.* 2021;p. 102505.

[26] Islam N, Rashid MM, Pasandideh F, *et al.* A review of applications and communication technologies for Internet of Things (IoT) and unmanned aerial vehicle (UAV) based sustainable smart farming. *Sustainability.* 2021;13(4):1821.

[27] Rauniyar A, Engelstad P, and Østerbø ON. Capacity enhancement of NOMA-SWIPT IoT relay system with direct links over Rayleigh fading channels. *Transactions on Emerging Telecommunications Technologies.* 2020;31(12):e3913.

[28] Mao S, Leng S, Hu J, *et al.* Power minimization resource allocation for underlay MISO-NOMA SWIPT systems. *IEEE Access.* 2019;7:17247–17255.

[29] Rauniyar A, Engelstad PE, and Østerbø ON. On the performance of bidirectional NOMA-SWIPT enabled iot relay networks. *IEEE Sensors Journal.* 2020;21(2):2299–2315.

[30] Kader MF. Interference aided cooperative SWIPT for cellular IoT networks towards 5G and beyond. *Physical Communication.* 2020;43:101223.

[31] Na H, and Lee C. A full-duplex SWIPT system with self-energy recycling to minimize energy consumption. *EURASIP Journal on Wireless Communications and Networking.* 2018;2018(1):1–12.

[32] Zhang S, Shi X, Song R, *et al.* Achievable rate analysis and maximisation for the SWIPT-based half-duplex quantise-and-forward cooperation in IoT networks. *IET Communications.* 2020;14(22):3989–3995.

[33] Huang Y, Liu M, and Liu Y. Energy-efficient SWIPT in IoT distributed antenna systems. *IEEE Internet of Things Journal.* 2018;5(4):2646–2656.

[34] Abbasizadeh H, Kim SY, Samadpoor Rikan B, *et al.* Design of a 900 MHz dual-mode swipt for low-power iot devices. *Sensors.* 2019;19(21):4676.

[35] Boshkovska E, Zlatanov N, Dai L, *et al.* Secure SWIPT networks based on a non-linear energy harvesting model. In: *2017 IEEE Wireless Communications and Networking Conference Workshops (WCNCW)*. IEEE; 2017. p. 1–6.

[36] Wang D, Liu H, Ma X, *et al.* Energy harvesting for internet of things with heterogeneous users. *Wireless Communications and Mobile Computing.* 2017;2017:1858532.

[37] Xu K, Zhang M, Liu J, *et al.* SWIPT in mMIMO system with non-linear energy-harvesting terminals: Protocol design and performance optimization. *EURASIP Journal on Wireless Communications and Networking.* 2019;2019 (1):1–15.

[38] Wang J, Wang G, Li B, *et al.* Massive MIMO two-way relaying systems with SWIPT in IoT networks. *IEEE Internet of Things Journal.* 2020.

[39] Mishra D, Alexandropoulos GC, and De S. Harvested power fairness optimization in MISO SWIPT multicasting IoT with individual constraints. In: *2018 IEEE International Conference on Communications (ICC)*. IEEE; 2018. p. 1–6.

[40] Li B, Dai Y, Dong Z, *et al.* Energy-efficient resources allocation with millimeter-wave massive MIMO in ultra dense HetNets by SWIPT and CoMP. *IEEE Transactions on Wireless Communications.* 2021.

[41] Schaible S. Fractional programming. II, On Dinkelbach's algorithm. *Management Science.* 1976;22(8):868–873.

[42] Feng W, Tang J, Yu Y, *et al.* UAV-enabled SWIPT in IoT networks for emergency communications. *IEEE Wireless Communications.* 2020;27 (5):140–147.

[43] Kang JM, and Chun CJ. Joint trajectory design, tx power allocation, and rx power splitting for UAV-enabled multicasting SWIPT systems. *IEEE Systems Journal.* 2020;14(3):3740–3743.

[44] Sharma S, Roy SD, and Kundu S. Physical layer security in bidirectional communication with SWIPT. *Procedia Computer Science.* 2020;171: 1093–1101.

[45] Sharma N, Magarini M, and Alam MM. Internet of drones-enabled smart cities. In: *IoT Architectures, Models, and Platforms for Smart City Applications*. IGI Global; 2020. p. 107–133.

[46] Ullah Z, Al-Turjman F, Moatasim U, *et al.* UAVs joint optimization problems and machine learning to improve the 5G and beyond communication. *Computer Networks.* 2020; p. 107478.

[47] Li X, Liu M, Deng C, *et al.* Joint effects of residual hardware impairments and channel estimation errors on SWIPT assisted cooperative NOMA networks. *IEEE Access.* 2019;7:135499–135513.

[48] Deka K, and Sharma S. Hybrid NOMA for future radio access: Design, potentials and limitations. arXiv preprint arXiv:201208106. 2020.

[49] Clerckx B, Huang K, Varshney LR, *et al.* Wireless power transfer for future networks: Signal processing, machine learning, computing, and sensing. arXiv preprint arXiv:210104810. 2021.

[50] An W, Zhang P, Xu J, *et al.* A novel machine learning aided antenna selection scheme for MIMO Internet of Things. *Sensors*. 2020;20(8):2250.

[51] Li L, Ma H, Ren H, *et al.* Learning-aided resource allocation for pattern division multiple access based SWIPT systems. *IEEE Wireless Communications Letters.* 2021;10(1):131–135.

Chapter 10
Energy-efficient computing for future IoT applications

Muhammad Wasif Shabir[1], *Muhammad Inyat Ullah Imran*[1], *Hira Tariq*[1], *Umar Farooq*[1], *and Muhammad Awais Javed*

Abstract

Energy management is a key challenge for future Internet of Things (IoT) applications. As the IoT nodes are energy-constrained devices, energy management for IoT networks is of paramount importance. A major component of future 6G-empowered IoT networks will be massive computing enabled by fog nodes. In this chapter, we present recent work carried out related to energy-efficient computing at the fog nodes and energy-efficient transmission between IoT devices and fog nodes. This chapter covers techniques and algorithms related to energy-efficient task offloading, device processing parameters control, content caching, and intelligent-surface-based transmissions. Finally, we present future opportunities and open challenges to improve energy efficiency for IoT applications.

10.1 Introduction

In the modern age, technology has made possible to realize concepts such as smart cities, smart devices, and smart industries. Internet of Things (IoT) will connect cyber-physical systems with the Internet, thus providing ubiquitous connectivity. With the accelerated expansion of IoT, a substantial number of gadgets such as tablets, smartphones, and wearable devices require access to the Internet for various applications [1]. According to the statistics in 2017, 20 billion IoT devices were connected to the Internet, which has increased to about 30 billion in 2020, and it will double once again by 2025. This rapid growth in devices creates a massive

[1]Department of Electrical and Computer Engineering, COMSATS University Islamabad, Islamabad, Pakistan

amount of data that needs to be stored and processed and hence brings a lot of new challenges to solve [2].

The devices in IoT are generally small sensors that have many resource constraints such as power, computation ability, and storage capability. Due to these constraints, cloud computing has been a popular technology of the last decade, which provides on-demand network access to IoT devices. Cloud computing allows IoT devices to store their data in the cloud and also transfer their computation task to the cloud using the mobile networks, thus resulting in reduced computational load [3]. Moreover, shifting data to the cloud brings advantages in terms of scalability and availability, which can be essential for the development of the Internet of Everything [4,5].

However, cloud nodes are far away from the IoT devices, and communication between these two nodes may not be cost effective. Fog computing allows computing nodes to be installed near the IoT devices, thus allowing IoT devices easier access to the computing nodes. While fog computing nodes are of less computational capacity as compared to the cloud, they can be installed in large numbers near IoT devices. As a result, IoT applications can easily benefit from their computational power and storage capacity.

Fog computing nodes provide two major benefits, first is the content caching, and second is the task offloading. Popular contents may be cached at strategic locations at the fog nodes, thus resulting in a lower load on the base stations. It will also allow users to access contents with a higher data rate. Similarly, IoT nodes will require help in terms of task computation. With the increased number of applications, IoT nodes will have to perform several tasks which may not be computationally possible and result in energy challenges. Thus, fog nodes will compute tasks on behalf of the IoT nodes.

Communication between the IoT nodes and fog computing nodes is also critical as cached content and computed tasks must be retrieved from the fog nodes. Reconfigurable intelligent surfaces (RIS) are a central building block of future 6G systems and can improve the signal-to-noise ratio at the receiver by providing increased diversity. RIS is a combination of large numbers of low cost and passive reflecting elements. With the help of these reflecting elements, the incoming signal is exploited and reflected toward the desire location. Each reflecting element, also called meta-surface, is composed of ferromagnetic materials and is used for the manipulation of impinging signals in a perfect manner. RIS is incapable of complex signal processing techniques; thus, a smart controller present between the transmitter and the RIS with a wired or wireless link helps in the establishment of RIS-assisted communication [2].

In this chapter, we present a review of the current techniques related to energy-efficient computing and transmissions for IoT applications. We first categorize the current work into four categories, namely, energy-efficient task offloading, energy-efficient device processing parameters control, energy-efficient caching, and energy-efficient intelligent-surface-enabled transmissions. A detailed review of work carried out in these four categories is presented in Section 10.2. We further highlight opportunities for future work and challenges that still need to be addressed related to energy-efficienct computing and transmissions for IoT applications in Section 10.3. Finally, conclusions are drawn in Section 10.4.

10.2 Recent work related to energy-efficient computing in IoT

In this section, we discuss the recent work related to energy-efficient computing in IoT. Figure 10.1 shows the architecture of future energy-efficient computing for IoT applications. The four major components that are related to the energy management of IoT include task offloading, device processing parameters control, content caching, and intelligent-surface-aided transmissions.

10.2.1 Energy-efficient task offloading

IoT devices have limited resources in terms of power, storage, and computation. Thus, local computation of complex tasks is not possible and can result in long delay and high energy consumption. Cloud computing is an efficient solution to this problem where cloud servers provide their computational resources for the execution of complex tasks. A major limitation of cloud nodes is that they are centralized nodes and are far away from the edge of the network. A longer transmission is required in cloud-assisted task offloading which consumes a larger bandwidth. Moreover, cloud-supported task offloading is also not efficient in terms of energy consumption and delay.

Fog computing brings computing servers nearer to the edge of the network. Several fog nodes are placed at different locations to serve IoT nodes in terms of computation and storage. Although these fog nodes have higher computation and energy capabilities as compared to the IoT nodes, they still are less capable devices as compared to the cloud servers. Hence, energy efficiency is a big challenge.

Several works in the literature have proposed solutions to enable energy-efficient task offloading, as shown in Table 10.1. In [3], the authors investigated the problem of computational offloading for resource allocation (RA) in fog computing collaborated with cloud computing-based heterogeneous network (HetNet).

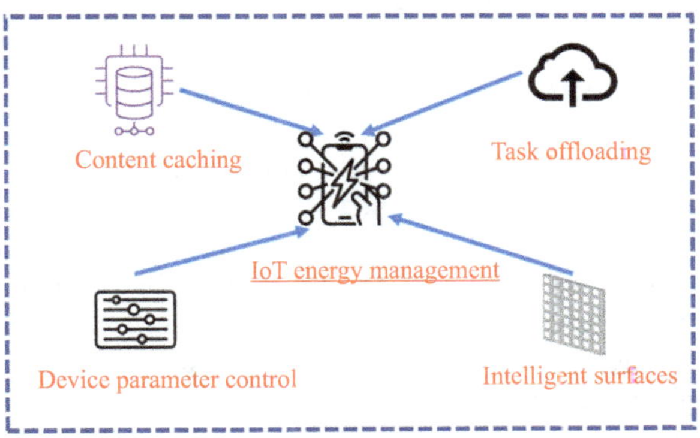

Figure 10.1 Architecture of energy-efficient computing for IoT

Table 10.1 Recent work done in energy-efficient task offloading in IoT networks

Ref.	Technique used	Objectives	Network
[3]	Constrained optimization	Reduction of system cost	MFC
[6]	DTMC and MDP	Minimize energy consumption	MCC
[7]	BIP and genetic algorithm	Improvement of energy consumption and QoS	MCC, MFC
[8]	MDP and dynamic programming	Efficient use of EVs as fog nodes	VFC
[9]	MILP	Minimize power consumption	MCC, MFC
[10]	MDP, VIA	Optimize computation time and energy	MEC
[11]	Fractional programming	Minimize maximal weightage cost	MFC, MCC

The authors presented a scheme named energy-efficient computation offloading and resource allocation (ECORA), which uses a constrained optimization model to manage the offloading of tasks among fog nodes, allocation of transmission power, and allocation of computation resources to different tasks. The ECORA scheme improves the task offloading performance in terms of system cost, i.e., offloading delay as compared to conventional schemes.

Terefe *et al.* [6] formulated an algorithm for multi-site offloading in mobile cloud computing by using the Markov decision process (MDP) for minimization of energy consumption. The authors modeled the channel fading and heterogeneity of the network using discrete-time Markov chain (DTMC) and formulated the energy optimization problem with a limited time utilization constraint. To solve this optimization problem, the authors proposed an energy-efficient multi-site offloading policy algorithm that uses the value iteration algorithm to make optimal task offloading decisions for multi-site execution application.

In [7], Mebrek *et al.* proposed an efficient solution to balance the trade-off between energy and delay in the IoT-Fog-cloud-based network. The main focus is to minimize the consumption of energy and latency constraints per the required QoS. An optimization problem is formulated for the energy-latency tradeoff and solved by using evolutionary algorithms including the broadcast incremental power (BIP) algorithm and an improved genetic algorithm (IGA). Results show that the energy efficiency and delay of the proposed algorithm are improved as compared to the traditional cloud networks.

Habtamu *et al.* [8] worked on the concept of vehicular fog computing (VFC) in which the electric vehicles (EVs) are used as fog nodes and used to save energy of the mobile devices (MDs). The key idea behind VFC is to exploit underutilized vehicle computational resources as fog nodes, thus providing an on-demand computation facility to the mobile devices that are constrained in terms of energy and processing capabilities. MDP is applied to formulate the problem of RA and solved by using dynamic programming.

In [9], the authors modeled the complete IoT-fog-cloud architecture and proposed a solution to minimize the consumption of power for networking and task processing. Task offloading decision to a particular fog/cloud node is taken using a mixed-integer linear programming (MILP)-based algorithm with a goal to reduce total power consumption. The authors also consider different offloading scenarios for IoT applications such as splitting of a task and offloading different parts of the task to multiple fog nodes.

In [10], the authors used a similar idea of multi-site offloading as given in [3], but here the authors have used edge servers and cloud servers as offloading sites. MDP is used to propose an efficient offloading technique to conserve energy within a limited amount of time. Performance evaluation of the proposed method shows that energy consumption in the edge server is less than the cloud server, and the computation time cost for multi-site execution on cloud and edge is also less than the time cost for local computation.

The work done in [11] identifies tasks that are computationally intensive and may consume high energy and offload these tasks to the fog and cloud nodes. The authors propose a computation offloading and resource allocation (CORA) algorithm to optimize the computational resource allocation, bandwidth, and transmission power. Results show that the algorithm converges within a limited time, and improvements are achieved in terms of energy efficiency. Besides, these algorithms [12–15] also present ideas related to energy-efficient task offloading in fog computing networks.

10.2.2 Energy-efficient device control

The computational processing parameters of fog nodes have an impact on energy, and hence, selection of these parameters becomes critical in an IoT network. Two important computational parameters are the voltage and frequency of the fog nodes. A higher voltage and frequency imply a higher computational processing speed; however, it results in higher energy consumption. Voltage and frequency of fog nodes can thus be adaptively varied per the task delay and fog node energy requirements. A few of literature related to device control as listed in Table 10.2 is reviewed one by one.

Table 10.2 Recent work done in energy efficient device control

Ref. no	Method used	Objectives	Network
[16]	MapReduce-DVFS	Increasing EE, decreasing average queue length	Fog-cloud computing network
[17]	Adaptive-BLE, Ski-rental optimization	Increasing EE of IoTs' battery	IoT-FOG communication networks
[18]	Adaptive-duty-cycle	Energy conservation	IoT-wireless sensor network
[19]	FL	Energy conservation	Fog network
[20]	AMO	Energy conservation	Fog network
[21]	EADCO, dynamic scaling method	Energy conservation	Fog-cloud computing network
[22]	Lyapunov optimization technique	Reducing load on power grid	Fog network

Two varied asymptotic data generated from fog-coordinated and fog-assisted cloud platforms were scaled and compared by using the MapReduce scheme in [16]. MapReduce is a standard distributed programming scheme for dispatching large data analytics that works in parallel to increase the energy efficiency of the model. In this context, MapReduce helps improve the energy efficiency of the dynamic-voltage–frequency scaling (DVFS)-aided cloud computing (CC) platform. Results accumulated from the asymptotic analytics showed that the proposed fog-assisted CC platforms can outperform fog-coordinated cloud computing platforms by approximately 32.2% in terms of time-average energy efficiency and approximately 37% in terms of time-averaged queue-length.

The power efficiency node discovery (PEND) technique is presented in [17] to overcome the issue of efficient node sensing and discovery in a fog environment. An improved Bluetooth low-energy (BLE) scheme was applied for the IoT nodes' communication with the fog network by adding a switching ON/OFF method and a ski-rental optimization scheme to control the rate of discovery message advertisement.

Wireless sensor networks (WSNs) in IoT platforms are energy-constrained devices, therefore to conserve the energy of WSNs' battery, an adaptive duty-cycle scheme is proposed in [18]. This scheme removes the idle listening mode of the sensor nodes and replaces it with the sleep mode. Subsequently, each time slot is divided into two modes, a wake-up mode and a sleep mode. Energy conservation is achieved by optimally selecting the two modes of operation.

The authors in [19] proposed a federated learning (FL) technique to efficiently control the CPU frequency. The goal of the FL technique is to reduce the learning complexity and only use local training at the IoT nodes for developing the model. A non-linear programming model is used to optimize the CPU frequency.

Fog computing supports low latency-constrained applications that are emerging in the modern world. The nature of the fog network is highly distributed and limited in resources, so one of the challenges is how efficiently fog computing can be utilized in delay-sensitive tasks. The authors in [20] proposed an ant mating optimization (AMO) to discover the best possible policy for task offloading while maintaining the stringent requirements of energy consumption.

The energy optimal computation offloading (EADCO) scheme is developed in [21] for fog computing-based industrial Internet of Things (IIoT). The system model of this paper consists of a central cloud server and fog nodes establishing communication links between IIoT devices and cloud servers. Different fog device parameters such as the number of offloaded tasks, delays, and other network constraints are included for the joint optimization problem formulation of the offloading ratio and transmission time. Moreover, for the improvement of the proposed solution, a dynamic scaling method is also used. Numerical results have shown that the proposed optimization framework can conserve energy consumption while reducing the task completion time.

The work in [22] introduced the concept of utilized renewable energy (RE) sources in fog computing. The power on the main grid is considerably reduced if RE resources are managed appropriately. Depending on workload and energy source availability, efficient DVFS-based request dispatching machines are designed that can help scale the frequency and modulation perfectly. The system is extended by investigating the time-varying non-linear state equation. An optimization technique called Lyapunov provides asymptomatic optimality while keeping the RE usage sufficient and balanced while minimizing the service time.

10.2.3 Energy-efficient caching

Content caching is another efficient technique that can enhance the energy efficiency of IoT nodes. Caching works by placing popular contents (that are likely to be downloaded extensively and requested by several IoT nodes) on the different fog nodes. By allowing data to be downloaded from the nearby fog nodes instead of cloud nodes or cellular base stations, IoT node's energy is conserved. Some of recent works affiliated with efficient caching are given in Table 10.3.

One of the works related to caching proposed by authors in [23] used an energy-aware caching technique for infotainment facilities in intelligent connected vehicles (ICVs). The goal of the technique is to conserve energy while maintaining the QoS. The proposed technique used optimal stopping theory to find the optimal set of content caching nodes such that the energy of the vehicles is conserved while keeping the latency within the allowed threshold.

In another work in [24], the energy-efficient node-association algorithm was proposed for content caching in heterogeneous fog networks. First, the decision of energy-efficient content caching and transfer in either fog nodes or the fog access point (FAP) was made while considering a fixed-node association scheme. An optimization problem is formulated that minimizes the energy consumption in both caching scenarios (at fog nodes or FAPs) under the restrictions of caching memory. Results show that the proposed scheme results in an enhanced battery lifetime of the IoT nodes.

Table 10.3 Recent work done in energy efficient caching

Ref.	Method used	Objectives	Network
[23]	Optimal stopping theory	Energy-efficient caching	Vehicular network
[24]	EE optimization/joint-node association algorithm	Maximizing ee	Heterogeneous fog networks
[25]	Numerical method	Improving EE	CDC
[26]	Convex optimization-AES scheme-dedicated powerlines based scheme	Minimizing energy consumption	Heterogeneous network
[27]	Optimization scheme	Minimizing energy consumption body text	UDN
[28]	ILP-heuristic algorithm	Minimizing energy consumption	Metro area networks
[29]	CRA optimization	Minimizing energy consumption	Information centric wireless network

Vu *et al.* [25] investigated the cache-aided content delivery-network (CDC) to improve the energy efficiency, where users requested the content from the data center (DC) through a base station (BS). In this model, both BS and users can cache the data (using either the coded or the uncoded technique) and an optimization problem is formulated to minimize energy utilization while maintaining the QoS. Results generated from the numerical method demonstrate that the uncoded caching technique was able to conserve energy better than the coded technique.

In an attempt to reduce the energy utilization of conventional smart-grid while maintaining the QoS, the authors in [26] proposed an alternative-iteration-based algorithm to jointly optimize the three layers of content caching, content delivery, and energy assistance in an HetNet. The proposed technique uses a convex-optimization scheme and matching theory. Furthermore, two different schemes of energy cooperation were also proposed, first is the aggregated energy-sharing (AES) scheme and the second one is the dedicated powerlines-based scheme.

Content caching in the mobility-aware scenario is presented in [27] to minimize the energy consumption in ultra-dense networks (DNS). The technique proposed clusters of small cells (SCs) that are grouped where mobile users (MUs) can roam between clusters. Energy consumption optimization in cache-aided UDNs was transformed into two sub-problems. In the first sub-problem, a polynomial-heuristic-algorithm was proposed to reduce the energy utilization of the inter-cluster caching, whereas the second sub-problem was formulated to place the contents in SCs within a cluster to reduce its energy utilization.

To minimize the energy utilization of the network, the on-and-off energy-efficient video-on-demand (VoD) content caching scheme is proposed in [28] for both fixed and dynamic traffic models. The integer-linear programming (ILP) method and heuristic algorithm were applied to turn the cache on and off. Cache nodes were resourcefully placed in the core-access and metro network topology and were beneficial to reduce the video load on the network.

In an order to conserve the energy of the users while reducing the time delay to obtain the requested cache, an optimization scheme of cache-resource allocation (CRA) was proposed in [29]. The CRA scheme can place the contents of a power-constrained user in the nearby cache node. To achieve the optimal CRA, the problem was further divided into two sub-problems, BS caching-capacity (BS-CC) and resource-matching (RMAT). Both problems are solved separately, and an optimal resource allocation is obtained by combining the two results.

10.2.4 Energy efficient RIS

Deployment of RIS in future wireless and IoT applications are proven to be highly energy efficient and the demand for massive connectivity can be attained by using these planner surfaces. Some of the latest papers for improving and making the existing communication efficient using RIS are summarized in Table 10.4. RIS while enabling line-of-sight (LoS) communication can support cell edge users, augmented reality and virtual reality (AR/VR), dead-end users, unmanned aerial vehicles (UAVs), smart transportation, and edge computing [30].

Table 10.4 Recent work done in energy efficient RIS

Ref.	Method used	Objective	System model
[31]	AO	Power minimization	MISO
[32]	DRL	Power minimization	MISO
[33]	AO	Power minimization	SISO
[34]	AO	SNR maximization, coverage extension	3D architecture, ARIS
[35]	Quadratic Transformation	Maximization of EE and spectral efficiency, ergodic capacity.	ARIS
[39]	AO	Sum power minimization	MISO, SWIPT
[41]	AO, penalty Method	Power minimization	MISO, SWIPT
[40]	AO, BCD	Sum rate maximization	MIMO, SWIPT
[43]	AO, Gaussian Randomization	Sum throughput maximization	MISO, WPCN, NOMA
[44]	AO	Information rate maximization	MISO, MEC
[46]	AO, gradient descent, majorize-minimization	Energy efficiency maximization	MISO
[47]	AO, Dinkelbach's Method	Energy efficiency maximization	MISO

RIS integrated with conventional multiple-input-single-output (MISO) was discussed in [31] whereby jointly optimizing the transmit beamforming at the transmitter and reflect beamforming at the RIS, the received signal power is maximized at the end-user. Two algorithms were applied as part of the work, first is a semi-definite technique-based centralized algorithm with channel state information (CSI) available globally, while the second is a low complexity distributed algorithm. Results show a significant gain in throughput as compared to traditional communication models.

Another learning-based algorithm was proposed for RIS in an MISO setup in [32]. The deep reinforcement learning (DRL) framework is used to tackle the non-convexity of the signal-to-noise (SNR) ratio optimization problem. The main aim is to maximize the SNR by using the deep deterministic policy gradient (DDPG) algorithm in the proposed DRL method. The DRL framework shows better results in terms of execution time as compared to the other techniques.

A power minimization problem for RIS-assisted communication was formulated in [33], where orthogonal multiple access (OMA) and non-orthogonal multiple access (NOMA) schemes are compared for downlink transmission. The conventional superiority of NOMA over OMA was established from the numerical results. Apart from the assisting transmitter and receiver while remaining at one fixed location, RIS can also support a variety of mobility-aware applications. One potential IoT application is UAVs. A new concept called arial RIS (ARIS) was introduced in [34] where RIS is coated or mounted over the UAV. RIS mounted UAVs flying at high elevations helps in expanding the coverage area and ultimately minimizing the blockage possibilities. The worst-case SNR maximization problem was formulated here while jointly optimizing the active and passive beamforming and placement of RIS.

In [35], a UAV-borne RIS was deployed and an extensive optimization framework was given for different transmission modes. These include UAV-only mode, RIS-only mode, and UAV-RIS mode for the comparison of different performance evaluation parameters such as outage probability, ergodic capacity, and energy efficiency. Aside from the introduction of ARIS, the RIS-integrated UAV can also be used for the strengthening of cellular ground communication with the help of passive beamforming at RIS [36–38].

A substantial amount of work was also done on the IoT-based wireless application called simultaneous information and wireless power transfer (SWIPT) and wireless power communication networks (WPCNs). The efficiency of SWIPT can be improved by deploying the RIS-assisted SWIPT system with multiple energy and information receivers [39–41]. RIS can notably recompense for SWIPT power loss via passive beamforming for smart home or smart office applications.

An MISO setup is given in [39] for assisting the energy-harvesting receivers (EHRs) and information decoding receivers (IDRs). The goal of this work was to maximize the weighted sum power at EHRs while optimizing the transmit precoding (TPC) vector at an AP and phase shifts at RIS. The nature of the formulated problem is non-convex, and it is subjected to IDRs signal-to-interference-noise ratio constraint. In the end, a new suboptimal algorithm is developed, and it was found that both IDRs and EHRs can be served while transmitting information signal only to AP.

Another investigation of SWIPT was given in [42] where the ultimate goal is to minimize the transmit power while optimizing the TPC vectors and phase shifts. The efficiency of multiple-input-multiple-output (MIMO)-based SWIPT was investigated in [40] where a weighted sum rate (WSR) problem was formulated. By using the block coordinate descent (BCD) method, the coupled TPC matrices and phase shifts were optimized iteratively

RIS is also being applied in traditional WPCN setups where RIS is also utilized for efficient wireless energy transfer (WET) and uplink wireless information transfer (WIT). Generally, the hybrid access point (H-AP) is considered here and works in two phases. In the first phase, the IoT user harvests energy from both the transmit signal at H-AP and the reflection from RIS. In the second phase, the user uplinks the information via a novel NOMA scheme to the AP, resulting in maximization of the throughput [43].

Another area where RIS can be used to improve the energy efficiency of IoT is MEC. The response, problems, and challenges in the area of RIS-assisted MEC are not fully exploited; however, the joint design of RIS-assisted communication, as well as computing, is presented in [44,45] in which the computational latency, energy, and spectral efficiency of the whole system are improved using RIS. In [46], energy efficiency of the RIS-based wireless network is maximized using alternate optimization. User fairness or minimum QoS was considered for the formulation of non-convex optimization problems. Two approaches are used to solve this problem, the first one was the power allocation using fractional programming, and the second one was the phase shifts optimization using the gradient descent algorithm. Furthermore, in the second technique, sequential fractional programming is used for optimizing the phase shifts.

Energy efficiency in terms of green communication for RIS-assisted wireless networks was studied in [47]. The reflect beamformer with a low overhead (in terms of a number of bits) was considered. A combined algorithm was designed for both power allocation to the users and the reflect beamformer. RIS with even 1-bit phase was shown to be more efficient as compared to the traditional relay facilitated communication.

Millimeters (mm) wave and terahertz (THz) communication models provide high data rates to applications such as AR/VR and high-quality graphics gaming. These models deliver a high data rate, but the propagation environment here is also more vulnerable to signal blockages and limited coverage. Thus, RIS can facilitate the coverage and service region extension in these scenarios [48,49]. Similarly, multiple RIS-aided cognitive radio for licensed and unlicensed users were discussed in [50,51] to improve energy efficiency.

10.3 Future opportunities and challenges

10.3.1 Integrated energy solution

Future IoT networks will require an integrated energy solution comprising all the components as described in this chapter. A big challenge will be how to develop protocols that address the energy issues of these different components. For example, energy-aware protocols must jointly consider resource allocation for caching and task offloading, and optimal device control. Moreover, RIS-based transmissions will also impact the caching and task offloading decisions. A content may be more popular and good candidate for caching, but there may be no RIS between the fog node and the IoT node, and hence, it may not be cached as content retrieval could be difficult. Therefore, future protocols must consider challenges across different technologies to come up with an intelligent energy-aware solution.

10.3.2 Federated learning

Machine learning-based solutions and artificial intelligence will be a key component of future IoT and 6G networks [52]. In this regard, it will be difficult to develop models based on global data. Hence, distributed learning models developed using data from a certain area will be implemented. This idea known as federated learning will be more secure and help in developing more accurate and quick solutions. In the context of energy management, federated learning will provide energy-efficient caching and task offloading solutions with lower complexity and quick convergence time.

10.3.3 Energy harvesting

Energy harvesting is an important technique that will enable future IoT devices to obtain energy from sources such as solar and RF [53]. Energy harvesting will impact caching and task offloading resource allocation of fog nodes. Computation load on fog nodes may be increased if it has higher energy harvesting capability.

Hence, efficient load-balancing techniques need to be developed that consider the energy level of the fog nodes including the amount of harvested energy.

10.4 Conclusion

In this chapter, we present an overview of energy-efficient computing techniques for future IoT networks. We discuss the recent energy efficiency-related work for four key components of computing in IoT networks, task offloading, device processing parameters control, content caching, and intelligent-surface-based transmissions. We also present important future opportunities and challenges that need to be considered in future IoT networks.

List of acronyms

Acronym	Definition
AES	Aggregated energy-sharing
AR	Augmented reality
ARIS	Arial RIS
AR/VR	Augmented and virtual reality
A2C	Advantage-actor-critic
BCD	Block coordinate descent
BIP	Broadcast incremental power
BLE	Bluetooth low energy
BS	Base station
BSCC	Base station caching-capacity
CC	Cloud computing
CDC	Content delivery-network
CORA	Computation offloading and resource allocation
CRA	Cache-resource allocation
CSI	Channel state information
DC	Data center
DDPG	Deep deterministic policy gradient
DRL	Deep reinforcement learning
DTMC	Discrete-time Markov chain
DVFS	Dynamic-voltage-frequency scaling
EA	Evolutionary algorithm
EADCO	Energy optimal computation offloading
ECORA	Energy-efficient computation offloading and resource allocation
EHRs	Energy harvesting receivers
EMOP	Energy-efficient multi-side offloading policy
EV	Electric vehicles
FAPs	Fog access points
FL	Federated learning
GNEP	Generalized Nash equilibrium problem
H-AP	Hybrid access point
HetNet	Heterogeneous network
HNs	Heterogeneous nodes

(Continues)

(*Continued*)

Acronym	Definition
ICV	Intelligent connected vehicles
IDRs	Information decoding receivers
IGA	Improved genetic algorithm
IIoT	Industrial IoT
ILP	Integer-linear programming
IoT	Internet of Things
IoE	Internet of Everything
IPM	Interior point method
KKT	Karush–Kuhn–Tucker
LoS	Line-of-sight
MCC	Mobile cloud computing
MD	Mobile devices
MDP	Markov decision process
MEC	Mobile edge computing
MFC	Mobile fog computing
MILP	Mixed integer linear programming
MIMO	Multiple-input-single-output
MISO	Multiple-input-single-output
ML	Machine learning
mm	Millimeters
MPCA	Module placement method by classification and regression tree algorithm
MPMCP	Module placement with Markov chain process
MUs	Mobile users
NOMA	Non-orthogonal multiple access
OFDMA	Orthogonal frequency division multiple access
OMA	Orthogonal multiple access
PEND	Power efficiency node discovery
RA	Resource allocation
RE	Renewable energy
RIS	Reconfigurable intelligent surfaces
RMAT	Resource-matching
SCs	Small cells
SNR	Signal-to-noise
SWIPT	Wireless power transfer
THz	Terahertz
TPC	Transmit precoding
UAVs	Unmanned aerial vehicles
UDN	Ultra-dense networks
VFC	Vehicular fog computing
VIA	Value iteration algorithm
VoD	Video-on-demand
VR	Virtual reality
WET	Wireless energy transfer
WIT	Wireless information transfer
WPCN	Wireless power communication networks
WSN	Wireless sensor networks
WSR	Weighted sum rate

References

[1] B. Omoniwa, R. Hussain, M. A. Javed, S. H. Bouk and S. A. Malik, "Fog/edge computing-based IoT (FECIoT): architecture, applications, and research issues," *IEEE Internet of Things Journal*, vol. 6, no. 3, pp. 4118–4149, June 2019.

[2] M. A. Javed and S. Zeadally, "AI empowered content caching in vehicular edge computing: opportunities and challenges," *IEEE Network*, 2021.

[3] Q. Li, J. Zhao, Y. Gong, and Q. Zhang, "Energy-efficient computation offloading and resource allocation in fog computing for internet of everything," *China Communications*, vol. 14, no. 3, pp. 32–41, 2019.

[4] M. A. Javed, S. Zeadally, and E. B. Hamida, "Data analytics for cooperative intelligent transport systems," *Vehicular Communications*, vol. 15, pp. 63–72, 2019.

[5] K. H. Abdulkareem, M. Abed Mohammed, S. Shamini Gunasekaran, *et al.*, "A review of fog computing and machine learning: concepts, applications, challenges, and open issues," *IEEE Access*, vol. 7, pp. 153123–153140, 2019.

[6] M. B. Terefe, H. Lee, N. Heo, G. C. Fox, and S. Oh, "Energy-efficient multisite offloading policy using Markov decision process for mobile cloud computing," *Pervasive and Mobile Computing*, vol. 27, pp. 75–89, 2016.

[7] A. Mebrek, L. Merghem-Boulahia, and M. Esseghir, "Efficient green solution for a balanced energy consumption and delay in the IoT-Fog-Cloud computing," in *2017 IEEE 16th International Symposium on Network Computing and Applications (NCA)*, 2017, IEEE, pp. 1–4.

[8] H. M. Birhanie, M. A. Messous, S.-M. Senouci, E.-H. Aglzim, and A. M. Ahmed, "MDP-based resource allocation scheme towards a vehicular fog computing with energy constraints," in *2018 IEEE Global Communications Conference (GLOBECOM)*, 2018, IEEE, pp. 1–6.

[9] B. A. Yosuf, M. Musa, T. Elgorashi, and J. Elmirghani, "Energy efficient distributed processing for IoT," *IEEE Access*, vol. 8, pp. 161080–161108, 2020.

[10] K. R. Alasmari, R. C. Green, and M. Alam, "Mobile edge offloading using Markov decision processes," in *International Conference on Edge Computing*, 2018, Springer, pp. 80–90.

[11] J. Du, L. Zhao, J. Feng, and X. Chu, "Computation offloading and resource allocation in mixed fog/cloud computing systems with min–max fairness guarantee," *IEEE Transactions on Communications*, vol. 66, no. 4, pp. 1594–1608, 2017.

[12] L. Liu, Z. Chang, and X. Guo, "Socially aware dynamic computation offloading scheme for fog computing system with energy harvesting devices," *IEEE Internet of Things Journal*, vol. 5, no. 3, pp. 1869–1879, 2018.

[13] L. Liu, Z. Chang, X. Guo, S. Mao, and T. Ristaniemi, "Multiobjective optimization for computation offloading in fog computing," *IEEE Internet of Things Journal*, vol. 5, no. 1, pp. 283–294, 2017.

[14] D. Rahbari and M. Nickray, "Task offloading in mobile fog computing by classification and regression tree," *Peer-to-Peer Networking and Applications*, vol. 13, no. 1, pp. 104–122, 2020.

[15] X. Gu, L. Jin, N. Zhao, and G. Zhang, "Energy-efficient computation offloading and transmit power allocation scheme for mobile edge computing,"*Mobile Information Systems*, vol. 2019, 2019.

[16] S. Hou, W. Ni, S. Zhao, B. Cheng, S. Chen, and J. Chen, "Frequency-reconfigurable cloud versus fog computing: an energy-efficiency aspect," *IEEE Transactions on Green Communications and Networking*, vol. 4, no. 1, pp. 221–235, 2019.

[17] R. Venanzi, L. Foschini, P. Bellavista, B. Kantarci, and C. Stefanelli, "Fog-driven context-aware architecture for node discovery and energy saving strategy for Internet of Things environments," *IEEE Access*, vol. 7, pp. 134173–134186, 2019.

[18] A. M. bin Baharudin, M. Saari, P. Sillberg, P. Rantanen, J. Soini, and T. Kuroda, "Low-energy algorithm for self-controlled wireless sensor nodes," in *2016 International Conference on Wireless Networks and Mobile Communications (WINCOM)*, 2016: IEEE, pp. 42–46.

[19] J. Yao and N. Ansari, "Enhancing federated learning in fog-aided IoT by CPU frequency and wireless power control," *IEEE Internet of Things Journal*, vol. 8, no. 5, pp. 3438–3445, 2021.

[20] S. Ghanavati, J. H. Abawajy, and D. Izadi, "An energy aware task scheduling model using ant-mating optimization in fog computing environment," *IEEE Transactions on Services Computing*, 2020.

[21] S. Chen, Y. Zheng, W. Lu, V. Varadarajan, and K. Wang, "Energy-optimal dynamic computation offloading for industrial IoT in fog computing," *IEEE Transactions on Green Communications and Networking*, vol. 4, no. 2, pp. 566–576, 2019.

[22] A. Karimiafshar, M. R. Hashemi, M. R. Heidarpour, and A. N. Toosi, "Effective utilization of renewable energy sources in fog computing environment via frequency and modulation level scaling," *IEEE Internet of Things Journal*, vol. 7, no. 11, pp. 10912–10921, 2020.

[23] H. Wu, J. Zhang, Z. Cai, F. Liu, Y. Li, and A. Liu, "Toward energy-aware caching for intelligent connected vehicles," *IEEE Internet of Things Journal*, vol. 7, no. 9, pp. 8157–8166, 2020.

[24] K. Wang, J. Li, Y. Yang, W. Chen, and L. Hanzo, "Content-centric heterogeneous fog networks relying on energy efficiency optimization," *IEEE Transactions on Vehicular Technology*, vol. 69, no. 11, pp. 13579–13592, 2020.

[25] T. X. Vu, S. Chatzinotas, B. Ottersten, and T. Q. Duong, "Energy minimization for cache-assisted content delivery networks with wireless backhaul," *IEEE Wireless Communications Letters*, vol. 7, no. 3, pp. 332–335, 2017.

[26] F. Yin, M. Zeng, Z. Zhang, and D. Liu, "Coded caching for smart grid enabled HetNets with resource allocation and energy cooperation," *IEEE*

Transactions on Vehicular Technology, vol. 69, no. 10, pp. 12058–12071, 2020.

[27] I. Keshavarzian, Z. Zeinalpour-Yazdi, and A. Tadaion, "Energy-efficient mobility-aware caching algorithms for clustered small cells in ultra-dense networks," *IEEE Transactions on Vehicular Technology*, vol. 68, no. 7, pp. 6833–6846, 2019.

[28] O. Ayoub, F. Musumeci, M. Tornatore, and A. Pattavina, "Energy-efficient video-on-demand content caching and distribution in metro area networks," *IEEE Transactions on Green Communications and Networking*, vol. 3, no. 1, pp. 159–169, 2018.

[29] Y. Tang, "Minimizing energy for caching resource allocation in information-centric networking with mobile edge computing," in *2019 IEEE Intl Conf on Dependable, Autonomic and Secure Computing, Intl Conf on Pervasive Intelligence and Computing, Intl Conf on Cloud and Big Data Computing, Intl Conf on Cyber Science and Technology Congress (DAsC/PiCom/CBDCom/CyberSciTech)*, 2019, IEEE, pp. 301–304.

[30] Q. Wu, S. Zhang, B. Zheng, C. You, and R. Zhang, "Intelligent reflecting surface aided wireless communications: a tutorial," *IEEE Transactions on Communications*, 2021.

[31] Q. Wu and R. Zhang, "Intelligent reflecting surface enhanced wireless network via joint active and passive beamforming," *IEEE Transactions on Wireless Communications*, vol. 18, no. 11, pp. 5394–5409, 2019.

[32] K. Feng, Q. Wang, X. Li, and C.-K. Wen, "Deep reinforcement learning based intelligent reflecting surface optimization for MISO communication systems," *IEEE Wireless Communications Letters*, vol. 9, no. 5, pp. 745–749, 2020.

[33] B. Zheng, Q. Wu, and R. Zhang, "Intelligent reflecting surface-assisted multiple access with user pairing: NOMA or OMA?," *IEEE Communications Letters*, vol. 24, no. 4, pp. 753–757, 2020.

[34] H. Lu, Y. Zeng, S. Jin, and R. Zhang, "Enabling panoramic full-angle reflection via aerial intelligent reflecting surface," in *2020 IEEE International Conference on Communications Workshops (ICC Workshops)*, 2020, IEEE, pp. 1–6.

[35] T. Shafique, H. Tabassum, and E. Hossain, "Optimization of wireless relaying with flexible UAV-borne reflecting surfaces," *IEEE Transactions on Communications*, 2020.

[36] S. Li, B. Duo, X. Yuan, Y.-C. Liang, and M. Di Renzo, "Reconfigurable intelligent surface assisted UAV communication: joint trajectory design and passive beamforming," *IEEE Wireless Communications Letters*, vol. 9, no. 5, pp. 716–720, 2020.

[37] D. Ma, M. Ding, and M. Hassan, "Enhancing cellular communications for UAVs via intelligent reflective surface," in *2020 IEEE Wireless Communications and Networking Conference (WCNC)*, 2020, IEEE, pp. 1–6.

[38] M. Hua, L. Yang, Q. Wu, C. Pan, C. Li, and A. L. Swindlehurst, "UAV-assisted intelligent reflecting surface symbiotic radio system," arXiv preprint arXiv:2007.14029, 2020.

[39] Q. Wu and R. Zhang, "Weighted sum power maximization for intelligent reflecting surface aided SWIPT," *IEEE Wireless Communications Letters*, vol. 9, no. 5, pp. 586–590, 2019.

[40] C. Pan, H. Ren, K. Wang, *et al.*, "Intelligent reflecting surface aided MIMO broadcasting for simultaneous wireless information and power transfer," *IEEE Journal on Selected Areas in Communications*, vol. 38, no. 8, pp. 1719–1734, 2020.

[41] Y. Tang, G. Ma, H. Xie, J. Xu, and X. Han, "Joint transmit and reflective beamforming design for IRS-assisted multiuser MISO SWIPT systems," in *ICC 2020 – 2020 IEEE International Conference on Communications (ICC)*, 2020: IEEE, pp. 1–6.

[42] Q. Wu and R. Zhang, "Joint active and passive beamforming optimization for intelligent reflecting surface assisted SWIPT under QoS constraints," *IEEE Journal on Selected Areas in Communications*, vol. 38, no. 8, pp. 1735–1748, 2020.

[43] D. Song, W. Shin, and J. Lee, "A maximum throughput design for wireless powered communications networks with IRS-NOMA," *IEEE Wireless Communications Letters*, 2020.

[44] Y. Cao and T. Lv, "Intelligent reflecting surface enhanced resilient design for MEC offloading over millimeter wave links," arXiv preprint arXiv: 1912.06361, 2019.

[45] Y. Liu, J. Zhao, Z. Xiong, *et al.*, "Intelligent reflecting surface meets mobile edge computing: enhancing wireless communications for computation offloading," arXiv preprint arXiv:2001.07449, 2020.

[46] C. Huang, A. Zappone, G. C. Alexandropoulos, M. Debbah, and C. Yuen, "Reconfigurable intelligent surfaces for energy efficiency in wireless communication," *IEEE Transactions on Wireless Communications*, vol. 18, no. 8, pp. 4157–4170, 2019.

[47] C. Huang, G. C. Alexandropoulos, A. Zappone, M. Debbah, and C. Yuen, "Energy efficient multi-user MISO communication using low resolution large intelligent surfaces," in *2018 IEEE Globecom Workshops (GC Wkshps)*, 2018, IEEE, pp. 1–6.

[48] N. S. Perović, M. Di Renzo, and M. F. Flanagan, "Channel capacity optimization using reconfigurable intelligent surfaces in indoor mmWave environments," in *ICC 2020 – 2020 IEEE International Conference on Communications (ICC)*, 2020, IEEE, pp. 1–7.

[49] W. Hao, G. Sun, M. Zeng, *et al.*, "Robust design for intelligent reflecting surface assisted MIMO-OFDMA terahertz communications," arXiv preprint arXiv:2009.
05893, 2020.

[50] J. Yuan, Y.-C. Liang, J. Joung, G. Feng, and E. G. Larsson, "Intelligent reflecting surface (IRS)-enhanced cognitive radio system," in *ICC 2020 – 2020 IEEE International Conference on Communications (ICC)*, 2020, IEEE, pp. 1–6.

[51] J. Yuan, Y.-C. Liang, J. Joung, G. Feng, and E. G. Larsson, “Intelligent reflecting surface-assisted cognitive radio system,” *IEEE Transactions on Communications*, vol. 69, no. 1, pp. 675–687, 2021.

[52] M. A. Javed and S. Zeadally, “RepGuide: reputation-based route guidance using internet of vehicles,” *IEEE Communications Standards Magazine*, vol. 2, no. 4, pp. 81–87, 2018.

[53] S. Khan, A. N. Alvi, M. A. Javed, Y. D. Al-Otaibi, and A. K. Bashir, “An efficient medium access control protocol for RF energy harvesting based IoT devices,” *Computer Communications*, vol. 171, pp. 28–38, 2021.

Chapter 11

Intelligent MapReduce technique for energy harvesting through IoT devices

Nawab Muhammad Faseeh Qureshi[1], *Isma Farah Siddiqui*[2], *Bhawani Shankar Chowdhry*[3], *and Dong Ryeol Shin*[4]

Abstract

MapReduce is a programming model that processes large-scale datasets in the distributed computing environment. It manages enormous data (big data) using system software such as Apache Hadoop. We recently found that MapReduce does not restrict itself for execution on data repositories only; devices are also using it for edge-node-based runtime efficient data computing. Thus, it also enables MapReduce effective on Internet of Things (IoT) devices for various reasons, such as producing data analytics within the device and preparing a complex analytics dataset. Therefore, we find it applicable that it could be programmed on an IoT device to regulate energy consumption and perform an energy harvesting technique based on the data's production pattern. This chapter presents a novel technique for energy harvesting through the IoT-enabled MapReduce paradigm. The Amazon Web Services platform's evaluation shows how a simple MapReduce paradigm harvests energy through an IoT device.

11.1 Introduction

Energy harvesting (EH) means an arrangement of transformed energy from an environment for a device [1]. It is a process that derives ambient energy from external environmental resources such as solar, thermal, and wind environment and transforms the pre-processed ambient energy into electrical energy. This electrical energy is captured well through precise small-scale reservoirs and stored for small and microdevices [2]. The EH process is inefficient because it involves a loss of

[1]Department of Computer Education, Sungkyunkwan University, Seoul, South Korea
[2]Department of Software Engineering, Mehran University of Engineering and Technology, Jamshoro, Pakistan
[3]Professor Emeritus and Advisor, Mehran University of Engineering and Technology, Jamshoro, Pakistan
[4]College of Software, Sungkyunkwan University, Suwon, South Korea

energy at each step. EH could be well understood by a simple example of harvesting solar energy. At first, we capture the light through panels and then use charge controllers to store the transformed direct current (DC) energy into batteries. The stored energy is further inverted into alternate current (AC) for the general domestic usage, as shown in Figure 11.1. The steps involved in solar energy harvesting lose energy at each step due to functional and capacity overflow problems. Hence, we do not observe that any single component of the solar energy harvesting system facilitates lossless energy management [3].

Internet of Things (IoT) device is a combination of sensors, information processing middleware, mechanical component to regulate a device, and input/output (I/O) handlers. It performs the function of networked things tied together to produce a group output [4]. EH uses an IoT device in a dual role such as (1) it is used as a source component to harvest energy from the environment and (2) it is a potential device to consume energy from harvested energy [5]. This chapter focuses on the first role that discusses how IoT devices are used to harvest energy from the environment. A generic IoT device consists of essential components such as amplifiers, front-end modules, RF chipsets, wireless technologies (wireless wide area network, wireless local area network, and wireless personal area network), sensors, an application processor, actuators, and modem processors [6]. The front-end module uses long-term evolution (LTE) and Wi-Fi bands. It receives signal data through multi-antenna operations and sensors, which are then further processed with the technical support of RF chipsets and modem processors to retrieve a shape of raw data. This dataset is further processed through an application processor that transforms raw data into meaningful functional data known as compressed data. Thus, an IoT device fulfills its purpose and bundles sensory data for the end node [7].

Big data analytics is a macro-process that examines the large-scale data and filters out the result sets through discovering hidden data patterns, trends, and correlations [8]. This giant system process is carried out through various system software such as Apache Hadoop, Cloudera, and MapR [9]. Apache Hadoop is open-source system software that processes large-scale datasets in the distributed computing environment cost-effectively [10]. Hadoop is designed and programmed to be used according to user specifications easily by reusing the open-source code and plugging-in new addons such as IoT known and unknown data formations [11]. It is further categorized into five main components, i.e., YARN, Hadoop Commons, HDFS, MapReduce, and Ozone [12]. Yet another resource negotiator (YARN) is the

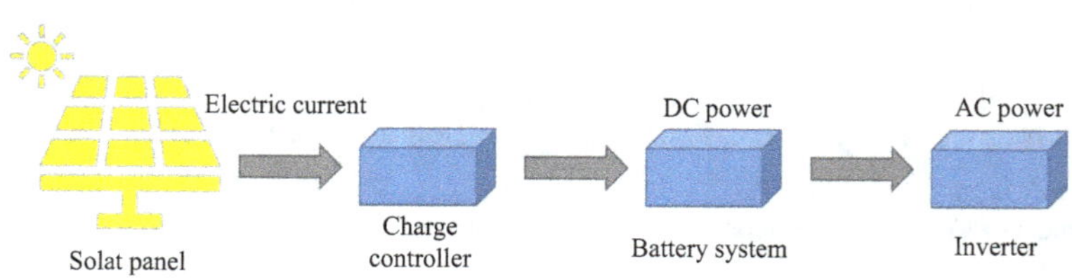

Figure 11.1 Solar energy harvesting

Hadoop brain and it facilitates the system by scheduling jobs and assigning resources to the tasks. Hadoop Commons is the system software environment that regulates all the I/O operations, manages interconnection between all the core components, and provides secure connectivity of external devices and storage mediums such as SAN. Hadoop distributed file system (HDFS) is Hadoop's backbone and regulates the I/O files through the file system's features. The distributed tag is associated with its name because it manages files across the network. MapReduce is the programming model that processes large-scale files in the Hadoop cluster. It is solely responsible for receiving the job and applying the available datasets to produce the cluster's output. Ozone is the storage-tier component that manages the cluster files by assigning them objects, so a file does not need to more instead, an object will move, and later a file has to move once the process gets finishes [13], as shown in Figure 11.2 [10].

MapReduce is a programming model that produces large-scale datasets in the networked environment. It processes datasets through input data, split, record reader, map, partition, sort, reduces, and output. Therefore, when a client requests to process a job, the server ensures that the required dataset on which the job is to be executed is available in the cluster. It then authorizes the client to interact with the system directly. MapReduce receives the job and the path of the dataset in the input data module and then uses a split function as per the number of slave nodes to divide the job. After that, the record reader captures all the indices of the dataset and reads the data file. Map function decides to execute the job on the assigned number of slaves and generate the individual ends. The partition function schedules the map results and generates file patterns to align the result. The sort function then assembles the job results and presents them to reduce function. The reduce function compiles all the job results available at multiple slaves, combines it to a single node, and generates the output, as shown in Figure 11.3 [10].

Recently, we observed that MapReduce is used to extract the resultant data from IoT nodes in the smart grid [14]. This extraction of datasets is carried out by applying analytics at the local edge node. However, we do not find it yet that an IoT device is harvesting energy from the environment; it may use data as a source. This chapter discusses this possibility and recommends the solution through an informal evaluation.

The main contributions of this chapter are as follows:

(a) Explore a novel point of view that EH is possible through an IoT device's data generation patterns.

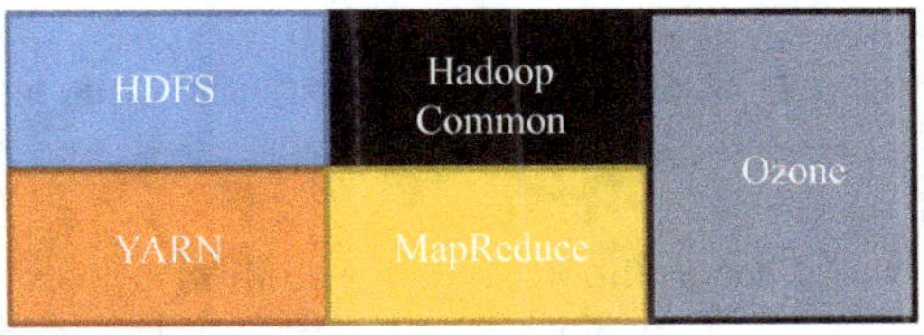

Figure 11.2 Apache Hadoop core components

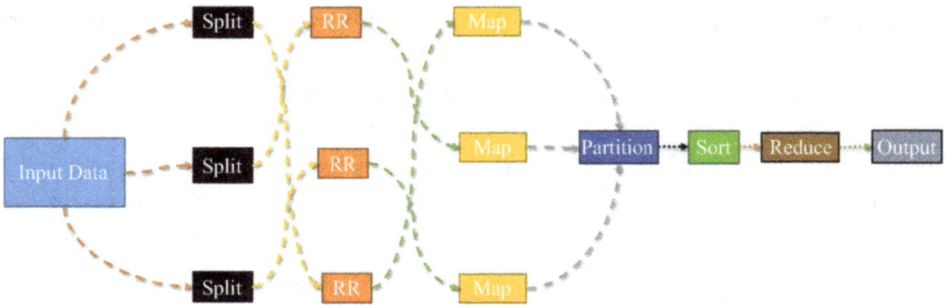

Figure 11.3 Apache Hadoop MapReduce

(b) Efficient programming models such as MapReduce helps devices to figure out an EH procedure.
(c) Group of IoT devices calculates an average gain of EH in a distributed computing environment.
(d) Usage of piezoelectrical buzzer for harvesting energy through data patterns.

The rest of the chapter is organized in such a way that we observe motivation for the need for MapReduce to harvest energy through IoT devices. After that, how energy is harvesting through an IoT device and the brief role of MapReduce in an IoT device. Followed by that, we find a simple Amazon Web Services (AWS) IoT analytics platform evaluation on an IoT device. Finally, ending the chapter with a conclusion.

11.2 Motivation

IoT devices are used to manage and store sensory data into respective distributed repositories. This workflow could be well understood through an example where a smart grid contains various sensory data units such as factory plants, electricity plants, wind power plants, and home appliance units. All the respective units collect sensory data through a type of IoT device known as "Smart Meter." This data is further bundled to the central collection point passing through the transmission channel and stored in the smart grid repository. The smart grid's core controls the data and performs user-defined data analytics, as shown in Figure 11.4. We find that IoT devices' utilization has drastically increased, which undoubtedly leads to either optimizing the energy resources or looking for new EH ways. At this time, we find various forms of EH through IoT devices; however, many of them use either a storage component to harvest the energy or a regulator to redirect the harvested energy. Thus, we could categorize them into three kinds of EH strategies through an IoT device such as storage-less and converter-less harvesting [15], minimizing energy consumption of an IoT network [16], and autonomous power manager for an IoT device [17].

Because we observe a continuous data generation process of an IoT device, this chapter motivates us to consider a novel approach to intelligent data-oriented EH techniques through an IoT device. The proposed technique uses data generation

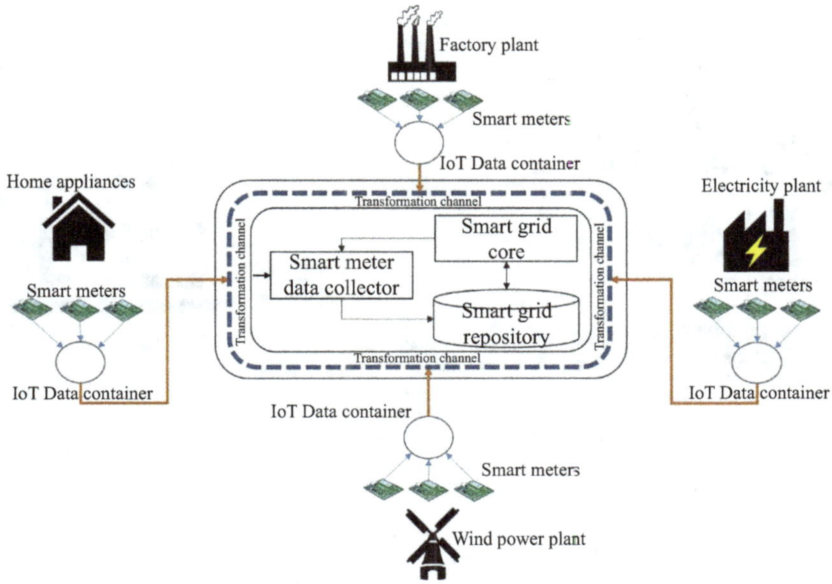

Figure 11.4 IoT-enabled smart grid case study

patterns to note down the workflow. It uses MapReduce to bundle the data batch compression and releases a piezo function that produces an electric buzzer. Thus, an IoT device generates datasets with a regulated amount of harvested energy for the other devices.

11.3 MapReduce-based energy harvesting through an IoT device

The proposed approach consists of an IoT device equipped with an addon framework (MapBuz) of In-place MapReduce processing through an edge node. This framework fetches the fresh arrival of data chunks to a partition-based large buffer and assembles them to a compressed data bundle through MapReduce. Soon after the bundle gets ready, a piezo sensor generates a buzzer that results in the formation of electric charge stored in a reservoir, as shown in Figure 11.5.

11.3.1 Partition-based buffer and bundle compression

The buffer is responsible for handling sensory data's arrival in two partitions, i.e., (1) data partition and (2) metadata partition. The data partition is receiving the sensory data files, and the metadata portion is storing the file information. Once a file reaches a threshold value of compression, such as $\text{Bundle}_i = \sum_{0\to i}\text{File}_{\text{EOF}(\Delta T)}$ equals to 64 MB size, it assembles them through enhanced HAR compression [18]. It presents an appropriate source form to process MapReduce operation on them, as shown in Figure 11.6.

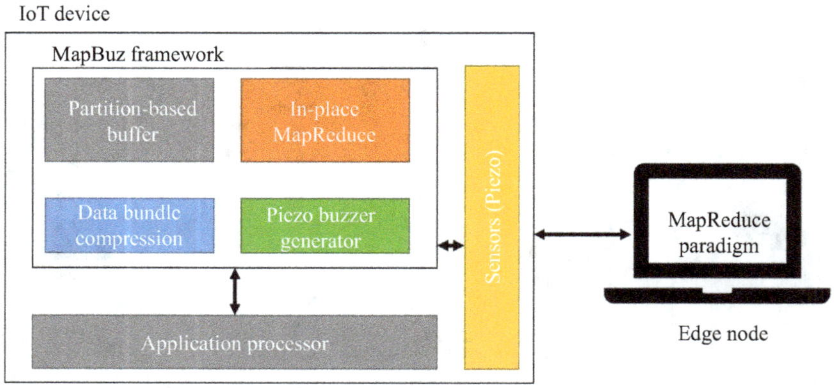

Figure 11.5 MapBuz framework

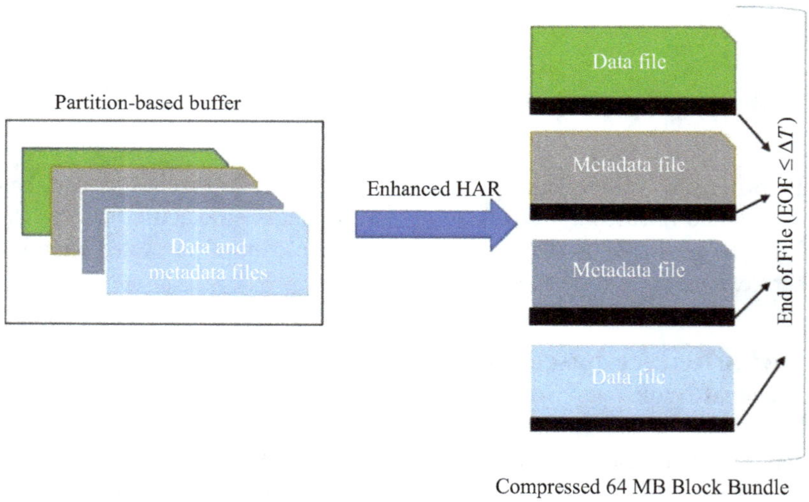

Figure 11.6 Partition buffer and bundle compression

11.3.2 In-place MapReduce

The in-place MapReduce uses compressed blocks for data block placement through the associated edge node [14]. This paradigm receives a compressed 64 MB block from an IoT device as a source for the job task and initializes job confirmation of replica exchanges to other associated edge nodes. Once the block replicas are exchanged at a minimum of three edge nodes, it generates a map function for the said edge nodes and uses a bloom filter [19] to ensure the bundle compression assembly. The bundle assembly processes activate the Piezo buzzer [20], as shown in Figure 11.7.

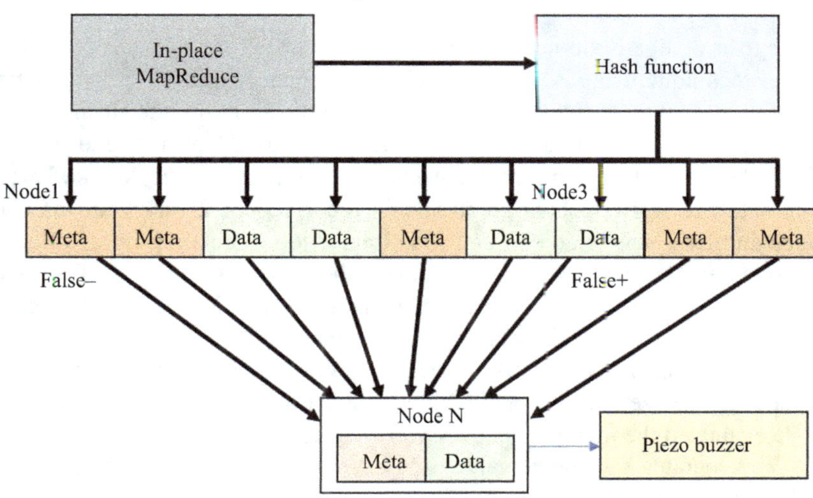

Figure 11.7 In-place MapReduce job operation and piezo buzzer

Table 11.1 MapBuz evaluation table

AWS IoT channel	Pipeline	Dataset used (64 MB)	Status MR job	Time (s)	No. of Buz
Channel-1	Pipeline-1	0.4 MB	Successful	8.19	1
Channel-1	Pipeline-1	0.9 MB	Successful	9.75	1
Channel-1	Pipeline-1	1.3 MB	Successful	8.89	1
Channel-1	Pipeline-1	1.8 MB	Successful	8.76	1
Channel-1	Pipeline-1	2.2 MB	Successful	9.45	1
Channel-2	Pipeline-2	2.8 MB	Successful	8.43	1
Channel-2	Pipeline-2	3.2 MB	Successful	8.84	1
Channel-2	Pipeline-2	0.8 MB	Successful	9.57	1
Channel-2	Pipeline-2	0.6 MB	Successful	8.63	1
Channel-2	Pipeline-2	1.1 MB	Successful	9.81	1
Channel-3	Pipeline-3	1.9 MB	Successful	8.66	1
Channel-3	Pipeline-3	2.9 MB	Successful	9.24	1
Channel-3	Pipeline-3	2.7 MB	Successful	9.19	1
Channel-3	Pipeline-3	3.9 MB	Successful	9.67	1
Channel-3	Pipeline-3	0.1 MB	Successful	9.41	1
Channel-4	Pipeline-4	0.8 MB	Successful	8.78	1
Channel-4	Pipeline-4	1.6 MB	Successful	8.99	1
Channel-4	Pipeline-4	1.1 MB	Successful	9.48	1
Channel-4	Pipeline-4	2.9 MB	Successful	8.82	1
Channel-4	Pipeline-4	2.1 MB	Successful	9.01	1

11.4 How much IoT devices' harvest energy

In order to investigate the number of buzzers generated for EH, we evaluate the proposed technique using AWS IoT device management and AWS IoT analytics paradigms. We set three things with a type of Smart meter with the attribute key as MapBuz and value as Evaluator with tag name as MapBuzT and value as EvaluatorT. After setting the things, we set the Amazon Elastic MapReduce job cluster with the software configuration of Core Hadoop: 2.10.1 with m5.xlarge instance and EC2 key-value pair and MapReduce job calculating mean value in a single bundle. The dataset used to evaluate the approach consists of an active channel of four states, four pipelines, smart-meter synthetic 512 MB dataset [21] with a custom datastore. This dataset includes smart meter sensory data Temp_file and Temp_meta on AWS IoT. We added an AWS Arduino that plays the role of Piezo speaker (buzzer) for harvesting energy situation.

We evaluated the mentioned dataset for MapReduce mean value job and found that AWS generates an average value of 9 s to sound Piezo speaker (buzzer) at the end of the job, as shown in Table 11.1.

11.5 Conclusion

This chapter discusses a novel technique of EH through in-place MapReduce data processing over IoT devices. The proposed technique explains how big data analytics paradigms help generate energy resources through continuous IoT device usage. We may say that a device uses a dataset to produce energy for IoT devices. The AWS platform evaluates our proposed prototype that consists of IoT devices, In-place MapReduce operation on a cluster, and Arduino buzzer. The performance evaluation shows that the proposed technique harvests energy through IoT devices and the prototype can be utilized well in the industry for harvesting energy in the distributed computing environment.

In the future, we plan to use synthetic sensors that could use bio-enabled datasets for the purpose of EH. The acronyms used in the chapter are shown in Table 11.2.

Table 11.2 List of acronyms

Acronyms	Descriptionh
AWS	Amazon Web Services
EH	Energy harvesting
IoT	Internet of Things
LTE	Long-term evolution
WPAN	Wireless personal area network
WWAN	Wireless wide area network

References

[1] S. Priya and D. J. Inman, eds. *Energy Harvesting Technologies*. Vol. 21. New York: Springer, 2009.

[2] A. Harb. "Energy harvesting: State-of-the-art." *Renewable Energy* 2011;36 (10):2641–2654.

[3] F. K. Shaikh and S. Zeadally. "Energy harvesting in wireless sensor networks: A comprehensive review." *Renewable and Sustainable Energy Reviews* 2016;55:1041–1054.

[4] L. Atzori, A. Iera, and G. Morabito. "The internet of things: A survey." *Computer networks* 2010;54(15):2787–2805.

[5] P. Kamalinejad, C. Mahapatra, Z. Sheng, S. Mirabbasi, V. C. M. Leung and Y. L. Guan. "Wireless energy harvesting for the Internet of Things." *IEEE Communications Magazine* 2015;53(6):102–108.

[6] Z. Abbas and W. Yoon. "A survey on energy conserving mechanisms for the internet of things: Wireless networking aspects." *Sensors* 2015;15(10): 24818–24847.

[7] A. P. Plageras, K. E. Psannis, C. Stergiou, H. Wang, and B. B. Gupta. "Efficient IoT-based sensor BIG Data collection–processing and analysis in smart buildings." *Future Generation Computer Systems* 2018;82:349–357.

[8] N. M. F. Qureshi and D. R. Shin. "RDP: A storage-tier-aware robust data placement strategy for Hadoop in a cloud-based heterogeneous environment." *KSII Transactions on Internet and Information Systems (TIIS)* 2016;10(9):4063–4086.

[9] N. M. F. Qureshi, D. R. Shin, I. F. Siddiqui, and B. S Chowdhry. "Storage-tag-aware scheduler for Hadoop cluster." *IEEE Access* 2017;5:13742–13755.

[10] J. Dean and S. Ghemawat. "MapReduce: Simplified data processing on large clusters." *Communications of the ACM* 2008;51(1):107–113.

[11] T. White. *Hadoop: The Definitive Guide*. Sebastopol, CA: O'Reilly Media, Inc., 2009

[12] J. Koo, N. M. F. Qureshi, I. F. Siddiqui, A. Abbas, and A. K. Bashir. "IoT-enabled directed acyclic graph in spark cluster." *Journal of Cloud Computing* 2020;9:1–15.

[13] J. Koo and N. M. F. Qureshi. "Fine-grained data processing framework for heterogeneous IoT devices in sub-aquatic edge computing environment." *Wireless Personal Communications* 2021;116:1407–1422.

[14] N. M. F. Qureshi, I. F. Siddiqui, M. A. Unar, *et al.* "An aggregate MapReduce data block placement strategy for wireless IoT edge nodes in smart grid." *Wireless Personal Communications* 2019;106(4):2225–2236.

[15] H. G. Lee and N. Chang "Powering the IoT: Storage-less and converter-less energy harvesting." *The 20th Asia and South Pacific Design Automation Conference*. IEEE, 2015, pp. 124–129.

[16] Z. Yang, W. Xu, Y. Pan, C. Pan and M. Chen. "Energy efficient resource allocation in machine-to-machine communications with multiple access and

energy harvesting for IoT." *IEEE Internet of Things Journal* 2018;5(1): 229–245.

[17] S. Carreon-Bautista, L. Huang, and E. Sanchez-Sinencio. "An autonomous energy harvesting power management unit with digital regulation for IoT applications." *IEEE Journal of Solid-State Circuits* 2016;51(6):1457–1474.

[18] I. F. Siddiqui, N. M. F. Qureshi, B. S. Chowdhry, and M. A. Uqaili. "Pseudo-cache-based IoT small files management framework in HDFS cluster." *Wireless Personal Communications* 2020;113(3):1495–1522.

[19] T. Lee, K. Kim, and H.-J. Kim. "Join processing using bloom filter in mapreduce." *Proceedings of the 2012 ACM Research in Applied Computation Symposium*. 2012, pp. 100–105.

[20] C. A. Howells. "Piezoelectric energy harvesting." *Energy Conversion and Management* 2009;50(7):1847–1850.

[21] AWS Synthetic dataset of Smart meter, configured according to custom MapBuz dataset https://aws.amazon.com/blogs/industries/tag/smart-meter-data/.

Part IV

Security and energy harvesting

Chapter 12

Hide-and-detect: forwarding misbehaviors, attacks, and countermeasures in energy harvesting-motivated networks

Sunho Lim[1], Cong Pu[2], Jinseok Chae[3], Manki Min[4], and Yi Liu[5]

Abstract

Multi-scale, heterogeneous, and battery-powered Internet-of-Things (IoT) sensors and devices (later in short, nodes) have been widely deployed in diverse applications and networks. Due to the limited amount of battery energy, energy harvesting-motivated networks (EHNets) powered by immediate environmental resources are increasingly popular and rapidly emerging as the next generation of ubiquitous communication infrastructure. However, EHNets are admittedly vulnerable to a denial-of-service (DoS) attack because of the shared medium, centralized coordination, and limited computing and communicating capabilities. Because of inherent resource constraints, EHNets seldom deploying conventional heavy-weight cryptographic techniques and secure algorithms and protocols. In light of these, we first investigate energy harvesting-based networking operations and applications. Second, we analyze the different types of forwarding misbehavior and attack caused by malicious nodes and their corresponding detection strategies. We introduce a set of adversarial scenarios and visualize its communication activities to capture vulnerable scenarios and potential malicious nodes. Here, single and multiple malicious nodes colluding together are considered. Lastly, we comprehensively compare the detection strategies of forwarding misbehavior by considering six perspectives and provide future research directions with interdisciplinary points of view.

[1]T[2]WISTOR: TTU Wireless Mobile Networking Laboratory, Department of Computer Science, Texas Tech University, Lubbock, USA
[2]Department of Computer Sciences and Electrical Engineering, College of Engineering and Computer Sciences, Marshall University, Huntington, USA
[3]Department of Computer Science and Engineering, Incheon National University, Incheon, South Korea
[4]Computer Science Program, Louisiana Tech University, Ruston, USA
[5]Department of Computer and Information Science, University of Massachusetts, Dartmouth, USA

12.1 Introduction

Recent advances in technology have fueled the development of a tiny and low-power node available to expedite fast deployment and improve portability, availability, and accessibility. Internet-of-Things (IoT) sensors and devices (later in short, nodes) have been used in diverse applications and networks, where nodes are often multi-scale, heterogeneous, and battery-powered. Nodes are seamlessly interconnected for actuation, sensing, and communication activities. IoT applications have been deployed in a variety of areas, such as smart homes, healthcare, infrastructure monitor, transportation and logistics, surveillance, and so on. As the demand for IoT applications is rapidly increasing globally, the IoT market is predicted to reach more than 2 trillion by 2023, which is three times higher than 2016 [1]. We envision that IoT-based networks will not only play an important role in realizing diverse applications ranging from civilian to military but also become the next generation of ubiquitous communication infrastructure.

Since nodes are primarily powered by batteries, it is unavoidable to replace or replenish batteries. This could be a critical issue if multiple nodes are deployed in a hard-reach area or a very wide area. It would be hard (if it is not impossible) to manually replace or replenish batteries. In light of these, we investigate energy harvesting-motivated networks (EHNets) to replenish or at least reduce the number of times in replacing batteries. In EHNets, each self-sustainable node is equipped with energy harvesting capabilities and powered by an immediate environment, e.g., solar, wind, or thermal. Nodes can communicate with others directly or indirectly through multi-hop relays.

Although a great research effort has been allocated to energy harvesting literature, we focus on a cybersecurity issue in the sense of forwarding misbehaviors, attacks, and countermeasures in EHNets. In this chapter, we summarize our contribution in threefold:

- First, we explore energy harvesting aided applications and research areas, consider system and adversarial models of energy harvesting capable nodes, and raise a denial-of-service (DoS) issue in EHNets.
- Second, we present a set of adversarial scenarios, analyze the forwarding interactions between legitimate and malicious nodes, and identify vulnerable scenarios and potential forwarding misbehavior.
- Third, we comprehensively compare and analyze the detection strategies of forwarding misbehavior with six major perspectives and provide future research directions with interdisciplinary insights.

The rest of this chapter is organized as follows. We review forwarding misbehavior in EHNets in Section 12.2. Both system and adversary models are introduced in Section 12.3. Energy harvesting-motivated attacks with adversarial scenarios and their detection strategies are discussed in Sections 12.4 and 12.5. Finally, we discuss future research directions with interdisciplinary aspects and insights and conclude the chapter in Sections 12.6 and 12.7, respectively.

12.2 Background and related work

We explore energy harvesting techniques and their applicable network operations and analyze forwarding detection strategies deployed in battery-powered networks.

Energy harvesting-motivated networks: Wireless communication could be responsible for more than half of total energy consumption in wireless and mobile networks [2]. In light of this, power-efficient and power-aware routing techniques have been developed [3–6]. However, it is hard to locate and replace low-power batteries because nodes often operate for a long period in an unattended environment. Researchers in academia and industry have been focusing on energy harvesting from various environmental sources [7–12] in which each node's battery can be rechargeable (or renewable).

In particular, energy harvesting from photovoltaic cells has been intensively investigated in the last two decades [8,9,13–26]. A variety of issues have been identified including solar-based routing and scheduling policy [8,9,15,21,27,28], resource allocation [19,23,29,30], energy synchronization [18], bounding communication delay [31], duty cycle [16], and data extraction [17] in multi-hop wireless networks. For example, a solar-based energy harvesting model is applied to scheduling and routing protocols [8,15]. The proposed energy harvesting-aware routing can increase network lifetime compared to that of traditional battery-based routing schemes. Several threshold policies are to maximize the communication performance in the network, where each node is assumed to be randomly recharged and changes its state into one of three states, active, passive, or ready [13]. A solar-aware routing scheme forwards packets to the nodes powered by solar energy [9].

Harvesting energy from ambient vibrations using a piezoelectric transducer has been investigated and applied to a wide range of civil and mechanical engineering applications for ease of battery replacement and energy replenishment [4,11,32–43]. Piezoelectric polymer patches are implanted into a living body to harvest energy from breathing [36]. Body heat, blood pressure, and even breath pressure have the potential to generate electric energy. Piezoelectric materials are used in the soles of shoes, where electrical power is generated through walking [4]. Researchers have also demonstrated the possibility of embedding a piezoelectric component in a textile [42]. Mechanical flow energy in oceans and rivers is utilized to convert electrical energy by using piezoelectric polymer actuators [39]. The piezo-based actuators can provide a large number of electrical power levels because of the vast size of the flowing water resource. Kinetic (motion)-based energy harvesting has received considerable attention [44–47].

Detection of forwarding misbehavior: A Watchdog technique and its variants [48,49] have been widely deployed to detect any communication misbehavior in infrastructure-based networks and infrastructure-less networks, such as mobile ad hoc networks (MANETs) and wireless sensor networks (WSNs). This technique often relies on overhearing the packets transmitted around neighbor nodes and checks whether the packets are heading to the right receivers. Nodes continuously monitor and observe communication activities in the network, and thus, they are required to stay in an active state for an extended period. Due to the non-negligible energy consumption, this technique cannot directly be applied to battery-powered networks.

Since nodes' communication activities can be spanned over multiple network layers, the algorithms and communication protocols embedded into the layers should not be conflicted. For example, IEEE 802.11 supports the power saving mechanism (PSM) in its medium access control (MAC) layer specification [50]. Each node can switch its state between active mode (AM) and power save (PS) mode. A node in AM stays awake all the time and conducts communication activities at any moment but wastes battery energy during idling. A node in PS periodically wakes up during the packet advertisement period and sees if there is any packet to receive. After staying awake and receiving any pending packet, the node puts itself to the low-power sleep state, PS, again for power saving. Thus, the Watchdog technique conflicts with the PSM embedded in the link layer. Although the Watchdog is not originally designed to work with the PSM, nodes are implicitly assumed to conduct communication activities in a resource-constrained environment.

We classify the detection strategies of forwarding misbehavior into three categories and briefly summarize their key ideas: (i) monitor, (ii) acknowledgment, and (iii) inducement. First, the basic idea of the monitor-based approach [51–55] is to check whether there is any forwarding misbehavior or network abnormality by observing the communication activities conducted among nodes, the amount of network traffic, or channel quality/condition. Second, the key operation of the acknowledgment-based approach [56–59] is that a set of designated nodes located between the source and the destination observes the forwarding operation of its very next node and sends an acknowledgment (*Ack*) packet to the source if an event or misbehavior is detected. Third, the basic idea of the inducement-based approach [60–62] is that nodes hide or fake their communication activities from malicious nodes to draw their forwarding misbehavior.

In summary, most detection approaches of forwarding misbehavior often require nodes not only to stay in an active state for an extended period but also to monitor/observe the communication activities via overhearing in a battery-supported network. Nodes are also supposed to generate a non-negligible number of control packets (i.e., *Ack*) to report any forwarding misbehavior that indeed consumes additional battery energy. Nevertheless, there is plenty of space to investigate EHNets with self-sustainable nodes that are under the charge-and-spend harvesting policy.

12.3 System and adversarial models

A system model mainly describes self-sustainable nodes, the energy harvesting process and policy, and initial network deployment. An adversarial model describes the potential misbehavior of malicious nodes in EHNets.

System model: First, a set of nodes is randomly distributed in a rectangle network, where each node can harvest energy. Nodes replenish their rechargeable battery [63] periodically or non-periodically, such as an event driven. A piezo-based device is feasible to harvest energy from an immediate environment, such as disturbances or body movements. This piezo-based device can generate at least sufficient power for IoT nodes to transceive packets [44–46]. For example, the IEEE 802.15.4-compliant Texas Instrument Chipcon CC2420 radio can support a set of different transmission power levels from 3 μW to 1 mW [64]. The Cisco Aironet 340 and 350 series can also support four or six different transmission power levels [65]. In [66], both piezo devices and integrated self-charging power cells (SCPCs) can be combined to improve the efficiency of energy harvesting.

Second, a two-state Markov process is deployed to model an energy harvesting process: *active* and *harvest* states. Each node initially selects one of two states, spends a certain period, and changes the current state to the other. An average period spent in each state may vary depending on the deployed energy harvesting device and environmental resources. If a node changes the states in a short period frequently, both energy consumption and operational delay increase. To manage the energy efficiently and strategically, a *charge-and-spend* energy harvesting policy [22,52,62,67] is deployed in EHNets. Under this policy, a node in the harvest state cannot receive an incoming packet before it harvests a certain level of energy for communication. Nodes minimize the communication activities during the harvest state and replenish battery energy quickly. More importantly, each node in the harvest state periodically broadcasts a one-hop *State* packet to prevent its adjacent nodes from forwarding packets, resulting in a packet loss.

Third, when a node senses an event or detects an abnormality, it generates and forwards a sensed data packet toward a sink. We deploy a simple broadcast-based forwarding scheme to quickly propagate the packet to the sink [68]. To initially conduct a network deployment process, a one-time *Hello* packet contained with a field (number of hops, initially set to zero) is broadcasted at the sink [68]. When a node receives the *Hello* packet, it rebroadcasts the packet after increasing the packet's number of hops by one. If the received packet contains a smaller number of hops, the node remembers the hop and rebroadcasts the packet. If not, the node discards the packet immediately. This procedure is repeated until all nodes receive and broadcast the packet. Finally, each node can identify its one-hop apart node(s) and how many hops are away from the sink. Then, the packet can be forwarded to single or multiple neighbor nodes that are located to the sink closer.

Fourth, we assume a reasonably dense network, where there are at least single or multiple nodes that can forward a packet. If two separate networks are connected solely by a single node, this node can be a single point of failure or a malicious node that may conduct forwarding misbehavior. Then, the network can easily be divided into two isolated sub-networks. Note that this network partition significantly affects the network performance in an infrastructure-less network, such as an EHNet, MANET, or WSN.

Adversarial model: First, single or multiple adversaries are to interfere with ongoing communications, intercept on-flying packets, and disrupt network algorithms and protocols. An adversary may physically capture a legitimate node and compromise it to behave maliciously, e.g., forwarding misbehavior. A malicious node is assumed to have no energy constraint and stay in an active state as long as it wants.

Second, we consider three types of misbehavior. (i) A single malicious node may blindly drop incoming packets (i.e., blackhole attack) or selectively/strategically drop/forward incoming packets (i.e., selective forwarding attack) to a sink. (ii) A single malicious node may overhear/eavesdrop on on-flying packets, inject fake information, or alter packet header information to lead network traffic to the wrong destination. If a sender applies an authentication technique to a packet, such as a lightweight digital signature [69], then a receiver can detect whether the packet has been modified during the transmission. In this chapter, we focus on a DoS attack under diverse forwarding misbehavior scenarios that cannot be detected by cryptographic techniques. Thus, cryptographic primitives are out of scope. (iii) Multiple malicious nodes may collude together to hide their forwarding misbehavior.

12.4 Energy harvesting-motivated adversarial scenarios and attacks

We first observe and analyze a set of forwarding misbehaviors and adversarial scenarios and then briefly introduce corresponding detection schemes in EHNets. Single or multiple malicious nodes with control packet exchanges are investigated using a simple network topology to clearly see the forwarding misbehaviors and attacks.

12.4.1 Single malicious node

Adversarial scenarios, AS1: These four adversarial scenarios are based on the overhearing of implicit acknowledgment in the network, where four energy harvesting enabled nodes interact as shown in Figure 12.1. Unlike an explicit acknowledgment by receiving an *Ack* packet, an implicit acknowledgment implies that a sender overhears if one-hop apart adjacent nodes have forwarded the received packet.

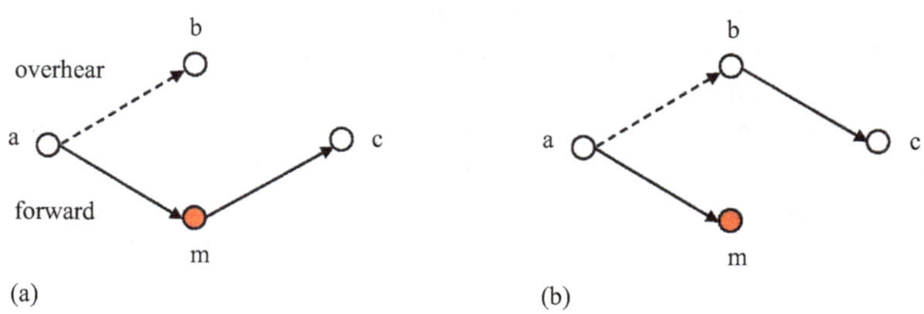

Figure 12.1 A single malicious node (n_m), shaded as red, in a network, where solid and dashed-arrow lines mark a packet forwarding and overhearing, respectively

First, suppose an active state sender (n_a) transmits a packet to one-hop apart nodes, n_b or n_m. If the sender is currently in a harvest state, it does not forward the packet until it becomes an active state. Whenever the state is changed, the sender broadcasts a one-hop *State* packet. Suppose n_a sends a packet to n_m in Figure 12.1 (a). n_b can overhear the packet and temporarily cache it in the local storage. If n_m is in the harvest state, n_a would send the packet to n_b. If n_m is in the active state and forwards the packet to n_c, both n_a and n_b can overhear the packet as an implicit acknowledgment. If n_m simply holds or discards the packet, both n_a and n_c cannot overhear the packet. Since n_b overheard the packet before, it can forward its cached packet to n_c directly after a timeout period, as depicted in Figure 12.1(b). n_a can now overhear the packet from n_b and may suspect the forwarding behavior of n_m. Second, suppose n_m receives the packet from n_a and forwards it to n_c, while n_c is in the harvest state. If n_b is in the harvest state, n_m would not be suspected because n_a can still overhear the packet from n_m. Third, if n_b is in the active state, it suspects n_m because n_b knows that n_c is in the harvest state. Thus, n_b forwards its cached packet to n_c after the timeout period. If n_a overhears the packet forwarded from n_b rather than the original forwarder n_m, it may suspect the forwarding behavior of n_m. Fourth, suppose n_m receives a packet from n_a, changes its state to harvest from active, and does not broadcast the *State* packet. n_b can forward its cached packet to n_c because it is in the active state. If n_b is in the harvest state, but n_a cannot overhear the packet forwarded from n_m, n_a considers n_m as a failure node and tries to find other forwarding candidate nodes.

In Figure 12.2, we highlight the vulnerable cases in which a malicious node can show forwarding misbehavior. The first misbehavior case is shown in Figure 12.2(a). When n_c is in the harvest state, n_m tries to forward the packet received from n_a to n_c. Similarly, n_m tries to forward the packet to n_c when n_b and n_c are in the harvest state concurrently, as shown in Figure 12.2(b). This is the second misbehavior case because n_a overhears the forwarded packet from n_m and considers the forwarding operation as valid.

Cooperative detection: A hop-by-hop cooperative detection (HCD) scheme [52] is proposed to discourage forwarding misbehavior by reducing the forwarding probability of malicious nodes in EHNets. The basic idea is that each node overhears the communication activities conducted around its neighbor nodes and

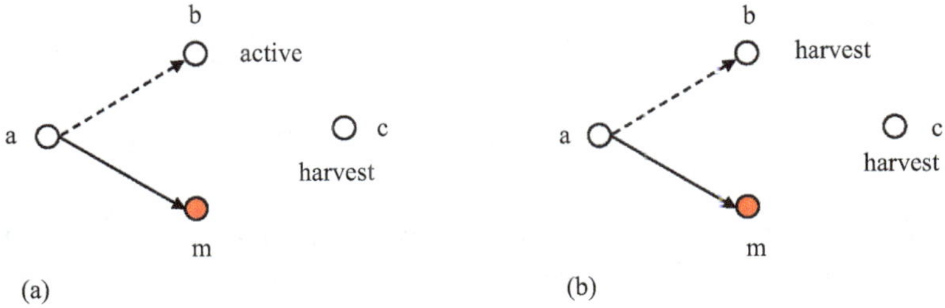

Figure 12.2 Vulnerable cases in adversarial scenarios, where a malicious node shows forwarding misbehavior in a network

records the trace of forwarding operations. This is different from the prior approach, in which each node reports a suspected malicious node either to a source node [48,56,57] or to a centralized server (i.e., credit clearance service) [70]. Then, the source node or server decides how to assign a credit/penalty and whether to isolate the malicious node from participating in the communication activities in the network, accordingly.

12.4.2 Single malicious node with an additional control packet

Adversarial scenarios, AS2: We enhance the adversarial scenarios of AS1 by including an additional control packet, *Wait*, in the network, as depicted in Figure 12.3.

The first scenario depicted in Figure 12.3(a) is the same as the first scenario shown in AS1. Here, since n_a and n_b are in the active state, n_m does not hold or drop the packet on purpose. This is because the forwarding misbehavior of n_m can be detected by either n_a or n_b. Thus, n_m forwards the packet just like a legitimate node. Second, as shown in Figure 12.3(b), suppose the harvest state n_c periodically broadcasts a *State* packet to its one-hop adjacent nodes. To avoid any forwarding misbehavior suspect, n_m simply holds the packet and waits until n_c changes the state back to active and broadcasts another *State* packet. Then, n_m sends a *Wait* packet back to n_a and behaves like a legitimate node. After receiving the *Wait* packet, n_a can select other forwarding candidate nodes, e.g., n_b. Third, suppose the harvest state n_b periodically broadcasts a *State* packet to one-hop neighbor nodes as shown in Figure 12.3(c). This is similar to the case when n_c is in the harvest state and broadcasts a *State* packet. n_m may send a *Wait* packet to n_a to intentionally

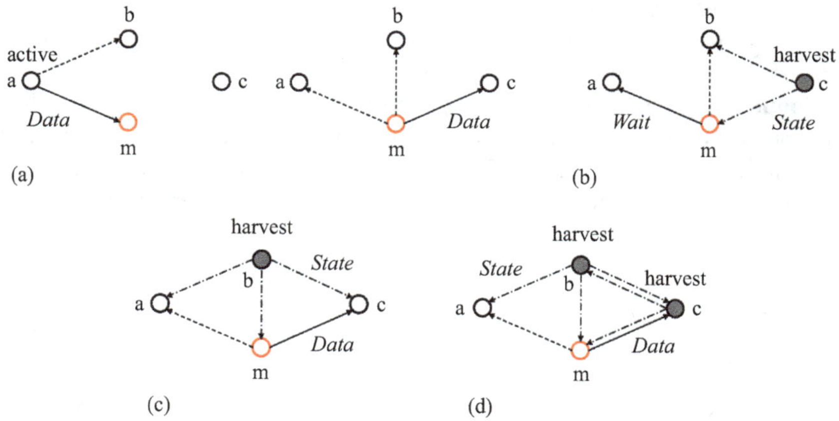

Figure 12.3 Adversarial scenarios with an additional control packet are depicted. Here, a malicious node (n_m) is marked as red, and nodes in the harvest state (n_b or n_c) are marked as shade. Forwarding, overhearing, and broadcasting operations are marked as solid, dotted, and dashed–dotted lines, respectively

delay the packet transmission. However, n_c can overhear the *Wait* packet and may suspect n_m of forwarding misbehavior. To avoid a forwarding misbehavior suspect, n_m behaves just like a legitimate node and forwards the packet to n_c rather than holding or dropping the packet on purpose. Fourth, when both n_b and n_c are in the harvest state, they periodically broadcast a *State* packet, as shown in Figure 12.3(d). Since the neighbor nodes of n_m (n_b and n_c) become blind, n_m can forward the packet to n_c and incur a packet loss intentionally. This is a vulnerable case, in which the forwarding misbehavior of malicious nodes cannot be detected even though n_a overhears the packet from n_m.

Camouflage-based active detection: A camouflage-based active detection (CAM) scheme [62] is proposed to detect the forwarding misbehavior of malicious nodes in EHNets. In the CAM, each node hides its current state and does not broadcast a *State* packet. A harvest state node pretends to conduct energy harvesting but, in fact, observes the communication activities of neighbor nodes to detect a lurking malicious node. The CAM is different from the prior approach [48,49,52,71,72], where each node *passively* observes the routing operations in the network.

12.4.3 Multiple malicious nodes

Adversarial scenarios, AS3: We further investigate eight adversarial scenarios, in which five energy harvesting enabled nodes are deployed in the network, as shown in Figure 12.4. In AS3, two malicious nodes (n_{m_A} and n_{m_B}) located along with the forwarding path observe communication activities and collude together for a selective forwarding attack. Suppose a sender (n_a) forwards a *Data* packet to n_c via intermediate nodes, n_b, n_{m_A}, and n_{m_B}.

The first scenario depicted in Figure 12.4(a) is similar to the aforementioned scenarios in AS1 and AS2 except for two malicious nodes in the network. When a sender (e.g., n_a, n_{m_A}, or n_{m_B}) forwards the received packet, a set of adjacent nodes (e.g., n_a, n_b, or n_{m_A}) can overhear and cache the packet in their local storage. If both n_a and n_b are in the active state, n_{m_A} and n_{m_B} do not conduct any forwarding misbehavior by holding or dropping the packet on purpose. This is because n_a and n_b can overhear any forwarded packet. Thus, n_{m_A} and n_{m_B} forward the packet just like a legitimate node.

Second, suppose n_{m_B} intentionally changes to the harvest state and broadcasts a *State* packet, as depicted in Figure 12.4(b). If n_b is in the active state, n_{m_A} may forward the received packet to n_{m_B} on purpose, resulting in a packet loss. Since the active state n_b can overhear the forwarded packet, this forwarding operation may be suspected as misbehavior. Thus, n_{m_A} replies a *Wait* packet to the sender, n_a, for delaying the packet transmission intentionally. Third, suppose harvest state n_c broadcasts a *State* packet as depicted in Figure 12.4(c). Both n_a and n_b are in active state and can overhear the communication activities. Both n_{m_A} and n_{m_B} behave as legitimate nodes and forward the received packet to the next-hop neighbor nodes. Note that n_{m_B} may conduct forwarding misbehavior without being detected by simply forwarding the packet to n_c, resulting in a packet loss. Fourth, if n_c is in the harvest state, as depicted in Figure 12.4(d), n_{m_A} should play as a legitimate node not to gain any suspect of forwarding misbehavior.

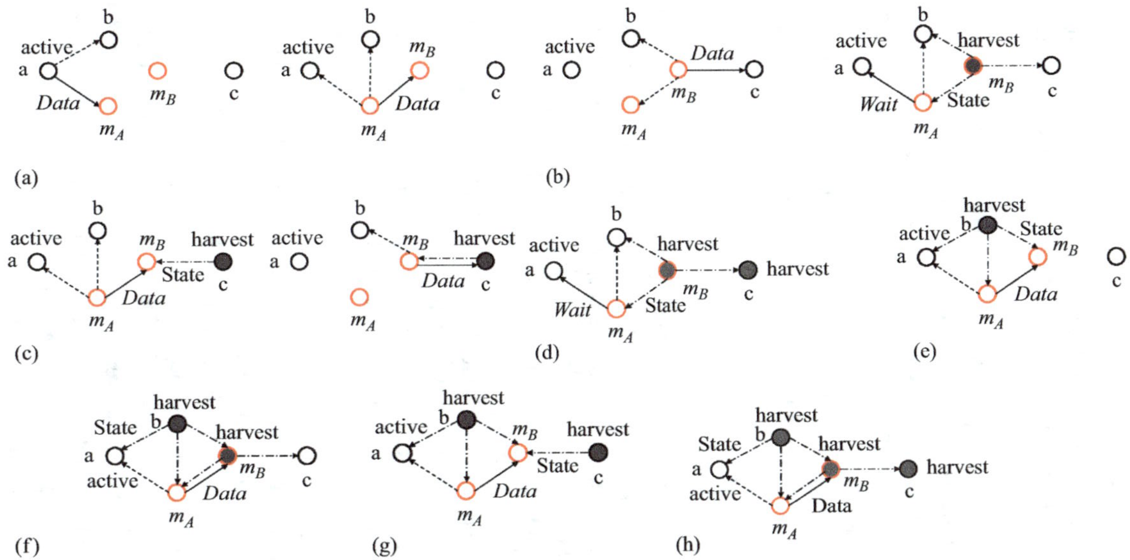

Figure 12.4 Adversarial scenarios with two colluding malicious nodes in a network. Here, a malicious node is marked as red and a harvest state node is shades, respectively. Forwarding, overhearing, and broadcast operations are marked as solid, dotted, and dashed–dotted lines, respectively

Fifth, suppose harvest state n_b broadcasts a *State* packet in Figure 12.4(e), where n_{m_A} may forward the received packet to n_{m_B}. Then, n_{m_B} may not forward the packet to the one-hop adjacent nodes but instead maliciously hold or drop it. It is hard for n_b and n_c to detect this forwarding misbehavior of n_{m_A} and n_{m_B}, because they cannot overhear any forwarded packet. Here, n_b is in the harvest state and n_c is far away from n_{m_A}. Sixth, as depicted in Figure 12.4(f), both legitimate and malicious nodes (i.e., n_b and n_{m_B}) are in the harvest state and broadcast a *State* packet, respectively. In this vulnerable case, n_a cannot detect forwarding misbehavior of n_{m_A} because n_b is in the harvest state and even cannot forward its cached packet after the timeout period. Thus, n_{m_A} may forward the received packet to n_{m_B} on purpose, resulting in a packet loss. Since the forwardee node, n_{m_B}, is also a malicious node, both n_{m_A} and n_{m_B} may collude together for the forwarding attack. Seventh, suppose both harvest state n_b and n_c broadcast a *State* packet as depicted in Figure 12.4(g). This is another vulnerable case because both n_{m_A} and n_{m_B} can collude together for forwarding misbehavior without being detected. Since both n_b and n_c are blind, they cannot overhear any communication activity in the network. n_{m_A} may keep quiet when it receives the forwarded packet from n_{m_B}, ultimately resulting in a packet loss. Lastly, suppose n_c is in the harvest state as depicted in Figure 12.4(h). n_{m_A} may still forward the received packet to n_{m_B} on purpose without being detected.

According to the analysis of adversarial scenarios, if more than one malicious node is located consecutively in a sparse network, it would be hard to detect their collusion of forwarding misbehavior. As aforementioned, the network should be dense enough not only to prevent network partition but also to discourage forwarding misbehavior.

Inducement- and monitor-based detection: The proposed countermeasure, called EYES, is to detect and discourage the forwarding misbehavior of colluding malicious nodes in EHNets [73]. The EYES is different from the prior approach [48,49,56–58,71,72,74–78], in which nodes *passively* monitor any forwarding misbehavior in the battery-powered networks. The EYES consists of inducement- and monitor-based sub-schemes, called SlyDog and LazyDog, respectively. The SlyDog is extended from the CAM [62]. Each node pretends to harvest energy without monitoring, but, in fact, it observes the forwarding operations conducted in one-hop neighbor nodes to efficiently detect shy malicious nodes and collusion of malicious nodes. In the LazyDog, each node counts the number of received/overheard packets and requests this information to its one-hop neighbor nodes. Then, each node can receive the information from its two-hop neighbor nodes and analyze the information to detect forwarding misbehavior.

12.5 Comparison and analysis of detection strategies

In Table 12.1, we summarize and categorize the detection strategies of forwarding misbehavior deployed in diverse networks. We analyze the strategies using six key perspectives.

- **Collusive attack**: We investigate whether multiple malicious nodes collude together to conduct a forwarding attack and achieve their attack goal(s) in the

Table 12.1 The comparison of detection strategies of forwarding misbehavior extended from [73]

Approach	Collusive attack	Computation overhead	Communication overhead	Detection latency	Punishment	Architecture
Watchdog [48]	N	Low	N	N	N	Stand-alone
CDS [51]	N	High	Low	Medium	N	Centralized
HED [54]	Y	Low	Low	Medium	Y	Distributed
HCD [52]	N	Medium	Low	High	Y	Distributed
CHEMAS [57]	N	Medium	High	Low	N	Centralized
CAD [79]	N	Medium	Medium	Medium	N	Centralized
SCAD [80]	Y	Medium	Medium	Low	N	Centralized
APS [59]	N	Medium	High	Medium	Y	Distributed
CBDS [60]	Y	Medium	Medium	High	N	Distributed
SNBDS [61]	Y	Medium	Medium	High	N	Distributed
CAM [62]	N	Low	N	N	Y	Stand-alone
ACIDS [81]	N	Medium	N	Medium	Y	Centralized
SCM [75]	N	Low	N	Medium	N	Stand-alone
EAACK [76]	N	Medium	High	Medium	N	Centralized
FADE [77]	Y	Medium	High	Low	N	Centralized
CRS [78]	Y	High	Medium	Medium	Y	Distributed
SlyDog	Y	Low	N	N	Y	Stand-alone
LazyDog	N	Low	Low	Medium	Y	Distributed

network. A single malicious node can selectively cooperate with a legitimate node or an infrastructure network component.

- **Computation overhead**: To detect forwarding behavior, either a sender, receiver, or intermediate nodes may record the history of communication activities that occurred in their adjacent nodes for a certain period. They can share and cross-check the history to detect any forwarding misbehavior. To realize this, each resource-constrained node is required a certain level of computing power to process.
- **Communication overhead**: In EHNets, whenever a node is harvest state, it broadcasts a *State* packet. The original intention of this periodic packet is for one-hop neighbor nodes not to mistakenly send a packet to the energy harvesting node. Depending on the detection purposes, one or multiple nodes have to generate and send a series of control packets to the packer sender or designated nodes, resulting in a communication overhead in such a resource-limited network.
- **Detection latency**: We consider how quickly each node suspects its one-hop neighbor nodes, detect their forwarding misbehavior, and isolate them from the network by excluding their participation in the communication activities in the network.
- **Punishment**: We also investigate whether there is a mechanism to discourage single or multiple malicious nodes for their forwarding misbehavior. For example, if a node suspects one of its neighbor nodes, it reduces the forwarding probability to the neighbor node. As more forwarding misbehaviors are suspected, the neighbor node gradually loses a chance to receive a packet to forward in the network. This is the same effect of network isolation. Legitimate nodes will not involve a malicious node for communication.
- **Architecture:** In this perspective, we consider three types of network operation in conducting detection strategies, *Centralized*, *Distributed*, or *Stand-alone* [58]. In *Centralized*, a set of designated nodes conduct most major detection operations, but the rest of the nodes have relatively simple operations, such as monitoring communication activities, and report any event or abnormality to the designated nodes. In *Distributed*, every node has an equal responsibility to monitor and detect forwarding misbehavior in the network. Nodes frequently exchange control packets or the history of communication activities for detection. *Stand-alone* is the same as *Distributed*, but each node does not exchange or share any information with others.

We also analyze major ideas and operations of the detection strategies. A centralized detection system (CDS) [51] is proposed to detect packet-dropping attacks in clustered IoT networks. The basic idea is that an uplink packet drop probability of IoT devices is calculated to monitor the behavior of the gateway, which is associated with IoT devices. A detection rule is provided by conducting a generalized likelihood ratio test, in which attack probabilities are approximated based on the maximum likelihood estimation. In a heuristic-based detection (HED) scheme [53,54], a suppression attack is discouraged in a multicast protocol for low power and lossy networks (LLNs). A malicious node multicasts a series of spoof data packets with continuous sequence numbers to prevent legitimate nodes from accepting valid data packets in the network. In the HCD [52], each node monitors its adjacent nodes' forwarding misbehavior by tracing a limited amount of forwarding history in EHNets. Each node also gradually reduces the forwarding probability of suspected malicious nodes to exclude them from participating in the routing operation. A monitor-based approach (CMD) [55] is to mitigate forwarding misbehavior in LLNs. Each node monitors the preferred parent node to observe its packet loss rate, compares this rate with the collected packet loss rates from one-hop neighbor nodes, and detects forwarding misbehavior.

In [56] and its extended approach, a proposed checkpoint-based multi-hop acknowledgment scheme (CHEMAS) [57] randomly selects checkpoint nodes to monitor ongoing forwarding operations and replies an *Ack* packet to the original packet sender in WSNs. Each intermediate node located along the forwarding path counts the number of received *Ack* packets corresponding to the number of *Data* transmissions. If the node receives the less number of *Ack* packets, it may suspect the next located neighbor node for forwarding misbehavior, e.g., dropping either *Data* or *Ack* packet. Then, the node generates an *Alarm* packet and transmits it to the original packet sender for reporting a malicious node that is potentially involved in the forwarding operation. In the CHEMAS, intermediate nodes often receive and forward many *Ack* and *Alarm* packets, resulting in high battery energy consumption. To efficiently detect a selective forwarding attack, a single checkpoint-based countermeasure (SCAD) [58] is proposed in resource-constrained WSNs. Unlike the CHEMAS, a single checkpoint node is randomly selected in the network and detect forwarding misbehavior. This approach can be combined with the timeout technique and hop-by-hop retransmission operation to mitigate the unexpected packet losses that are primarily caused by forwarding attacks or fluctuating channel qualities. To discourage a malicious node dropping data packet, an acknowledgment-based punishment and stimulation scheme (APS) [59] is proposed in MANETs. In the APS, each node estimates the reputation of neighbor nodes based on routing reliability and shares its recommendation to identify a malicious node.

To detect both selective forwarding and blackhole attacks, a cooperative bait detection scheme (CBDS) [60] is proposed based on the dynamic source routing (DSR) in MANETs. In the CBDS, a source node virtually creates a destination address to monitor the reaction of a potential malicious node. Since the malicious node does not know whether the destination address is real, it may reply a fake route reply (RREP) packet to the source. Upon receiving the RREP packet, the source can trace back the route and identify the malicious node. Based on the ad hoc on-demand distance vector routing (AODV), a sequence number-based bait detection scheme (SNBDS) [61] is proposed in MANETs. A series of sequence numbers piggybacked in the RREP packet is used to see if there is any packet drop over the transmission. Each node examines whether there is any gap between sequence numbers in the receiving packets. The next-hop neighbor node may be suspected of forwarding misbehavior if the gap is greater than a predefined threshold value. A CAM scheme [62] is deployed in EHNets, and its operational summary is presented in Section 12.4.2.

An accurate and cognitive intrusion detection system (ACIDS) [81] is proposed to defend against blackhole attacks in MANETs. In the ACIDS, each node monitors key parameters (e.g., destination sequence number and route reply) and checks the amount of deviation from the normal to detect an intruder. An attack detection framework [82] is proposed to defend IoT cyber-attacks using a deep learning technique, in which an attack detector is implemented and embedded into fog nodes. In [83], a dependence estimator-based scheme is proposed in IoT sensor networks, where a deep analysis of network traffic is conducted. This scheme can identify key IoT network traffic parameters and help in detecting any malicious network activity and traffic accurately.

12.6 Discussion and future research directions

We envision that energy harvesting-motivated computing and networking under security awareness are essential to support future IoT networks. To see the full potential of research introduced in this chapter, we discuss promising research issues and directions with the interdisciplinary points of view.

Vibration sensitive medium access control: In vibration-motivated energy harvesting, a disturbance event initiates the direct piezoelectric effect actively or passively. A passive event can be caused by surrounding environmental resources (e.g., ground disturbance or wind) in a static EHNet, where the nodes located nearby the event sense and transform it into mechanical vibration energy for communication. Since multiple nodes can respond to the same event, they may initiate the transmission simultaneously that may result in packet contention, collision, and retransmission. On the other hand, an active event can be caused by immediate environmental resources (e.g., the kinetic motion of walking or running) in a mobile EHNet, where each node responds to the event. Note that each node must maximize the utilization of harvested energy for communication.

The prior energy harvesting-aware MAC protocols have concentrated on solar- [22,84] or thermal-based [85] energy harvesting. However, there is plenty of space to extend by deploying vibration-motivated energy harvesting from intermittent kinetic movements and their integration with the IEEE 802.11 MAC protocol. This research approach newly considers underlying properties of ambient vibrations and practical obstacles in terms of the medium access technique that will significantly affect the design of algorithms and communication protocols embedded in upper layers, such as the network and application layers.

Energy harvesting-motivated lower power and lossy networks: IoT-based networks equipped with smart sensors and objects are expected to play an important role in building a future communication paradigm, such as minimizing or without human intervention for communication activities [86]. In the realm of IoT, IPv6-based LLNs consisting of a myriad of resource-constrained devices endowed with the capabilities of sensing, computing, and wireless communicating represent a key enabler for IoT applications. To overcome limited battery power, energy harvesting-motivated LLNs (EH-LLNs) are rapidly emerging and will be a major part of IoT-based networks, where energy harvesting-motivated nodes use a routing protocol for LLNs (RPL) [87]. Since the RPL was not originally designed for the energy harvesting features, we plan to develop an energy harvesting module to seamlessly integrate with RPL and conduct different simulation scenarios by using Contiki Cooja network simulation [88]. In addition, we plan to investigate the dissipation of harvested energy and traffic load, and design a traffic load and energy balancing RPL to further extend the network lifetime and improve the network performance.

Authentication with lightweight cryptography: We can find the presented scheme to apply to IoT device authentication, especially collaborative authentication on a group of IoT nodes using threshold cryptography [89–91]. The proposed countermeasure will allow us to further filter out the honest behaving nodes or generating the honestness weights, so when it is combined with some other security measures, we can further strengthen the threshold and improve the security level of the IoT network nodes. We can imagine that this level of work may happen at a much more powerful node such as a base station which possibly possesses the entire (or at least majority of) the IoT node topology, and the honestness of each node can aid the more accurate computation of group security measures.

Another research direction related to the proposed scheme is the authentication itself of each IoT device using very lightweight cryptographic functions such as cryptographic hash functions. Traditionally hash chain was found to be useful to balance off the computational overhead and the security level [92–95], but it does not work well to handle more complicated hierarchical structures. Hierarchically structured authentication has extensively been studied [96–99], and we can use more complicated hash structures such as hash Merkle tree, multidimensional hash chains, and hash vine. By designing a lightweight hash function to sacrifice the collision attack security, we can make it work with low computing-powered IoT devices for lower security but shorter lifetime protection of broadcast communications. This is a plausible direction in the sense that the lifetime of each broadcast is very short, so each hash computation needs to be safe against collision attacks for the short period that can be achieved with the lightweight hash design.

IoT software architecture: IoT, as a rapidly growing field, has applications in various domains such as healthcare, automated home services, smart energy and smart grid, food and water tracking, and transportation [100,101]. "Software engineering for the IoT poses challenges in light of new applications, devices, and services" [102]. The IoT adds additional complexity to software development as its nature of distribution and inclusion of heterogeneous devices, such as sensors and actuators [102]. One of the areas of research in the IoT from the Software Engineering perspective is software architecture.

Several reference architectures have been proposed to standardize the design of IoT systems, in which some reference architectures are more generic on industry scale implementation [102] while some are more specific [103] to the resources or environment, such as cloud computing. Some research targets specific software architecture for the IoT applications in different domains, for example, [104] presents a service-oriented software architecture for a data-driven smart city utility application and [105] did a mapping study on using microservice architecture as the building blocks for IoT systems and cloud computing solutions. The research work [106,107] have done mapping studies on exploiting software architecture models to develop IoT systems.

Although reference architectures give the software developers a general guide and the specific software architectures proposed for different domains and resources allow the developers to adapt the methodologies while developing the IoT systems similar settings, there are still scenarios in which these architectures are not applicable. The energy harvesting-based wireless sensor networks systems [108] involve the special requirements that need to be addressed in the software architectural design. The existing proposed IoT software architectures may need to be extended and expanded with the unique aspects of wireless sensor networks and energy harvesting-based computing involved. Thus, as one of the future works, we plan to exploit the software architectural styles that work as the best practice in EHNets powered by harvesting environmental resources.

12.7 Conclusion

Seamlessly interconnected IoT sensors and devices have been deployed in diverse applications and networks ranging from civil to military. Since IoT nodes are powered by batteries, they should be replaced or replenished ultimately but often hard, if it is not impossible. Due to the limited battery energy, energy harvesting from immediate environmental resources would be the best candidate to efficiently replenish or significantly reduce the frequency of replacing batteries.

This chapter introduces a DoS attack that must be considered in rapidly emerging EHNets. Depending on single or multiple malicious nodes, three sets of adversarial scenarios and their corresponding forwarding misbehaviors are observed and analyzed to find vulnerable scenarios and malicious nodes. Detection strategies of forwarding misbehavior are also compared and analyzed comprehensively in terms of six perspectives. In addition, potential future research directions for IoT and its variants are provided, including energy harvesting-aware MAC, energy harvesting-motivated LLNs, IoT lightweight authentication, and IoT software architecture.

We envision that this chapter will open many interesting research directions to pursue and enable the research community to quickly follow up the proposed energy harvesting-motivated networking research.

List of acronyms

Acronym	Description
ACIDS	Accurate and cognitive intrusion detection system
AODV	Ad hoc on-demand distance vector routing
APS	Acknowledgment-based punishment and stimulation scheme
CAD	Channel-aware detection
CAM	Camouflage-based active detection scheme

(Continues)

(*Continued*)

Acronym	Description
CBDS	Cooperative bait detection scheme
CDS	Centralized detection system
CHEMAS	Checkpoint-based multi-hop acknowledgment scheme
CMD	Monitor-based approach
CRS	Channel-aware reputation system
DSR	Dynamic source routing
EAACK	Enhanced adaptive acknowledgment
EHNet	Energy harvesting-motivated network
EH-LLN	Energy harvesting-motivated lower power and lossy network
FADE	Forwarding assessment based detection
HCD	Hop-by-hop cooperative detection
HED	Heuristic-based detection
IoTSN	Internet-of-Things sensor network
LazyDog	Proposed monitor-based detection scheme
LLN	Low power and lossy network
MAC	Medium access control
MANET	Mobile ad hoc network
PFCB	Fiber composite bi-morph
SCAD	Single checkpoint-based countermeasure
SCM	Side channel monitoring
SCPC	Integrated self-charging power cell
SlyDog	Proposed inducement-based detection scheme
SNBDS	Sequence number based bait detection scheme
WatchDog	Observation-based detection scheme
WSN	Wireless sensor network

References

[1] Internet of Things (IoT) Market – Share, Industry Trends, Development, Revenue, Demand and Forecast, to 2023. https://www.marketwatch.com/.

[2] Kravets R, and Krishnan P. Power Management Techniques for Mobile Communication. In: Proceedings of ACM MOBICOM; 1998. p. 157–168.

[3] Singh S, Woo M, and Raghavendra CS. Power-Aware Routing in Mobile Ad Hoc Networks. In: Proceedings of ACM MOBICOM; 1998. p. 181–190.

[4] Chang JH, and Tassiulas L. Energy Conserving Routing in Wireless Ad-Hoc Networks. In: Proceedings of IEEE INFOCOM; 2000. p. 22–31.

[5] Li Q, Aslam J, and Rus D. Online Power-Aware Routing in Wireless Ad-Hoc Networks. In: Proceedings of ACM MOBICOM; 2001.

[6] Lim S, Yu C, and Das CR. RandomCast: An Energy Efficient Communication Scheme for Mobile Ad Hoc Networks. *IEEE Transactions on Mobile Computing* 2009;8(8):1039–1051.

[7] Stordeur M, and Stark I. Low Power Thermoelectric Generator – Self-sufficient Energy Supply for Micro Systems. In: Proceedings of 16th International Conference on Thermoelectrics; 1997. p. 575–577.

[8] Kansal A, Potter D, and Srivastava MB. Performance Aware Tasking for Environmentally Powered Sensor Networks. In: Proceedings of ACM SIGMETRICS/Performance; 2004. p. 223–234.

[9] Voigt T, Ritter H, and Schiller J. Utilizing Solar Power in Wireless Sensor Networks. In: Proceedings of IEEE Local Computer Networks; 2003. p. 416–422.

[10] Raghunathan V, Kansal A, Hsu J, *et al.* Design Considerations for Solar Energy Harvesting Wireless Embedded Systems In: Proceedings of 4th International Symposium on Information Processing in Sensor Networks (IPSN); 2005. p. 457–462.

[11] Nathan SS, and Joseph PA. Energy Scavenging with Shoe-Mounted Piezoelectrics. *IEEE Micro.* 2001;21(3):30–42.

[12] Kim H, Priya S, Stephanou H, *et al.* Consideration of Impedance Matching Techniques for Efficient Piezoelectric Energy Harvesting. *IEEE Transactions on Ultrasonics, Ferroelectrics, and Frequency Control.* 2007;54(9):1851–1859.

[13] Kar K, Krishnamurthy A, and Jaggi N. Dynamic Node Activation in Networks of Rechargeable Sensors. In: Proceedings of IEEE INFOCOM; 2005. p. 15–26.

[14] Hsu J, Kansal A, Friedman J, *et al.* Energy Harvesting Support for Sensor Network. In: Proceedings of IEEE IPSN Demo; 2005.

[15] Kansal A, Hsu J, Srivastava M, *et al.* Harvesting Aware Power Management for Sensor Networks. In: Proceedings of Design Automation Conference (DAC); 2006. p. 651–656.

[16] Vigorito CM, Ganesan D, and Barto AG. Adaptive Control of Duty Cycling in Energy-Harvesting Wireless Sensor Networks. In: Proceedings of IEEE SECON; 2007. p. 21–30.

[17] Fan K, Zheng Z, and Sinha P. Steady and Fair Rate Allocation for Rechargeable Sensors in Perpetual Sensor Networks. In: Proceedings of ACM SenSys; 2008. p. 239–252.

[18] Zhu T, Zhong Z, Gu Y, *et al.* Leakage-Aware Energy Synchronization for Wireless Sensor Networks. In: Proceedings of MobiSys; 2009. p. 319–332.

[19] Liu R, Sinha P, and Koksal CE. Joint Energy Management and Resource Allocation in Rechargeable Sensor Networks. In: Proceedings of IEEE INFOCOM; 2010. p. 1–9.

[20] Gummeson J, Clark SS, Fu K, *et al.* On the Limits of Effective Hybrid Micro-Energy Harvesting on Mobile CRFID Sensors. In: Proceedings of MobiSys; 2010. p. 319–332.

[21] Eu ZA, Tan H, and Seah WKG. Opportunistic Routing for Wireless Sensor Networks Powered by Ambient Energy Harvesting. *Journal of Computer Networks.* 2010;54(17):2943–2966.

[22] Eu ZA, Tan H, and Seah WKG. Design and Performance Analysis of MAC Schemes for Wireless Sensor Networks Powered by Ambient Energy Harvesting. *Ad Hoc Networks.* 2011;9(3):300–323.

[23] Chen S, Sinha P, Shroff NB, *et al.* Finite-Horizon Energy Allocation and Routing Scheme in Rechargeable Sensor Networks. In: Proceedings of IEEE INFOCOM; 2011. p. 2273–2281.

[24] Gu Y, Zhu T, and He T. ESC: Energy Synchronized Communication in Sustainable Sensor Networks. In: Proceedings of 17th International Conference on Network Protocols; 2009. p. 52–62.

[25] Meraki solar. http://meraki.com.

[26] Proxim Wireless. The Solar-Power Alternative in Broadband Wireless Networks.

[27] Lei J, Yates R, and Greenstein L. Optimal Transmission Policy for Renewable Sensor Networks. In: Proceedings of 40th Annual Conference on Information Sciences and Systems; 2006. p. 81–86.

[28] Khouzani M, Sarkar S, and Kar K. Optimal Routing and Scheduling in Multihop Wireless Renewable Energy Networks. In: Proceedings of Sixth Information Theory and Applications Workshop; 2011.

[29] Niyato D, Hossain E, Rashid MM, *et al.* Wireless Sensor Networks with Energy Harvesting Technologies: A Game-Theoretic Approach to Optimal Energy Management. *IEEE Wireless Communications.* 2007;90–96.

[30] Gatzianas M, Georgiadis L, and Tassiulas L. Control of Wireless Networks with Rechargeable Batteries. *IEEE Transactions on Wireless Communications.* 2010;9(2):581–593.

[31] Gu Y and He T. Bounding Communication Delay in Energy Harvesting Sensor Networks. In: Proceedings of IEEE ICDCS; 2010. p. 837–847.

[32] Hauseler E, Stein L, and Harbauer G. Implantable Physiological Power Supply with PVDF Film. *Ferroelectrics.* 1984;60:277–282.

[33] Kymissis J, Kendall C, Paradiso J, *et al.* Parasitic Power Harvesting in Shoes. In: Proceedings of IEEE Wearable Computing; 1998.

[34] Taylor GW, Burns JR, Kammann SM, *et al.* The Energy Harvesting Eel: A Small Subsurface Ocean/River Power Generator. *IEEE of Oceanic Engineering.* 2001;26(4):539–547.

[35] Xu N, Rangwala S, Chintalapudi KK, *et al.* A Wireless Sensor Network for Structural Monitoring. In: Proceedings of Sensys; 2004. p. 13–24.

[36] Biswas S and Morris R. ExOR: Opportunistic Multi-Hop Routing for Wireless Networks. In: Proceedings of ACM SIGCOMM; 2005. p. 133–144.

[37] Roundy S, Leland ES, Baker J, *et al.* Improving Power Output for Vibration-based Energy Scavengers. *Pervasive Computing.* 2005;p. 28–36.

[38] Paek J, Chintalapudi K, Cafferey J, *et al.* A Wireless Sensor Network for Structural Health Monitoring: Performance and Experience. In: Proceedings of EmNetS-II; 2005. p. 1–10.

[39] Chen C, Aksoy D, and Demir T. Processed Data Collection using Opportunistic Routing in Location Aware Wireless Sensor Networks. In: Proceedings of MDM; 2006. p. 150–157.

[40] Beeby SP, Tudor MJ, and White NM. Energy Harvesting Vibration Sources for Microsystems Applications. *Measurement Science and Technology.* 2006; p. 175–195.

[41] Kim S, Pakzad S, Culler D, *et al.* Health Monitoring of Civil Infrastructures Using Wireless Sensor Networks. In: Proceedings of 6th International

Symposium on Information Processing in Sensor Networks (IPSN); 2007. p. 254–263.
[42] Chebrolu K, Raman B, Mishra N, *et al.* BriMon: A Sensor Network System for Railway Bridge Monitoring. In: Proceedings of Sensys; 2008. p. 2–14.
[43] Swallow LM, Luo JK, Siores E, *et al.* A Piezoelectric Fibre Composite Based Energy Harvesting Device for Potential Wearable Applications. Measurement Science and Technology. 2008;p. 1–7.
[44] Starner T. Human-Powered Wearable Computing. *IBM Systems Journal.* 1996;35(3 & 4):618–629.
[45] Starner T and Paradiso JA. *Human Generated Power for Mobile Electronics.* CRC Press; 2004. p. 1–35.
[46] Wang ZL. *Nanogenerators for Self-Powered Devices and Systems*. Georgia Institute of Technology, Atlanta, USA; 2011.
[47] Gorlatova M, Sarik J, Grebla G, *et al.* Movers and Shakers: Kinetic Energy Harvesting for the Internet of Things. In: Proceedings of ACM SIGMETRICS; 2014.
[48] Marti S, Giuli TJ, Lai K, *et al.* Mitigating Routing Misbehavior in Mobile Ad hoc networks. In: Proceedings of ACM MOBICOM; 2000. p. 255–265.
[49] Raymond DR and Midkiff SF. Denial-of-Service in Wireless Sensor Networks: Attacks and Defense. *IEEE Pervasive Computing.* 2008;7(1):74–81.
[50] IEEE Std 802.11-1999, Local and Metropolitan Area Network, Specific Requirements, Part 11: Wireless LAN Medium Access Control (MAC) and Physical Layer (PHY) Specifications. http:// standards.ieee.org/getieee 802/download/802-11-1999.pdf; 1999.
[51] Abhishek NV, Tandon A, Lim T, *et al.* Detecting Forwarding Misbehavior in Clustered IoT Networks. In: Proceedings of ACM International Symposium on QoS and Security for Wireless and Mobile Networks; 2018. p. 1–6.
[52] Lim S and Huie L. Hop-by-Hop Cooperative Detection of Selective Forwarding Attacks in Energy Harvesting Wireless Sensor Networks. In: Proceedings of International Conference on Computing, Networking and Communications (ICNC); 2015. p. 315–319.
[53] Pu C, Zhou X, and Lim S. Mitigating Suppression Attack in Multicast Protocol for Low Power and Lossy Networks. In: Proceedings of IEEE LCN; 2018. p. 251–254.
[54] Pu C and Zhou X. Suppression Attack Against Multicast Protocol in Low Power and Lossy Networks: Analysis and Defenses. *Sensors.* 2018;18 (10):3236.
[55] Pu C and Hajjar S. Mitigating Forwarding Misbehaviors in RPL-based Low Power and Lossy Networks. In: Proceedings of IEEE CCNC; 2018.
[56] Yu B and Xiao B. Detecting Selective Forwarding Attacks in Wireless Sensor Networks. In: IEEE IPDPS; 2006. p. 1–8.
[57] Xiao B, Yu B, and Gao C. CHEMAS: Identify Suspect Nodes in Selective Forwarding Attacks. *Journal of Parallel and Distributed Computing.* 2007;67(11):1218–1230.

[58] Pu C and Lim S. A Light-Weight Countermeasure to Forwarding Misbehavior in Wireless Sensor Networks: Design, Analysis, and Evaluation. *IEEE Systems Journal.* 2018;12(1):834–842.

[59] Bounouni M and Bouallouche-Medjkoune L. Acknowledgment-Based Punishment and Stimulation Scheme for Mobile Ad Hoc Network. *Journal of Supercomputing.* 2018;74(10):5373–5398.

[60] Chang J, Tsou P, Woungang I, *et al.* Defending against Collaborative Attacks by Malicious Nodes in MANETs: A Cooperative Bait Detection Approach. *IEEE Systems Journal.* 2014;9(1):65–75.

[61] Jhaveri R and Patel N. A Sequence Number Based Bait Detection Scheme to Thwart Grayhole Attack in Mobile Ad Hoc Networks. *Wireless Networks.* 2015;21(8):2781–2798.

[62] Pu C and Lim S. Spy vs. Spy: Camouflage-based Active Detection in Energy Harvesting Motivated Networks. In: Proceedings of Military Communications Conference (MILCOM) – Track 3. Cyber Security and Trusted Computing; 2015.

[63] Lim S, Kimn J, and Kim H. Analysis of Energy Harvesting for Vibration-Motivated Wireless Sensor Networks. In: Proceedings of International Conference on Wireless Networks; 2010. p. 391–397.

[64] 2.4 GHz IEEE 802.15.4/ZigBee-ready RF Transceiver. http://www.ti.com/lit/ds/symlink/cc2420.pdf (Last accessed at Feb 2018).

[65] Cisco Aironet 802.11a/b/g wireless CardBus adapter;. https://www.cisco/com/c/en/us/products/collateral/wireless/aironet-802-11a-b-g-cardbus-wireless-lan-client-adapter-cb21ag=productdatasheet09186a00801ebc29.html (Last accessed at Feb 2018).

[66] Xue X, Wang S, Guo W, *et al.* Hybridizing Energy Conversion and Storage in a Mechanical-to-Electrochemical Process for Self-Charging Power Cell. *Nano Letter*. 2012;12(9):5048–5054.

[67] Fujii C and Seah WKG. Multi-Tier Probabilistic Polling for Wireless Sensor Networks Powered by Energy Harvesting. In: Proc. IEEE ISSNIP; 2011. p. 383–388.

[68] Pu C, Gade T, Lim S, *et al.* Light-Weight Forwarding Protocols in Energy Harvesting Wireless Sensor Networks. In: Proceedings of MILCOM; 2014. p. 1053–1059.

[69] Stallings W. *Cryptography and Network Security - Principles and Practices*, 6th Edition. Prentice-Hall; 2013.

[70] Zhong S, Chen J, and Yang YR. Sprite: A Simple, Cheat-Proof Credit-Based System for Mobile Ad Hoc Networks. In: Proceedings of IEEE INFOCOM; 2003.

[71] Hai TH and Huh E. Detecting Selective Forwarding Attacks in Wireless Sensor Networks Using Two-hops Neighbor Knowledge. In: Proceedings of IEEE NCA; 2008. p. 325–331.

[72] Shila DM, Yu C, and Anjali T. Mitigating Selective Forwarding Attacks with a Channel-Aware Approach in WMNs. *IEEE Trans on Wireless Communications.* 2010;9(5):1661–1675.

[73] Pu C, Lim S, Jung B, *et al.* EYES: Mitigating Forwarding Misbehavior in Energy Harvesting Motivated Networks. *Computer Communications.* 2018;124(2018):17–30.

[74] Liu K, Deng J, Varshney PK, *et al.* An Acknowledgment-Based Approach for the Detection of Routing Misbehavior in MANETs. *IEEE Trans on Mobile Computing.* 2007;6(5):536–550.

[75] Li X, Lu R, Liang X, *et al.* Side Channel Monitoring: Packet Drop Attack Detection in Wireless Ad Hoc Networks. In: Proceedings of IEEE ICC; 2011. p. 1–5.

[76] Shakshuki EM, Kang N, and Sheltami TR. EAACK: A Secure Intrusion-Detection System for MANETs. *IEEE Transactions on Industrial Electronics.* 2013;60(3):1089–1098.

[77] Liu Q, Yin J, Leung V, *et al.* FADE: Forwarding Assessment Based Detection of Collaborative Grey Hole Attacks in WMNs. *IEEE Transactions on Wireless Communications.* 2013;12(10):5124–5137.

[78] Ren J, Zhang Y, Zhang K, *et al.* Exploiting Channel-Aware Reputation System against Selective Forwarding Attacks in WSNs. In: Proceedings of IEEE Global Communications Conference; 2014. p. 330–335.

[79] Shila D, Cheng Y, and Anjali T. Mitigating Selective Forwarding Attacks with a Channel-Aware Approach in WMNs. *IEEE Transactions on Wireless Communications.* 2010;9(5):1661–1675.

[80] Pu C and Lim S. A Light-Weight Countermeasure to Forwarding Misbehavior in Wireless Sensor Networks: Design, Analysis, and Evaluation. *IEEE Systems Journal.* 2018;12(1):834–842.

[81] Sivanesh S and Dhulipala V. Accurate and Cognitive Intrusion Detection System (ACIDS): A Novel Black Hole Detection Mechanism in Mobile Ad Hoc Networks. *Mobile Networks and Applications.* 2020;p. 1–9.

[82] Samy A, Yu H, and Zhang H. Fog-Based Attack Detection Framework for Internet of Things Using Deep Learning. *IEEE Access.* 2020;8:74571–74585.

[83] Baig ZA, Sanguanpong S, Firdous SN, Vo VN, Nguyen TG, and So-In C. Averaged Dependence Estimators for DoS attack detection in IoT networks. *Future Generation Computer Systems.* 2020;102:198–209.

[84] Fafoutis X and Dragoni N. ODMAC: An On-Demand MAC Protocol for Energy Harvesting – Wireless Sensor Networks. In: Proc. PE-WASUN; 2011. p. 49–56.

[85] Vithanage MD, Fafoutis X, Andersen CB, *et al.* Medium Access Control for Thermal Energy Harvesting in Advanced Metering Infrastructures. In: Proceedings of EuroCon – Internet Services and Applications; 2013. p. 291–298.

[86] Pu C. Sybil Attack in RPL-Based Internet of Things: Analysis and Defenses. IEEE Internet of Things *Journal.* 2020;7(6):4937–4949.

[87] Winter T and Thubert P. RPL: IPv6 Routing Protocol for Low-Power and Lossy Networks; 2012. *RFC Standard* 6550.

[88] Romdhani I, Qasem M, Al-Dubai A, *et al. Cooja Simulator Manual*; 2016. Edinburgh Napier University.

[89] Abidin A, Aly A, and Mustafa MA. Collaborative Authentication Using Threshold Cryptography. In: *International Workshop on Emerging Technologies for Authorization and Authentication.* Springer; 2019. p. 122–137.

[90] Feng Q, He D, Wang H, *et al.* Lightweight Collaborative Authentication with Key Protection for Smart Electronic Health Record System. *IEEE Sensors Journal.* 2019;20(4):2181–2196.

[91] Rimmer V, Preuveneers D, Joosen W, *et al.* Frictionless Authentication Systems: Emerging Trends, Research Challenges and Opportunities. arXiv preprint arXiv:180207233. 2018.

[92] Bailey DV, Duane WM, and Katz A. Protected Resource Access Control Utilizing Credentials based on Message Authentication Codes and Hash Chain Values. Google Patents; 2015. US Patent 8,984,602.

[93] Bailey DV, Duane WM, and Young E. Protected Resource Access Control Utilizing Intermediate Values of a Hash Chain. Google Patents; 2015. US Patent 8,990,905.

[94] Alshahrani M, and Traore I. Secure Mutual Authentication and Automated Access Control for IoT Smart Home Using Cumulative Keyed-Hash Chain. *Journal of Information Security and Applications.* 2019;45:156–175.

[95] Pinto A, and Costa RF. Hash-Chain-Based Authentication for IoT; 2016.

[96] He M, Fan P, Kaderali F, *et al.* Access Key Distribution Scheme for Level-Based Hierarchy. In: Proceedings of the Fourth International Conference on Parallel and Distributed Computing, Applications and Technologies; 2003. p. 942–945.

[97] Shehab M, Bertino E, and Ghafoor A. Efficient Hierarchical Key Generation and Key Diffusion for Sensor Networks. In: 2005 Second Annual IEEE Communications Society Conference on Sensor and Ad Hoc Communications and Networks, 2005. IEEE SECON 2005; 2005. p. 76–84.

[98] Castiglione A, Santis AD, Masucci B, *et al.* Hierarchical and Shared Access Control. *IEEE Transactions on Information Forensics and Security.* 2016 April;11(4):850–865.

[99] Zaman MU, Shen T, and Min M. Hash Vine: ANew Hash Structure for Scalable Generation of Hierarchical Hash Codes. In: 2019 IEEE International Systems Conference (SysCon); 2019. p. 1–6.

[100] Borgia E. The Internet of Things Vision: Key Features, Applications and Open Issues. *Journal of Computer Communications*. 2014.

[101] Al-Fuqaha A, Guizani M, Mohammadi M, *et al.* Internet of Things: A Survey on Enabling Technologies, Protocols, and Applications. *IEEE Communications Surveys & Tutorials*. 2015;(17):2347–2376.

[102] Weyrich M, and Ebert C. Reference Architectures for the Internet of Things. *IEEE Software.* 2016;33(1):112–116.

[103] Breivold H. A Survey and Analysis of Reference Architectures for the Internet-of-Things. In: *Proc. Software Engineering Advances (ICSEA)*; 2017.

[104] Simmhan Y, Ravindra P, Chaturvedi S, *et al.* Towards a Data-driven IoT Software Architecture for Smart City Utilities. *Software: Practice and Experience*. 2018;48(7):1390–1416.

[105] Campeanu G. A Mapping Study on Microservice Architectures of Internet of Things and Cloud Computing Solutions. In: Proceedings of Mediterranean Conf. on Embedded Computing (MECO); 2018.

[106] Alreshidi A, and Ahmad A. Architecting Software for the Internet of Thing Based Systems. *Future Internet.* 2019;11(7).

[107] Mucchini H, and Moghaddam MT. IoT Architectural Styles: A Systematic Mapping Study. In European Conference on Software Architecture. In: Proceedings of European Conf. on SoftwareArchitecture; 2018.

[108] Gaglione A, Rodenas-Herraiz D, Jia Y, *et al.* A. Energy Neutral Operation of Vibration Energy-Harvesting Sensor Networks for Bridge Applications. In: Proceedings of Embedded Wireless Systems and Networks (EWSN); 2018.

Index

www.ingramcontent.com/pod-product-compliance
Lightning Source LLC
LaVergne TN
LVHW011847200926
848484LV00002B/412